Laser Precision Microprocessing of Materials

Laser Precision Microprocessing of Materials

A.G. Grigor'yants
M.A. Kazaryan
N.A. Lyabin

CISP

CRC Press is an imprint of the
Taylor & Francis Group, an **informa** business

Translated from Russian by V.E. Riecansky

CRC Press
Taylor & Francis Group
6000 Broken Sound Parkway NW, Suite 300
Boca Raton, FL 33487-2742

First issued in paperback 2022

ISBN 13: 978-1-03-240169-0 (pbk)
ISBN 13: 978-1-138-59454-8 (hbk)

DOI: 10.1201/9780429488771

Publisher's Note
The publisher has gone to great lengths to ensure the quality of this reprint but points out that some imperfections in the original copies may be apparent.

Visit the Taylor & Francis Web site at
http://www.taylorandfrancis.com

and the CRC Press Web site at
http://www.crcpress.com

Contents

Symbols and abbreviations **ix**
Introduction **xi**

1. Overview of the present state and the development of copper vapour lasers and copper vapour laser systems **1**
1.1. Discovery and first investigations and design of copper vapour lasers 1
1.2. The condition and development of CVL in Russia 3
1.3. The condition and development of CVL and CVLS in foreign countries 18
1.4. The current state and development of the CVL and CVLS in the Istok company 26
1.5. Conclusions and results for chapter 1 36

2. Possibilities of pulsed copper vapour lasers and copper vapour laser systems for microprocessing of materials **41**
2.1. The current state of the modern laser processing equipment for the processing of materials and the place in it of pulsed copper vapour lasers 41
2.2. Analysis of the capabilities of pulsed CVL for microprocessing of metallic and non-metallic materials 43
2.3. Equipment MP200X of Oxford Laser for microprocessing 51
2.4. The main results of the first domestic studies on microprocessing at the Kareliya CVLS and installations EM-5029 54
2.5. The first domestic experimental laser installation (ELI) Karavella 55
2.6. Conclusions and results for Chapter 2 66

3. A new generation of highly efficient and long-term industrial sealed-off active elements of pulsed copper vapour lasers of the Kulon series with a radiation power of 1–20 W and Kristall series with a power of 30–100 W 69
3.1. Analysis of the first designs of self-heating AE pulsed CVLs and the reasons for their low durability and efficiency 70
3.2. Investigation of ways to increase the efficiency, power and stability of the output radiation parameters of CVL 73
3.3. Choice of directions for the development of a new generation of industrial sealed-off self-heating AE of the CVLs 76
3.4. Appearance and weight and dimensions of industrial sealed-off AEs of the pulsed CVL of the Kulon and Kristall series 78
3.5. Construction, manufacturing and training technology, basic parameters and characteristics of industrial sealed-off AEs of the Kulon and Kristall CVL series 80
3.6. Conclusions and results for chapter 3 161

4. Highly selective optical systems for the formation of single-beam radiation of diffraction quality with stable parameters in copper vapour lasers and copper vapour laser systems 166
4.1. Distinctive properties and features of the formation of radiation in a pulsed CVL 167
4.2. Experimental settings and research methods 169
4.3. Structure and characteristics of radiation of CVL in single-mirror mode. Conditions for the formation of single-beam radiation with high quality 174
4.4. Structure and characteristics of the laser radiation in the regime with an unstable resonator with two convex mirrors 187
Conditions for the formation of single-beam radiation with diffraction divergence and stable parameters 187
4.5. Structure and characteristics of the radiation of CVL in the regime with telescopic UR. Conditions for the formation and separation of a radiation beam with diffraction divergence 196
4.6. Investigation of the conditions for the formation of a

powerful single-beam radiation with a diffraction divergence in a CVLS of the MO–PA type 204
4.7. Investigation of the properties of the active medium of a pulsed CVL using CVLS 231
4.8. Conclusions and results for chapter 4 235

5. Industrial copper vapour lasers and copper vapour laser systems based on the new generation of sealed-off active elements and new optical systems 243
5.1. The first generation of industrial CVLs 243
5.2. A new generation of industrial CVLs of the Kulon series 255
5.3. Two-channel Karelia CVLS with high quality of radiation 274
5.4. Two-channel lamp-pumped laser CVLS Kulon-15 289
5.5. Three-channel CVLS Karelia-M 292
5.6. Powerful CVLS 293
5.7. Conclusions and results for chapter 5 296

6 Modern automated laser technological installation Karavella (ALTI) 299
6.1. Requirements for pulsed CVL and CVLS in modern technological equipment 299
6.2. Industrial ALTI Karavella-1 and Karavella-1M on the basis of two-channel CVLS 300
6.2.1. *Composition, construction and principle of operation* 304
6.2.2. *Principle of construction and structure of the motion and control system* 318
6.2.3. *Main technical parameters and characteristics* 323
6.3. Industrial ALTIs Karavella-2 and Karavella-2M on the basis of single-channel CVL 326
6.3.1. *Basics of creating industrial ALTIs Karavella-2 and Karavella-2M* 326
6.3.2. *Composition, design and operation principle of ALTI* 330
6.3.3. *Main technical parameters and characteristics* 335
6.4. Conclusions and results for Chapter 6 339

7. Laser technologies of precision microprocessing of foil and thin sheet materials for components for electronic devices 342
7.1. The threshold densities of the peak and average radiation

power of CVL for evaporation of heat-conducting and refractory materials, silicon and polycrystalline diamond 343
7.2. Effect of the thickness of the material on the speed and quality of the laser treatment 347
7.3. Development of the technology of chemical cleaning of metal parts from slag after laser micromachining 350
7.4. Investigation of the surface quality of laser cutting and the structure of the heat-affected zone 355
7.5. Development of microprocessing technology in the production of LTCC multi-layer ceramic boards for microwave electronics products 363
7.6. Conclusions and results for Chapter 7 371

8. Using industrial automatic laser technological installations Karavella-1, Karavella-1M, Karavella-2 and Karavella-2M for the fabrication of precision parts for electronic devices 373
8.1. The possibilities of application of ALTI Karavella for the manufacture of precision parts 373
8.2. Examples of the manufacture of precision parts for electronic components at ALTI Karavella 378
8.3. Advantages of the laser microprocessing of materials on ALTI Karavella in comparison with traditional processing methods 389
8.4. Perspective directions of application of ALTI Karavella 390
8.5. Conclusions and results for Chapter 8 395

Conclusion 397
References 401
Index 416

Symbols and abbreviations

ALTI – automated laser technological installation
AM – active medium
AE – active element
MPG – master pulses generator
NPG – nanosecond pulse generator
MO – master oscillator
HAZ – heat-affected zone
PS – power supply
GVL – gold vapour laser
CVL – copper vapour laser
LPMET – metal vapour lasers
DSL – dye solution laser
CVLS – copper vapor laser system
NC – nonlinear crystal
UR – an unstable resonator
OE – optical element
SFC – spatial filter collimator
PA – power amplifier
PRF – pulse repetition frequency
ELTI – experimental laser technological installation
EOT – electroerosion treatment
c – speed of light
C_{cap} – capacity of the storage capacitor
C_{sc} – capacity of the sharpening condenser
D_{chan} – diameter of the discharge channel
n – number of double passes of radiation in the resonator
p_{Ne} – neon buffer gas pressure
P_{EA} – power input into the AE

P_{ext} – power consumed from the rectifier of the power source
P_{rad} – average output power
λ – wavelength of the radiation
l_{AM} – length of the active medium
l_{chan} – length of the discharge channel
L – length of the optical resonator
F – focal length of the lens
M – gain of the optical resonator
R – radius of curvature of the mirror
T_c – temperature of the discharge channel
τ – time of population inversion existence
τ_{time} – pulse duration
θ – radiation divergence
θ_{dif} – diffraction divergence of radiation
ρ – peak radiation power density
V_{AM} – volume of the active medium
U – voltage
W – energy in the radiation pulse

Introduction

The development of the electronic industry, with the further miniaturization of electronic components and the use of new materials, puts ever-increasing demands on the quality, reliability and competitiveness of manufactured products. That, in turn, makes higher demands on the parameters of the components, thus dictating the creation of new technologies and technological processes. A special recognition was given to laser technologies for microprocessing. In this case, the function of the processing tool is performed by a high-intensity focused light spot. To ensure high quality of machining, the tool should provide the following parameters – micron width of cut (1–20 μm), minimal heat-affected zone (≤3–5 μm) and roughness (≤1–2 μm). The radiation sources in the microprocessing equipment include short-pulsed, high-frequency lasers with a low energy in the pulse and a small reflection coefficient of the visible and ultraviolet radiation spectrum: solid-state, excimer, nitrogen and, in particular, lasers and laser systems on copper vapour (copper vapour lasers (CVL) and copper vapour laser systems (CVLS)). CVLs belong to the class of gas lasers on self-terminated transitions of metal atoms that generate on transitions from resonant to metastable levels [1–8].

CVL and CVLS with emission wavelengths λ = 510.6 and 578.2 nm, short pulses duration (τ_{pulse} = 20–40 ns), and high amplification of the active medium (AM) (k = 10^1–10^2 dB/m), the removal of medium power from one active element (AE) to 750 W, high pulse repetition rates (f = 5-30 kHz) and low pulse energy (W = 0.1–10 mJ) remain today the most powerful pulsed coherent radiation sources in the visible region of the spectrum. With these parameters and provided that the structure of the output radiation is single-beam and has a diffraction quality, the peak power density in the focused spot (d = 5–20 μm), even at relatively small values of the average power (P_{rad} = 1–20 W), reaches very high values – ρ = 10^9–10^{12} W/cm^2, sufficient for effective microprocessing of metallic materials and a large number of dielectrics and semiconductors [8–62]. The spectrum of processed materials includes: heat conducting – Cu, Al, Ag, Au; refractory – W, Mo, Ta, Re and other metals – Ni, Ti, Zr, Fe and their alloys, steel, dielectrics and semiconductors – silicon, polycrystalline diamond, sapphire, graphite, carbides and nitrides and transparent materials [15, 16, 20, 26].

More than four decades after the first generation of the CVLs was by the efforts of a number of scientific teams, primarily Russia, the USA, England, Australia and Bulgaria, these lasers were established both with the basic physical principles of work and design, and specific applications in science, technology and medicine. The bulk of the research was devoted to the 'pure' CVL operating in a mixture of the neon buffer gas and copper vapour at a discharge temperature of 1500–1600°C. In the last 10–15 years, researchers and developers have increased interest in its varieties operating at the same *r–m* junctions, but at relatively low temperatures (300–600°C) and higher repetition rates (up to hundreds kilohertz) to lasers on copper halides (CuCl, CuBr, and CuI) and 'hybrid' (with the pumping of a mixture of HBr, HCl, Br_2 or Cl_2 and Ne), and also with 'enhanced kinetics' (with the addition of H_2 or its compounds) [19, 35]. But on copper halides and 'hybrid' CVLs, without appreciable performance, the service life and stability of the output parameters remain relatively low for today, which is due to the instability in time of the composition and properties of the multicomponent gas mixture of the active medium (AM). Therefore, today, from the point of view of industrial production and practical application, the advantage remains on the side of 'pure' CVLs and with 'enhanced kinetics'.

At low radiation power levels (1–20 W), the CVL is designed constructively as a separate generator (monoblock) with one low-power active element (AE) and an optical resonator. To obtain the average (20–100 W) and especially high (unit or tens of kW) radiation power levels, CVLSs operating according to the master oscillator–power amplifier (MO–PA) scheme with one or several powerful AEs as PAs are used with a preamplifier (PRA), located in front of the PA. In the CVL of the MO–PA type, in comparison with the CVL operating in the single-generator mode, higher efficiency and the quality of the output beam are achieved [16, 20].

The CVL remains the most efficient (30–50% efficiency) source for pumping lasers on solutions of organic dyes (DSL) tunable along wavelengths in the near infrared region of the spectrum, non-linear crystals of the BBO type, KDP, DKDP (efficiency 10–25 %), transforming the generation of CVL into the second harmonic – λ = 255.3, 289.1 and 272.2 nm, i.e., into the ultraviolet region of the spectrum and titanium sapphire (Al_2O_3: Ti^{3+}), which converts generation to the near-IR region spectrum, and then with the help of the non-linear crystal – from the IR region to blue. The use of CVL with DSL and the non-linear crystal allows us to practically cover the wavelength range from the near UV to the near IR spectral region and, accordingly, to expand the laser's functional capabilities. Such tunable pulsed laser systems are unique and preferable for both practical and scientific spectroscopic studies and microprocessing by UV radiation [15, 16, 20, 21, 31].

A special place is occupied by the use of CVLs in combination with DSL tunable in wavelengths in high-power laser systems of the MO–PA type. Powerful CVLs of the MO–PA type are used mainly in the isotope separation system according to AVLIS technology, which uses a difference in the absorption spectra of atoms of different isotopic composition. This progressive optical technology makes it possible to produce substances with the necessary level of enrichment and high purity for use, primarily in nuclear power engineering and medicine [15, 16, 20–23, 32, 37].

A promising area of development for CVL is also medicine. Multifunctional modern medical devices such as Yakhroma-Med and Kulon-Med for use in oncology, low-intensity therapy, dermatology and cosmetology, microsurgery, etc. are created on its basis. This class of equipment is the leader in laser non-ablative technologies. Laser pulses act on the body's defects selectively, without damaging the surrounding tissue and without causing pain (anesthesia is not required) [43–56].

In addition, CVL is used as an intensifier for the brightness of the image of microobjects, in nanotechnology, high-speed photography, for analyzing the composition of substances, in laser projection systems for imaging on large screens and in open space, in lidar installations for probing the atmosphere and sea depths, in navigation systems, water treatment, gas flow visualization, laser acceleration of microparticles, holography, forensics and entertainment industry, etc. [12, 15, 16, 20, 21, 24, 27–30, 38–42, 57].

In the technology of material processing, industrial CO_2 lasers with λ = 10.6 μm are widely used, but such heat-conducting metals as Cu, Al, Au, and Ag are not efficiently treated with CO_2 laser radiation and other infrared lasers as the reflection coefficient exceeds 95%.

Powerful IR lasers are mainly used for high-speed cutting, cutting and welding of ferrous metals and stainless steel up to 20 mm thick [21, 25].

A widely distributed solid-state laser based on yttrium–aluminum garnet with neodymium (Nd:YAG laser) with λ = 1064 nm and frequency doubling with λ = 532 nm is close to the CVL in terms of spectrum, power, and efficiency, because of the appearance of thermal deformations in the active element, has relatively large divergences. Nd:YAG-lasers are widely used for marking and engraving parts and assemblies in the production process, for welding metals, including aluminum, in medicine and location [21, 25, 26, 58–61].

Lasers similar to the CVL are the solid state (SS) disk o the yttrium–aluminum garnet (Yb:YAG laser) at λ = 1030 nm and other SSs with pico- and femtosecond durations, for example, the German company Rofin-Sinar Laser and TRUMPF [21, 26], designed for drilling microholes in stainless steel up to 1 mm thick for injectors. Development of ultrashort pulse of the solid-state lasers was successfully conducted in a number of advanced countries (France, England, Latvia, etc.). The main distinguishing feature of laser systems with ultrashort pulses is that due to the low thermal impact on the base material and without the formation of a melt, the best microprocessing quality and high resolution are achieved. These lasers are used where it is impossible to achieve high quality of microprocessing by other lasers, for example, when drilling nozzles for injections, manufacturing medical stands and display glasses, etc. To their disadvantage today is a low average radiation power (1–10 W) and high cost.

At the stage of rapid development, highly efficient ytterbium (Yb) fiber lasers with wavelengths of 1060–1070 nm and an average radiation power of 10–50000 W continue to be developed and produced by the international research and production group IPG Photonics Corporation and IRE–Polyus [25, 62–64]. However, in these lasers, when operating in a single-mode pulsed regime with nanosecond duration, such high peak power densities as in CVL are not achieved, since non-linear effects and material destruction centres arise in the light-conducting fiber. Continuous fiber lasers activated by erbium and thulium with a power of 5–50 W at λ = 1530–1620 nm and λ = 1800–2100 nm, respectively, are produced. The main areas of application: precision cutting, cutting and welding of ferrous and non-ferrous metals, hardening and surfacing, marking and engraving, telecommunications and medicine.

Excimer gas lasers on halogen compounds on an inert gas (ArF, KrF, XeCl, XeF) and inert gas dimers (Ar, Kr) operate like a CVL in a pulsed mode with nanosecond duration, but have shorter radiation wavelengths – λ = 157; 193; 248; 282; 308; 351 nm [21, 25, 65, 66], i.e., they are generated in the near-UV range. This is their advantage for wide application in lithography processes in semiconductor manufacturing, in eye surgery, as well as in dermatology. However, due to the relatively large divergence and the smaller working pulse repetition frequencies (not more than 1–5 kHz), the quality and productivity of material processing is reduced, and this class of lasers is used mainly for the processing of plastics, ceramics, crystals, biological tissues.

Diode (semiconductor) lasers are small in size and can be produced in large batches at relatively low costs. Most diode lasers generate in the near-IR region – λ = 800–1000 nm. They are reliable and durable, but the output power of a single element is limited and have a high radiation divergence. The diode lasers are used in many spheres of human activity, mainly in the telecommunications and optical memory sectors [21, 25, 67, 68], and are also used in large quantities as pumping sources for solid-state and fiber lasers. The developed technology of adding single diodes to diode lines allows to increase the average laser power to 1–3 kW, which is enough for high-performance and high-quality welding, for example, aluminium parts.

The above comparative analysis of the characteristics of CVL with other known types of technological lasers confirms that the CVL remains a promising quantum device in terms of the set of radiation output parameters, primarily for the microprocessing of materials of electronic equipment and selective technologies for isotope separation, and also in spectroscopy, image brightness amplifiers, medicine and other fields of science and technology.

Figure I1 represents the global market for laser sales from 2007 to 2015, Fig. I2 shows structure of the world market of laser sources radiation for 2014 [69].

Figure I.3 presents the dynamics of the global sales volume of all types of lasers by years for the acquisition of technological equipment [69].

Technological equipment uses CO_2, solid-state, fiber, and excimer lasers. The use of CVL in specialized equipment, in spite of the unique combination of its output parameters, is extremely limited, due to the small number of commercial models on the market with

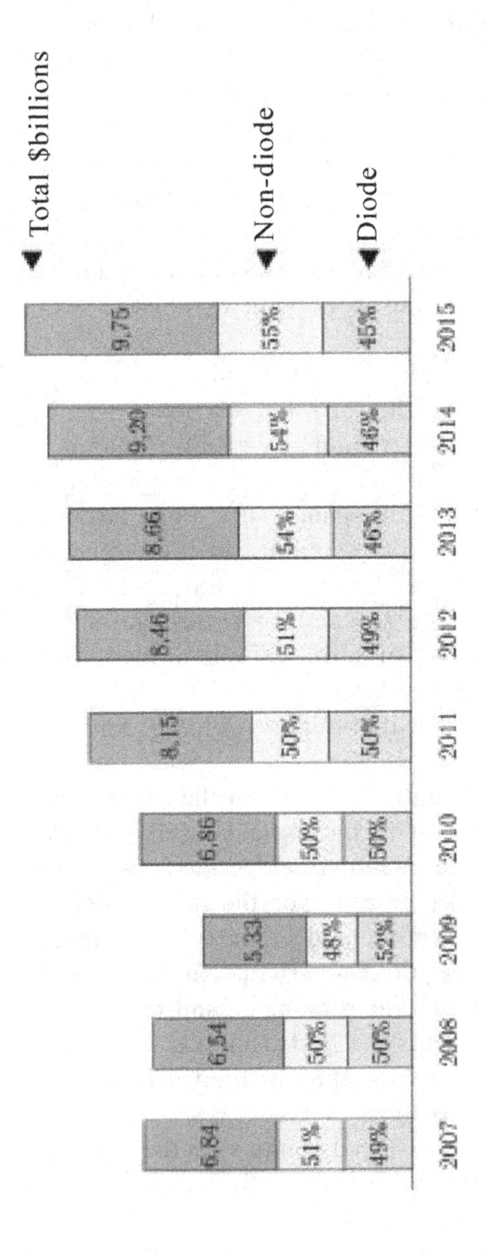

Fig. I1. Worldwide sales of lasers from 2007 to 2015.

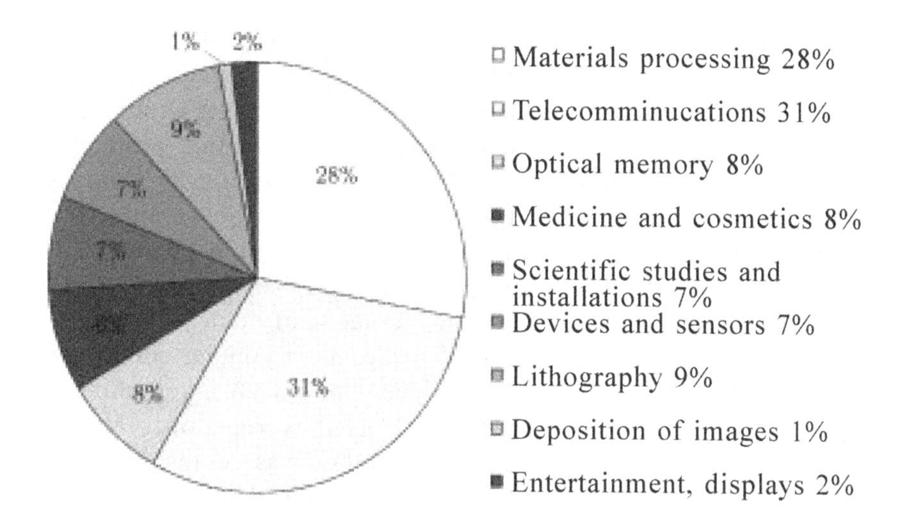

Fig. I2. The structure of the world market of laser radiation sources by fields of application for 2014.

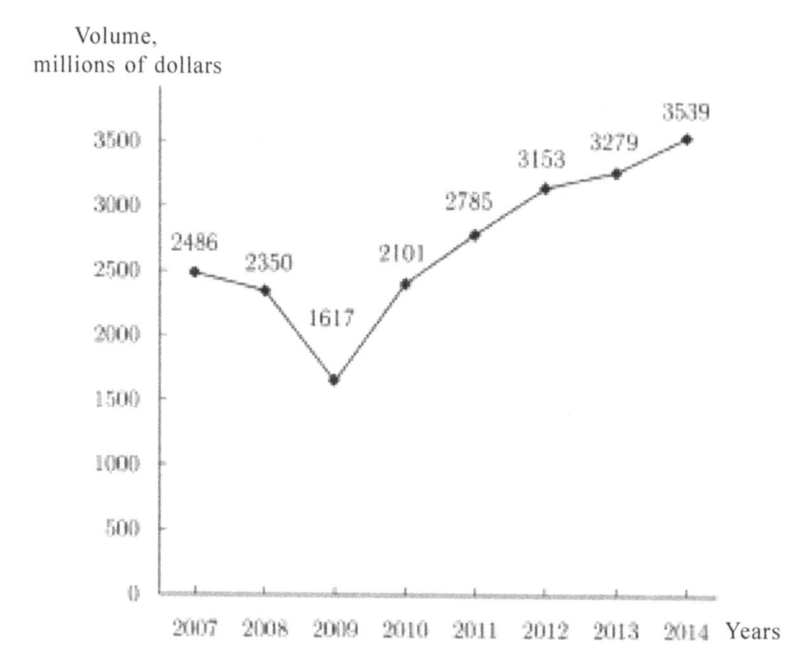

Fig. I.3. The volume of sales of lasers by years for the acquisition of technological equipment.

high reliability and radiation quality. This situation seems to have taken shape for a number of reasons. Firstly, in many research institutes of Russia (USSR), large-scale theoretical and experimental studies of physical processes in CVL were carried out, rather than industrial developments [8–14, 17–19]. Secondly, in the advanced countries abroad (USA, Britain, France, Japan), the main efforts were directed to research and development of high-power CVLS of the type MO–PA in providing laser isotope separation programs according to AVLIS technology for the needs of nuclear power [15, 20, 22, 23]. Thus, the development of the most popular commercial CVLs, which include lasers of small (1–20 W) and medium (30–100 W) power levels, remained, as it were, aloof. Thirdly, over the last 10–15 years, the laser market was represented by a relatively large number of CVLs and its varieties with a low level of reliability and radiation quality, which reduced the user demand for this type of laser. Nevertheless, today it is possible to single out several organizations and firms that continue to improve the old and create new commercial models of CVLs and CVLSs and, on their basis, modern technological equipment for microprocessing of materials and isotope separation, as well as medical facilities and other equipment. They include, first of all, Istok Co. (Fryazino, Moscow region) together with VELIT (Istra, Moscow region) and Chistye Tekhnologii (Izhevsk) with the scientific support of the P.N. Lebedev Physical Institute of the Russian Academy of Sciences, Oxford Lasers (England), Macquarie University (Australia) and Pulse Light (Bulgaria). The Lawrence Livermore National Laboratory is the leader in powerful CVLSs, intended for isotope separation technology, where the average power is brought to 72 kW and the Kurchatov Institute. Studies continue at the A.N. Prokhorov Institute of General Physics, Russian Academy of Sciences, TSU and the Institute of Optics and Atmosphere of the SB RAS (Tomsk), the Institute of Semiconductor Physics (Novosibirsk), Mekhatron, St. Petersburg National Research University of Information Technologies, Mechanics and Optics, Bauman Moscow State Technical University and the Joint Institute for High Temperatures of the Russian Academy of Sciences (Moscow) [15, 16, 20, 33, 34, 41, 42, 52, 53, 70–72].

To realize the advantages of pulsed radiation from CVL in the technology of precision microprocessing of materials and other modern technologies, it is necessary to create a new generation of highly efficient industrial CVLs and CVLSs. The introduction of the technology of laser microprocessing in the production of electronic components allows, in comparison with traditional methods of processing, including electroerosion mashining, to shorten the cycle of preparation of production, to order and increase the productivity, to exclude the mechanical pressure of the tool and the thermal influence, to improve the quality and resource of products.

This monograph is devoted to research and development aimed at creating a new generation of industrial pulsed CVLs and CVLSs with high quality and stability of radiation parameters and on their basis technological equipment and technologies for precision microprocessing of materials. The first chapter of the monograph presents a review of foreign and domestic literature on the state, development, and applications of CVLs operating in the regime of an individual oscillator and powerful CVLSs operating according to the effective MO–PA scheme, chapter 2 describes the possibilities of pulsed CVLs and CVLSs for microprocessing materials and the first technological installations created on their basis. The chapters 3–7 outline the results of the latest developments and research in these areas:

– creation of a new generation of high-performance, durable and stable parameters of industrial sealed-off laser AEs on copper vapour with an average radiation power of 1–100 W. Optimization of AE in terms of radiation power and efficiency from power consumption, buffer gas pressure and hydrogen additives, pulse repetition frequency and pump current pulse characteristics;

– development and research of highly efficient and reliable circuits for the execution of a high-voltage power source modulator with the nanosecond duration of the pump pulses;

– research and development of highly selective optical resonators and systems for the formation of single-beam radiation of diffraction quality in CVL and powerful CVLSs and with stable parameters for achieving high peak power densities (10^9–10^{12} W/cm^2);

– investigation of the properties of the active medium of the pulsed CVL and development on their basis of methods and electronic devices for on-line power control and pulse repetition frequency of radiation;

– development on the basis of a new generation of sealed-off AEs of new high-selective optical systems, electrical circuits and methods control of radiation parameters of industrial technological CVLs and CVLSs with the radiation power up to 100 W with high reliability, efficiency, quality and stability of the radiation parameters;

– creation of modern automated laser technological installation (ALTI) of the Karavella type on the basis of industrial CVLs and CVLSs and modern precision three-axis XYZ tables for efficient and high-quality precision laser microprocessing of materials;

– determination of the optimum density of peak and average radiation power of the CVL for efficient microprocessing of foil (0.01–0.2 mm) and thin-sheet (0.3–1 mm) materials. Investigation of the dependence of the processing speed on the thickness of the material;

– study of the quality of the laser cut surface and the structure of the heat-affected zone for refractory and high-heat conducting materials from processing parameters: speed, number of passes and power of radiation.

Chapter 8 presents concrete examples of manufacturing precision parts at ALTI Karavella and analysis of their quality, certain perspective technological directions for laser microprocessing in the field of electronic engineering.

The appendix contains multifunctional laser medical devices of the type Yakhroma-Med and Kulon-Med for practical and scientific medicine.

1

Overview of the present state and the development of copper vapour lasers and copper vapour laser systems

1.1. Discovery and first investigations and design of copper vapour lasers

The first generation on self-contained *r–m* transitions of metal vapour lasers (MVL) was obtained in 1965 by the American scientists G.R. Fowles and W.I. Silfast on atomic lead vapours in the red region of the visible spectrum at a wavelength $\lambda = 722.9$ nm [1]. In the same year, M. Piltch, V.T. Walter, N. Solimen, G. Gould and V.R. Bennet produced lasing on manganese vapours [2], in 1966 – on copper and gold vapours [3]. The best results were achieved with the transitions of copper atoms with emission wavelengths $\lambda = 510.6$ and 578.2 nm (Fig. 1.1).

Large financial and material resources were spent on the research and development of effective copper vapour lasers (CVL). The first design of copper and gold vapour lasers (CVL and GVL) was a ceramic tube made of alumina (Al_2O_3) with an external electric heater heated to 1500°C and filled with a buffer gas of helium [4]. The peak power on the green line ($\lambda = 510.6$ nm) with a pulse duration of 20 ns (at half-height) reached 1.2 kW with an efficiency of only 0.1%. The gain on the green line was 58 dB/m, on the yellow line – 42 dB/m. When a thyratron commutator (instead of a spark gap) and a gas discharge tube with a diameter of 10 mm and a length of 800 mm

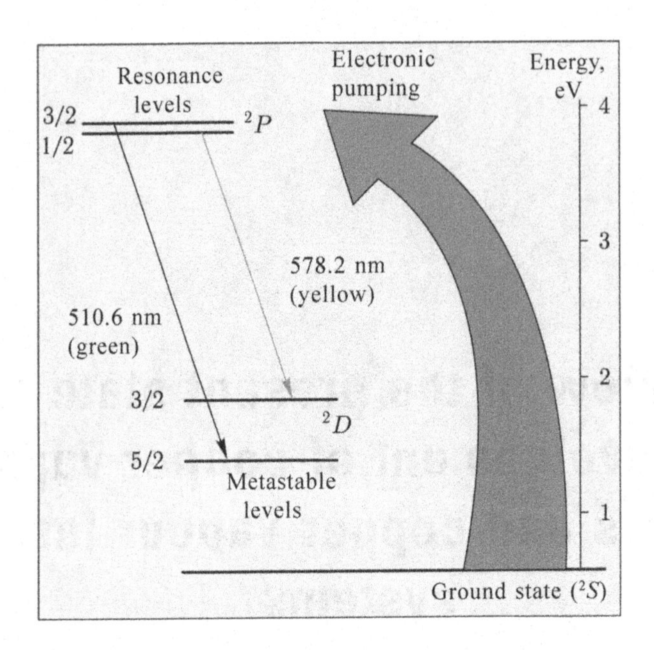

Fig. 1.1. Scheme of energy levels of pulsed copper vapour lasers.

were used as the power supply, the average output power of the CVL increased to 20 mW at 660 Hz pulse repetition frequency (PRF). In CVLs with a large channel diameter (50 mm), unprecedented peak power and efficiency were obtained at that time: 40 kW and 1.2%, respectively [5, 6]. At 1 kHz, average power of radiation reached 0.5 W. The first theoretical analysis of CVL was published in 1967 [7]. The small values of the lasing power were explained by the fundamental limitation of the increase in the PRF. It was believed that the relaxation of the lower laser (metastable) levels of metal atoms occurs only on the walls of the discharge tube, where they fall due to diffusion. Even with channel diameters of 20–40 mm, and even in the case of low buffer gas pressures (10–30 mm Hg), the PRF can not exceed 1 kHz. The presence in the laser design of an external high-temperature heating furnace also reduced the efficiency and depreciated the advantages of high-efficiency lasing at *r–m* transitions of metal atoms.

At the very beginning, CVL was used as an image brightness amplifier [73]. In order to expand the range of applications, it was required to improve the quality of its output radiation. Since 1971, searches have been made for optical systems forming in the CVLs beams with low divergence [74], as a result, the divergence of radiation has been reduced by approximately 30 times.

These 'pioneering' fundamental works of American scientists prompted an intensive study of MetVL (metal vapour lasers) in many countries of the world: the USA, Russia (USSR), England, France, Australia, Israel, Bulgaria, Japan, China, India, etc.

1.2. The condition and development of CVL in Russia

P.N. Lebedev Institute of Physics, Russian Academy of Sciences
In Russia (USSR), the first successes in the study of MetVL were obtained at the P.N. Lebedev Institute of Physics, in the Optics laboratory. In the initial period, mainly by the efforts of the staff of this laboratory, extensive studies of vapour lasers of various metals [8–11, 35, 75–90] – lead [80–83], gold [81, 82, 84], barium [85–89], manganese [81, 90], copper [75, 77, 79, 81, 82] were conducted. In the designs of self-heating active elements (AE), ceramic tubes made of Al_2O_3 were used as the gas-discharge channel, quartz tubes were used as an outer shell, cathode and anode were electrodes from flash lamp-flashes. Between the discharge channel and the shell there was placed a finely dispersed heat insulator of zirconium oxide (ZrO_2). Optical windows were glued to the ends of the shell, through which laser radiation was emitted. In 1971–1972 A.A. Isaev, M.A. Kazaryan and G.G. Petrash demonstrated the world's first pulsed CVL (Fig. 1.2) with a self-heating AE (without an external furnace) and a thyratron commutator of high-voltage pump pulses [9], which in fact predetermined in the main all the further development of this important type of laser, the most powerful in the visible region

Fig. 1.2. The first self-heating AE of the pulsed CVL.

of the spectrum [9–20]. In a self-heating CVL, the heating of the discharge channel with the metallic active substance to the operating temperature (T_{chan}~1500°C) occurs due to the energy of a pulsed-periodic arc discharge, which follows with a high pulse repetition frequency, which also excites copper atoms. The self-heating mode made it possible to simplify the design of CVL and to increase its radiation power and efficiency. With a discharge tube diameter of 15 mm, a length of its heated portion of 700 mm, and a PRF of 15–19 kHz, an average lasing power of up to 15 W and a peak power of 200 kW with a practical efficiency of ~1% are obtained. (Practical efficiency is defined as the ratio of the average output power to the electric power consumed from the power source rectifier.) With a discharge tube diameter of 4 mm, a record specific power output of 0.4 W/cm^3 was achieved. In the work [77], the possibility of obtaining high values of radiation power in CVL was demonstrated. At the PRF up to 20 kHz, the average radiation power in the non-stationary thermal regime of the AE reached 43.5 W with a practical efficiency of ~1%. In the years 1974–1975, an unstable resonator of the telescopic type was studied at the Institute of Physics of the Russian Academy of Sciences in order to reduce the divergence of the radiation of the CVL [91–94]. In the case of using such a resonator, radiation beams with a diffraction quality were formed at a magnification hundreds of times higher. At present, the main attention and efforts of the Institute of Physics of the Russian Academy of Sciences researchers are aimed at studying various varieties of CVL, in which lasing occurs in the same *r–m* transitions of copper atoms. These include lasers based on copper halides (CuCl, CuBr, CuI) with additives of hydrogen (H_2), a hybrid laser (Cu–Ne–HBr) and CVL with H_2, HBr, HCl additives, also called CVL with enhanced kinetics [35, 95].

For lasers of the first two types an efficiency of up to 3% was achieved with an average lasing power of 100–200 W [96]. In [97], an average power of 280 W and an AE efficiency of 3.8% was obtained for a 'hybrid' laser with an active medium volume of 19.5 l (AE length of 3 m) and a 17 kHz PRF. The main role in the kinetics of the active medium of these lasers is played by HBr or HCl molecules, which have relatively large dissociative attachment cross sections [97]. The authors believe that the future lies behind lasers with additives, and copper bromide lasers are more promising in industrial terms in this group of lasers.

The paper [35] describes their advantages over 'pure' CVLs. The 'pure' self-heating CVLs, which are widely used, operate at the discharge tube wall temperatures of 1500–1600°C, which reduces the life of the AE due to the limited choice of structural elements, and have a long heating time (about 1 h).

The most studied copper bromide laser has a number of potential advantages: its discharge tube temperature is about 1000°C lower, which makes it possible to use fused quartz. This simplifies and lowers the design of the AE, makes it possible to place the working substance in the extension and regulate its concentration in the active medium, regardless of the input power, and also significantly shortens the heating time. In principle, it is possible to practically create a fully heated AE, in which there will be no limitation of the service life associated with the removal of the working substance into the 'cold' zones. The addition of hydrogen leads to a significant increase in both the radiation power and the efficiency (up to 3% or more).

However, one can not agree with all the conclusions drawn in [19, 35]. Despite the combination of the above positive properties, the problem associated with the lifetime of lasers on copper halides and the preservation of high stability of the output radiation parameters remain open. In these lasers there is a more intensive consumption of the working substance, which can be due to several reasons. First, there is a deposition of copper atoms from the gas-discharge medium directly to the walls with respect to the 'cold' discharge tube; secondly, the copper atoms and their molecular compounds are diffused to the even 'cooler' AE end sections, and thirdly, the low pressure The buffer gas increases the diffusion rate of the working substance. High chemical activity of chlorine and bromine leads to an intensive (premature) destruction of the elements of the electrode assemblies and instability of the combustion of the discharge. The processes of the physico-chemical interaction of the gaseous medium with quartz and the gas evolution of quartz have also not been studied. In addition, for long-term preservation of the output radiation parameters, stabilization at the optimum level of the multicomponent composition of the active gas medium is required, in which a large number of physical processes and chemical reactions occur. For 'clean' CVLs, many problems related to the longevity and stability of the parameters have already been successfully solved [16, 42]. The efficiency in industrial 'clean' CVLs is 0.5–1%, and the average power output from one AE has reached the level of 500–750 W [20].

It should be emphasized that the optical laboratory under the leadership of M.A. Kazaryan carried out a series of theoretical and experimental studies of active optical systems with image brightness amplifiers for microobjects based on CVL, GVL (gold vapour laser), BaVL (barium vapour laser) etc. [24, 75, 78, 98, 358]. This direction is unique and requires further practical development. At present, ways are being considered to increase the efficiency of CVL and the quality of its output radiation beam by using new configurations of optical systems. Much attention is paid to the use of pulsed CVL radiation for microprocessing both metallic and non-metallic materials [99–116].

In the Raman scattering laboratopy under the leadership of V.S. Gorelik, high-intensity pulsed MetVLs were first used to excite the processes of Raman, hyper-Rayleigh, and hyper-Raman scattering of light in matter [117]. High average and peak power and PRF, nanosecond pulses of MetVL turned out to be crucial for the analysis of inelastic light scattering spectra of many substances previously unavailable for analysis. A new unique device – a molecular analyzer characterized by high sensitivity [39] – was created nn the basis of CVL and GVL with the use of sealed-off AE of low power of the series 'Kulon' produced by the Istok company [16, 42] . Theoretical and experimental studies of this laboratory offer great opportunities for the effective use of modern laser technology as a means of molecular monitoring of technological processes and for non-destructive testing of water, liquid and solid fuel, food, pharmaceutical and biological objects, determination of the molecular composition of gasoline and oil, biologically dangerous objects in food and environment, detection of especially dangerous and toxic substances.

The A.M. Prokhorov General Physics Institute, the Russian Academy of Sciences

In the Laboratory of macrokinetics of non-equilibrium processes under the scientific supervision of G.A. Shafeev a large amount of research has been carried out over the past 10 years on the formation of metal nanoparticles in liquids using pulsed laser radiation [118–120]. The process of formation of noble metal nanoparticles – gold (Au) and silver (Ag) in the ablation of metal targets in liquids – water (H_2O) and ethanol (C_2H_5OH) – using CVL radiation was experimentally studied in Ref. [118]. In the work [119] the dynamics of formation of an alloy of gold and silver nanoparticles under laser irradiation of a mixture of individual nanoparticles was studied and the factors influencing the formation of an alloy were determined. Individual nanoparticles were obtained by ablating metals in a liquid with radiation with a wavelength of 510.6 nm. It is shown that the size of the nanostructures essentially depends on the duration of the laser pulse and the power density in the beam focusing spot. The authors of the works assert that the investigated processes can undoubtedly find wide application in photonics, medicine and biology.

In the Scientific Centre of Laser Materials and Technologies, under the guidance of A.A. Sobol, it was established that the technique for recording Raman spectra at high temperatures is also applicable to the study of the luminescence of atoms and individual molecular groupings in vapours over superheated melts of certain compounds [121]. This technique, with the use of laser radiation, is effective for studying the synthesis of laser materials, for example, non-linear optical crystals based on barium borate (BaB_2O_4). Synthesis of low-intensity β-phase BaB_2O_4 is carried out from a multicomponent melt of the Na_2O–BaO–Ba_2O_3 system. The luminescence of atomic sodium excited by the yellow (578.2 nm) emission line of the atomic lithium in vapour over superheated melts of the Na_2O–BaO–Ba_2O_3 system and the Na_2WO_4 melt was detected. It is shown that the spectrum of the sodium doublet can be recorded upon excitation by the 578.2 nm line of CVL in vapours of any superheated melts containing sodium.

In all experimental studies in the pulsed CVLs the industrial sealed AEs of the Kulon series produced by the Istok company were used[16, 42].

Tomsk State University (TSU) and Institute of Atmospheric Optics of the Siberian Branch of the Russian Academy of Sciences

In parallel with the P.N. Lebedev Institute of Physics, The Tomsk State University (TSU) and the Institute of Atmospheric Optics (IAP, Tomsk) made a great contribution to the development of self-heating pulsed MetVLs at *r–m* junctions. The works of these collectives, under the scientific supervision of A.N. Soldatov and G.S. Evtushenko, are presented in [13, 14, 19, 36, 38, 122]. In the years 1975–1980 self-heating CVLs of the Milan series, and then Malachite type CuBr lasers with sealed AEs and on their basis medical, lidar and navigation systems, systems for show entertainment [13, 14, 19, 36, 38, 56, 122–125] were constructed. The design of the AE of the CVL and GVL is a quartz tube as an external vacuum-tight shell, inside of which a ceramic discharge channel made of beryllium oxide (BeO) or alumina (Al_2O_3) is coaxially installed. The space between the channel and the shell is filled with a thermal insulator of ZrO_2. In a CuBr laser, discharge tubes are also made of quartz. The average radiation power of the developed CVL is 0.5–15 W, GVL – up to 4 W with PRF 5–15 kHz, power consumption 0.7–3 kW (efficiency ~0.5%), readiness time 20–50 min, the service life is 300–500 hs. In works [19, 36 38, 122, 126, 127] the achievements and records in MetVL are presented and they are the following. Addition of molecular hydrogen (H_2) to CVL leads to a significant increase in the lasing efficiency – up to 3%. In the controlled ionization mode of the active medium, the CVL has an efficiency of 9%. In the 'pure' CVL, the PRF is 235 kHz, in the GVL it is 150 kHz. A record specific average radiation power was achieved in the PRF range of 30–60 kHz for PPI: 1–2 W/cm^3 for CVL and 0.2–0.3 W/cm^3 for GVL. The maximum PRFs were obtained in a CuBr laser with hydrogen and independent preheated CuBr vapour generators that amounted to 270–300 kHz and 400 kHz. It is predicted to receive PRF more than 500 kHz [19, 127, 128]. In lasers with hydrogen, high efficiency values are achieved, and the duration of the radiation pulses increases. The increase in the pulse duration, in the case of an unstable resonator of the telescopic type, in turn leads to an increase in the power in the beam of diffraction quality. The AE designs with spatially separated active media are proposed and implemented, allowing lasing simultaneously on several metals. Multicolour radiation is obtained in three to seven *r–m* transitions for various combinations of atoms: Cu + Au, Cu + Ba + Pb, etc. In the process of accomplishing the tasks of atmospheric optics, specific requirements for MetVL were developed and lasers for atmospheric probing were developed. The small-sized CVL and GVL, were used as a basis in the development of visual navigation systems for wiring and landing aircraft in conditions of reduced visibility such as Liman and Raduga [38].

Despite a number of advantages of the CuBr laser in comparison with the pure CVLs, nevertheless, today they are inferior to CVLs both in reliability and in the stability of output radiation parameters, which limits their application in practical medicine and in the technology of microprocessing of materials.

Establishment of the Russian Academy of Sciences 'Joint Institute for High Temperatures of the Russian Academy of Sciences'

In 2009 a monograph 'Lasers on self-terminating transitions of metal atoms' was published in two volumes, edited by V.M. Batenin, where a group of scientists of the Joint Institute for High Temperatures of the Russian Academy of Sciences provide extensive data on the entire class of pulsed MetVLs, of which a large proportion falls on the CVL [18, 19].[1]

In the first volume [18], methods for creating active MetVL media and their various excitation conditions are considered, and much attention is paid to the description of AE structures operating at high temperatures of the discharge channel (500–1500°C). The results of both theoretical and experimental research and development are presented in detail. The basic physical processes responsible for creating population inversion in lasers on the transitions of metal atoms are discussed. The book occupies a special place in the study of plasma parameters and discharge characteristics, analyzing their interrelation with the energy, time, and other characteristics of laser radiation [129–136].

The second volume [19] deals not only with pure metal vapour lasers, but with their chemical compounds (halides), as well as the possibility of achieving the limiting parameters of laser radiation [17, 18, 134, 135]. The results of experimental studies of radiation characteristics at high PRFs are presented. In the CVL with the diameter and length of the discharge channel of 4.5 and 300 mm, respectively, 70 kHz PRF the specific average power output of 1.3 W/cm^3 (efficiency ~1%) was obtained. The physics of the excitation of metal vapours by electron beams generated by a free discharge is discussed.

The studies [19, 137, 138] describe a method for increasing the peak power in a radiation pulse using multipass amplifiers. Increase in peak power is achieved due to a decrease in the duration of the pulses of radiation of the master oscillator. The master oscillator and power amplifier use sealed AEs on copper vapour GL-201 from the Kristall series produced by the Istok Company. Efforts are also being made to efficiently convert the visible radiation of CVL to ultraviolet using non-linear crystals such as DKDP and BBO, and high results have been achieved [19, 138–145]. The CVLs, supplemented by a non-linear frequency converter, have an extended emission range λ = 510, 578, 255, 271 and 289 nm. With a BBO crystal, an average

[1]The two volumes are available in English from CRC Press.........data to be added

radiation power of 3.6 W and an optical efficiency of 24% were obtained with the generation of the total frequency with a wavelength λ = 271 nm, with a DKDP of 2.1 W and 14%, respectively. When generating second harmonics using BBO, the following best results were achieved: 3.4 W and 44% at λ = 289 nm and 2.1 W and 27% at λ = 255 nm.

The results of the cycle of experimental studies of non-linear frequency transformations of the laser radiation with an unstable resonator were used in the Velit Company when developing a prototype of the Kulon-10Cu-UV industrial laser, generating pulsed radiation in the visible and ultraviolet ranges [143–145]. The use of lasers and amplifiers on copper vapour in combination with non-linear crystals is of great practical interest, since their capabilities in the field of spectroscopic research, microprocessing and nanotechnology are greatly expanded.

The Kurchatov Institute Russian Scientific Centre, Medical Sterilization Systems Organisation, Institute of Semiconductor Physics

At the Kurchatov Institute (Moscow), Medical Sterilization Systems Organisation (Khimki, Moscow Region) and the Institute of Semiconductor Physics (IFP, Novosibirsk), powerful technological systems have been developed for laser isotope separation. The systems are used mostly for the needs of nuclear energy and medicine. This class of complexes employs highly efficient copper vapour laser systems (CVLS) operating according to the master oscillator–power amplifier scheme, which are designed for optical pumping of wavelength tunable dye solution lasers (DSL). The DSL also operate according to the oscillator–power amplifier scheme, and the composition of its organic dye is determined by the necessary wavelength tuning range. The electrical circuit of the high-voltage modulators of nanosecond pulses of the pump source of the power source (PS) of the CVLS is made and optimized according to the effective scheme of capacitive voltage doubling and with two links of magnetic compression in which high-power pulsed hydrogen thyratrons of the type TG1-5000/50 are used as commutators [146]. Emitters in the CVLS are industrial sealed self-heating AEs of the series Kristall GL-205 A, GL-205B and GL-205C designation according to the Technical Instruction (TI)) with an average radiation power of 30, 40 and 50 W, respectively, produced by the Istok Company (Fryazino, Moscow Region) [16, 41, 42]. The master oscillator in the CVLS is usually represented by the AE models GL-205A or GL-205B with an unstable resonator of the telescopic type or with one convex mirror, the power amplifier in the form of GL-205B and GL-205C. For example, in the system of the Kurchatov Institute the master oscillator is AE GL-205B with one convex mirror, and the power amplifier has two AG-205C AEs, the average power level is 120 W, and the optical conversion coefficient of the DSL of the yellow–green pump emission spectrum (λ = 510 and 578 nm) in the near-IR radiation reaches up to 50%.

The results of their studies of the last period have been presented by these teams mainly at All-Russian (international) scientific conferences 'Physical and chemical processes in the selection of atoms and molecules' [22, 23, 32, 150–152].

Design Bureau of the P.N. Lebedev Institute Physics of the Russian Academy of Sciences. The Design Bureau (DB) of the P.N. Lebedev Institute of Physics, the Russian Academy of Sciences (Troitsk, Moscow Region), with the technical support of the Istok Company under the scientific supervision of I.V. Ponomarev, designed and constructed a modern, highly efficient and reliable medical device Yakhroma-Med based on a pulsed CVL with air cooling [48–51]. The commutator in the high-voltage power source of the CVL is a pulsed hydrogen thyratron TGI-500/20 with a guaranteed lifetime of more than 2000 hours, the gas discharge tube – an industrial sealed-off self-heating AE of the Kulon GL-206C series (designation according to TI) with an average radiation power of 3–5 W and a minimum operating time of at least 1500 hours produced by the Istok Company [16, 41, 42]. The medical facility Yakhroma-Med is the leader in non-ablative technologies and is optimal for removing vascular, pigmented and unpainted skin defects, treating acne, smoothing out wrinkles, i.e. in dermatology and cosmetology. Pulsed radiation of the CVL with wavelengths λ = 510 and 582 nm, PRF of 14–16 kHz and pulse duration of 20–25 ns selectively coagulates skin defects without damaging the surrounding tissue and without causing pain (without anesthesia). The device is used in more than 100 clinics in Russia and abroad.

Research Institute of Precision Engineering (PE) and Mekhatron Scientific and Production Organisation

In the 1980s, the PE studied extensively the use of pulsed CVL radiation in the field of projective microscopy, thin film processing, and obtaining of microcircuits for a specified program, repair and fine-tuning the passive elements of microcircuits and local annealing of implanted plate surfaces for integrated circuits. On the basis of these studies, technological equipment of the Luch-30 type was developed using the first generation of industrial sealed-off AEs of the pulsed CVL produced by the Istok Company (AE Kriogen UL-101, Kvant UL-102 and Kristall GL-201) [147–149]. But since the 1990s, for objective reasons, this direction was not further developed.

At that time, The Mekhatron Scientific and Production Organisation was established on the basis of the TM Research Institute, and today this enterprise has many years of experience in creating metal vapour lasers: CVL and GVL, dye lasers with CVL pumping. One of the main activities of Mekhatron is the development and production of laser medical systems of the Metalaz-M series for photodynamic therapy. The systems have successfully passed technical and clinical trials, which resulted in the permission of the Russian Ministry of Health for their serial production and clinical use in oncology. The second area of activity is the advertising and information systems Metalaz-R. The settings of this series are intended for laser show and advertising. Due to the combination of their energy characteristics, they allow to generate dynamic (graphic and text) information of high brightness and sizes on various scattering surfaces (http: //18236/ru.all.biz).

St. Petersburg State Polytechnic University

A group of experts at the Polytechnic University of St. Petersburg, under the scientific and technical guidance of Yu.M. Mokrushin, developed CVLs with an average radiation power of 30 W and higher at 10 kHz PRF. The commutators in the high-voltage power source of the laser are pulsed hydrogen thyratrons of the type TGI1-2000/35, TGI1-5000/50 and TGI1–1000/25 [146]; in the radiator with an unstable telescopic resonator there are industrial sealed-off AEs of the Kristall GL-205A series with a power of 30 W and GL-205C with a power of 40 watts produced by the Istok Company [16, 41, 42].

For many applications, in particular, the tasks of transmitting and converting optical information, when creating full-colour images it is necessary to have three primary colours. In the red region of the spectrum a gold vapour laser can be used successfully. As for the blue region of the spectrum, on the basis of the CVL, this group used an CVL to construct a powerful tunable laser system that can fill the existing gap, and it seems very promising [31, 104, 153,

154, 156]. Two-step transformation of yellow–green radiation of a pulse-periodic CVL using an α-Al_2O_3:Ti^{+3} laser and a non-linear BBO crystal in the blue region of the optical spectrum was experimentally realized. With an average mean radiation power of CVL of 6.9 W, an average generation power of 0.4 W at a wavelength λ = 450 nm was attained at a PRF of 15625 Hz and a pulse duration of 50 ns [31, 154]. This result allows one to expect to obtain large power levels in the blue region of the spectrum when using more powerful CVLs to pump radiation. In the framework of this work, preliminary studies were carried out to create a colour television projection system for large screens [104]. The proposed laser system makes it possible to obtain a higher spatial resolution than in the case of a solid-state Nd:YAG laser.

Employees also implemented a source of ultraviolet radiation based on CVL with acousto-optical control of the spectral and temporal characteristics [155]. An analysis of the results of experimental and theoretical studies shows that in order to ensure the conversion efficiency of the visible radiation of CVL into the UV region of the spectrum (efficiency 30–35%), it is necessary:

— use in an emitter of an unstable resonator of a telescopic type with an increase of 50–100 times;

— application of AE on copper vapour with a uniform distribution of the gain over the cross section of the active medium;

— use as a non-linear medium of BBO and c-LBO crystals;

— the use of anamorphic optical systems that make it possible to form a radiation beam with a divergence less than the crystal synchronization angle, which makes it possible to provide a higher radiation power density in this crystal and a high efficiency of non-linear frequency conversion.

The fulfillment of these conditions made it possible to obtain in the laser system UV radiation with a power of 1.8 W at a wavelength λ = 255.3 nm, operating according to the scheme with one AE GL-205 A and 4.7 W according to the master oscillator–power amplifier scheme. In both cases, the efficiency of non-linear frequency conversion was about 35%.

One of the applications of the created UV system with control of the radiation of CVL is the treatment of various materials. Another promising direction for the yellow-green emission spectrum of CVL, which was realized by the scientific and technical staff of the university, is the formation and controlled transfer of images into the external space, for example, to clouds and large objects.

VELIT Research and Production Enterprise (Istra, Moscow Region) is one of the leaders in the field of high-voltage engineering, in particular, in the field of development and serial production of a wide range of power supplies for X-ray systems. Over the past 10 years, the highly reliable high-voltage pulse power sources (PS) with high technical expertise have been created and investigated by the group of high-class specialists composed of V.G. Filippov, Yu.S. Priseko and N.M. Lepekhin in the VELIT Enterprise – generators of nanosecond pulses for pumping laser AEs on metal vapours. High-voltage pulsed hydrogen thyratrons such as TGI1-1000/25, TGI2-1000/25k and TGI1-500/20 in the 10–16 kHz PRF mode produced by NPO Plasma (Ryazan') are used as switches in the power sources. The electrical circuit of the generator is made and optimized according to the effective scheme of capacitive voltage doubling and with two links of magnetic compression of current pulses [146,157, 160–163, 175, 177, 178].

In the VELIT Enterprise the generators of nanosecond pulses developed by the VELIT and industrial sealed-off self-heating AEs GL-205 V, D, G, I (KPPG.433 757.004–007 TU) produced by the Istok Company [16,41,42] were used to develope high-efficiency industrial MetVL of the Kulon series (YuVIE.433 713.001 IU) with a frequency range of 10–16 kHz: Kulon-01 on copper vapour with wavelengths λ = 510.6 and 578.2 nm and an average power of 10 W; Kulon-02 gold vapours with λ = 627.8 nm and a power of 1.5 W; Kulon-03 on a mixture of copper and gold vapour with λ = 510.6; 578.2 and 627.8 nm and a total power of 4.5 W; Kulon-04 on copper vapor with a power of 15 W; Kulon-05 on copper vapor with the radiation modulation channel and Kulon-06 on copper vapour, structurally separately made by the emitter and the PS unit [158, 176, 183]. The lasers Kulon 01, 02, 03, 04, 05 are structurally executed in the form of a monoblock. To date, several tens of these systems have been produced.

The Kulon-05 CVL has two modifications – Kulon-10Cu-M with sealed self-heating AE GL-205E with average power of 10 W and Kulon-15Cu-M with AE GL-205I with a power of 15 W [16, 157, 1,]. The main advantage of these CVLs is the possibility of high-speed packet and pulse modulation, which allows to control the power characteristics by any given algorithm from an IBM PC. Modulation is achieved due to time delays in the AE of an additional current pulse, generated by a special low-power generator, relative to the main pump current pulse [159, 163,170, 171, 174, 180, 181]. This advantage provides ample opportunities for the use of the CVL in technological and medical equipment and scientific research [165, 168, 166, 167, 169, 179]. The CVL Kulon-10Cu-M was used in in the Lasers and apparatus TM (Zelenograd, Moscow) in constructing two laser models ML1-2 for high-quality drilling of microholes. In 2009–2011 the Istok Company used the Kulon-15Cu-M CVL to construct automated laser technological installations (ALTU) of the type Karavella-2 and Karavella-2M for efficient precision microprocessing of thin sheet metal and non-metallic materials of ET products.

A multifunctional laser medical device Kulon-Med was developed and certified. The Kulon-Med unit combines Kulon-10Cu-M CVL with wavelengths of 510.6 and 578.2 nm and DSL tunable over wavelengths in the range of 620–750 nm. It is intended for the treatment of oncological and non-oncological diseases by the method of photodynamic therapy, low-intensive therapy, in surgery, dermatology, cosmetology, etc. [166, 167, 172, 173, 182].

The Kulon-10Cu-M CVL and non-linear crystals such as DKDP and BBO were used to create a device with the generation of second harmonics and total frequency in the ultraviolet region of the spectrum – Kulon-10Cu-UV with wavelengths λ = 255, 271 and 289 nm and an average radiation power of 0.3–0.4 W, which expands the practical capabilities of CVL [141–144].

At present, an air-cooled compact low-power CVL with 14–17 kHz PRF and an average power of 1–1.5 W is being developed, which is the basis of a spectral molecular analyzer for operational diagnostics of the composition of liquid, solid, and gaseous substances. The principal electrical circuit and design of a high-voltage current of nanosecond current pulses for pumping a two-channel CVLS on the basis of sealed-off AE GL-205A and 205B (Kristall series) operating according to the master oscillator–power amplifier scheme, with a channel synchronization accuracy not worse than ±2 ns, is being worked out. The average radiation power of the CVLS should be 40–60 W with a PRF of 9–11 kHz. The purpose of the laser system of this power level is to complete the Karavella process units for precision microprocessing of metallic materials 1–2 mm thick.

Chistye tekhnologii (Clean Technologies) Company (Izhevsk) developed since 2001 modern two-channel high-voltage impulse transistor power source, intended for pumping CVLS, working under the master oscillator (MO) – power amplifier (PA) scheme. The MO and PA, AE use the series Kulon AEs: GL-206E and GL-206I with an average radiation power of 10 and 15 W, respectively, produced by the Istok Company [16, 33, 34, 41, 42, 190]. The two-channel power source consists of a control rack and two high-voltage magnetic modulators with bias current sources. The modulators provide the generation of current pulses with an amplitude of 220–250 A and a duration of no more than 150 ns with a pulse voltage of $\cong$15 kV and a PRF of 12–14 kHz and synchronization of the MO and PA channels with an accuracy of ±2 ns. The maximum output power of the laser beam in the laser is maintained in the automatic mode. On the basis of CVLS with two-channel transistor of the PS the Istok Company developed two models used by ALTU: Karavella-1 and Karavella-1M for precision microprocessing of metallic and non-metallic materials up to 0.5 and 1 mm in thickness for electronic equipment.

Lazery i apparatura TM (Laser and apparatus TM) (Zelenograd, Moscow) has more than 20 years of experience in the market and is today the acknowledged leader among Russian manufacturers of laser processing equipment. Under the scientific and technical guidance of L.G. Saprykin, the centre develops modern technological systems for precision machining, marking and engraving, cutting and cutting, welding, fitting resistors, etc., that successfully work at electronic, nuclear, aviation, space, instrument-making, defense and other industries. Recently, complexes developed on the newest solid-state and fiber lasers with diode pumping and the output power of hundreds and thousands of watts and kinematic systems on linear motors have been mastered in mass production for high-performance and high-quality precision processing of various materials with a thickness of 1–10 mm.

To date, a laser model ML1-2 for precision microprocessing of foil and thin materials has been developed at the NPC on the basis of pulsed CVL with nanosecond duration 'Kulon-10Cu-M' with radiation modulation. It is recommended, if necessary, to form a treatment spot of 5–15 microns in size and achieve a minimum thermal exposure zone.

Research in the Bauman Moscow State Technical University. The MSTU practically from the moment of the appearance of the first domestic lasers, carried out a large volume of theoretical and experimental studies of physical processes on the interaction of laser radiation with various materials and determination of areas of technological applications for various types of lasers [25, 184–190].

The Department of Laser Equipment under the scientific guidance of A.G. Grigor'yants, used gas-discharge CO_2 (λ = 10.6 μm), solid-state Nd:YAG (λ = 1.064 μm) and ytterbium fiber (Yb, λ = 1.06–1.07 μm) lasers and CVLs (λ = 0.51 and 0.58 μm) to study the following technological processes: surface hardening of alloys, surface coatings, laser welding of metals, separation of structural materials, dimensional processing, technologies in microelectronics, soldering of metals with ceramics, rapid prototyping techniques etc. [25].

The main field of application of pulsed CVLs with radiation in the visible range is precision microprocessing of materials for electronics, accurate instrumentation [16]. A special place is occupied by the technology of applying images in the volume of transparent media with the help of pulsed laser radiation of the CVL. This technology is unique, because during processing laser radiation is focused into the volume of the material (in glass or quartz), causing internal microfractures [25, 191–193]. Moving the focused radiation spot along the vertical Z axis, and a transparent object in the perpendicular plane along the two XY axes, one can obtain any three-dimensional image without disturbing its surface. The divergence of the laser radiation is several times smaller than that of solid-state Nd:YAG lasers, which determines its sharper focusing. The latter determines the smaller size of the microfractures and, correspondingly, the higher resolution of the 3D image.

At present, much attention is paid to the practical development of the technology of welding and cutting of various metallic materials using fiber lasers with a high level of average radiation power (up to 30 kW).

It should be noted that the successful development of lasers using copper and other metals and their compounds in Russia (and the former USSR) was facilitated by the work of other groups in various cities of Russia and the CIS. Here we confined ourselves to a brief description of the directions of the work of those collectives that not only conducted physical research, but also paid much attention to the technical and practical side of the question, thereby in many cases predetermining the current basic level of laser technology in this field.

1.3. The condition and development of CVL and CVLS in foreign countries

In the USA, the first self-heating pulsed CVL with wavelengths λ = 510.6 and 578.2 nm was a laser that was designed by T.S. Fahlen (1974), on a hollow copper cathode with an average radiation power of 270 mW, 12 kHz PRF, and an efficiency of 0.025 % [194]. In 1975, this CVL was first used for pumping a dye laser [195]. The usual self-heating pulsed CVL was reported in 1975 in the works of T.W. Karras and co-workers [196, 197]. One of the created CVLs with a diameter and length of the discharge channel of the AE of 8.5 mm and 350 mm, generated radiation with an average power of 1.3 W with a PRF of 6.8 kHz.

In 1973, in the Lawrence Livermore National Laboratory (LLNL) began the program AVLIS on laser isotope separation (LIS) of uranium by the method of selective photoionization in atomic vapour. The purpose of the AVLIS program is to provide an economically advantageous process of large-scale uranium enrichment with the ^{235}U isotope for the purpose of using it as fuel for nuclear reactors. To perform the three-step selective photoionization in the isolation of the ^{235}U isotope, a wavelength-tunable dye laser with pumping by CVL radiation was chosen because in combination these lasers have potentially high total power and efficiency [20]. Historically, the AVLIS program has served as a stimulus for the development of high-power CVLSs working on an effective optical master oscillator–power amplifier scheme. Already in 1976 in the LLNL on a Venus installation for the dye laser pumping used a laser system of eight self-heating CVLSs with an average output power of 5 W (diameter of a discharge tube of 25 mm), in 1979 from 21 CVLSs with a total output power of 260 W, in 1980 – from 32 CVLSs with a power up to 400 W. And in 1984, a separate laser AE with a discharge tube diameter of 80 mm, used as a power amplifier in an CVLS, generated a power of 200 W with a 5 kHz PRF. Due to the use of several components of magnetic compression of current pulses in the modulator of the power supply, the power of the radiation from individual amplifiers was raised from 200 to 300 W [198]. At the same time there appeared dye lasers tunable along the length with a total output power of 50–100 W and a line width of ±30 MHz. In 1985, an CVLS from 6 circuits with a radiation power of 2 kW was constructed, in 1991 – from 12 circuits with a power of 8–9 kW. Each circuit consisted of one master oscillator and three power amplifiers; the power take-off from one power amplifier reached 220–250 W. The use of a new generation of the power amplifier in the CVL made it possible to obtain a power of 1.5 kW [199]. In this case, the removal of the radiation power from a separate amplifier was more than 750 W at an efficiency of ~1%. The system consisting of dye lasers, operating according to the master oscillator–power amplifier scheme, generated radiation power above 2.5 kW (1992).

In 1992–1993, new generation equipment was created and demonstration complex tests were carried out, including powerful laser systems and separators of industrial scale, allowing to ensure the required productivity of the uranium enrichment process and the necessary production volumes. Production capacity can provide hundreds of kilograms of the required isotope. The CVLS developed here with the total average radiation power was brought to 72 kW, and a complex of dye laser tunable along the wavelengths, up to 24 kW [23, 200]. Progress in the development of selective laser technology allowed the United States to begin in the early 1990s the work on obtaining isotopically enriched plutonium for military purposes. In 1998, the US Isotope Enrichment Corporation submitted a plan for the creation of an AVLIS installation of a commercial level (early 2001) with full production capacity in 2005. The average radiation power of the CVL in the AVLIS system of the optical plant will be hundreds, and the dye lasers – tens of kilowatts [201, 202]. According to some information, it became known that the project of laser enrichment of uranium in the LLNL was first frozen, and then rolled up. And this is explained by the fact that in connection with disarmament processes in the world, enough uranium has been accumulated. However, it is already known today that in 2000, research on the laser separation of uranium was continued by the US uranium company USEC and Australian Silex Systems Ltd. To date, all major problems in uranium enrichment technology have been overcome and information has appeared that Global Laser Enrichment was planning to launch a pilot plant for laser enrichment of uranium in 2009. In case of conducting and successful experiments on obtaining commercially competitive technology of enrichment of fuel uranium, in relation to traditional technologies, it is planned to create an industrial plant with commissioning in 2012 with a design capacity of 3.5–6 million units of separation work units (EPP) [23].

The main promising areas for the use of isotopes (except uranium), obtained by the selective AVLIS technology, are nuclear power engineering, medicine, biology, and scientific and practical research.

The qualitative development of laser technology over the past decade allows us to put on a new generation of productive industrial isotope separation plants.

In connection with the large reserve of already enriched uranium, pulsed CVLs began to be applied in other fields of science and technology. These include the current laser micromachining of materials, medicine, spectral studies on the analysis of the composition of substances, increasing the brightness of the image, speed photography, etc. Therefore, some US firms have begun to develop low-power commercial air-cooled CVLs. For example, the company Laser Now advertises CVL models CVL-5W and CVL-10W. The average radiation power of the model CVL-5W with a diameter of the discharge channel of 14 mm is about 5 W, CVL-10W with a channel diameter of 20 mm has a power of 10 W with a PRF of 20 kHz. The minimum lifetime of soldered AE CVL-5W is 800 h, CVL-10W is 500 h. A group of experienced scientists from the USA and Korea from BISON MEDIKAL has developed a new generation medical device Cooper Bromide for innovative technologies in dermatology and cosmetology. The basis of the medical device is a pulsed low-power laser with sealed AE on copper bromide. The firm was established in 2002 in the USA, California. The device Cooper Bromide effectively treats vascular and pigmentary pathologies and other skin diseases. Vascular pathologies include teleantiectasias, 'wine spots', hemangiomas, senile angioma, pigmentosa – freckles, lentigo, melasma and other diseases – warts, acne, scars, wrinkles and senile skin defects. Another well-known company in the US for the development and production of commercial LMOs for applications in science, technology and medicine is Metalaser Technologies Inc.

The development of CVL in France, Japan, Britain, China, Israel, India and other countries. Programs similar to the US AVLIS programs, with the use of CVL and dye lasers in multicomponent master oscillator–power amplifier systems (MO–PA system), have also been developed in other countries. These include, first of all, France, Japan, Britain, China, Israel, India, Brazil, Korea.

In France, the development of the AVLIS program began in 1985 as part of the development of the next-generation isotope enrichment technology. CVL is developed by the Ministry of Atomic Energy and CILAS company [20, 203]. In 1989–1991 years the power amplifiers with an average output power of 100 W have already been used, and in 1993 CVLSs with three PAs with a total output power of 330 W were demonstrated. The MO had a power of 30 W with a PRF of 4–6 kHz. In 1996, along with a thyratron power source, solid-state sources with transistor switches, in particular, for CVL with a power of 400 W, began to be developed. And in the same year the total output capacity of the CVL in the system with six circuits was 2 kW. In 1997, the ASTER installation was demonstrated, which obviously became the basis for future industrial developments for AVLIS technology [204]. For today, there is information that in the French laser systems for separating isotopes, high-power Nd:YAG lasers with frequency doubling (λ = 532 nm) and a high service life have also been used as pumping sources for the dye lasers.

In Japan, the AVLIS technology, namely, uranium enrichment, has also attracted considerable interest, as nuclear power plants form the backbone of national energy. The implementation of the Japanese AVLIS program, called LASER-J, began in 1987 with the conclusion of contracts with state companies for the development of powerful CVLs [20, 205]. Hitachi, Toshiba, Mitsubishi, Kansai, etc., were involved in development of CVL. In 1990, for individual CVLs, the output average power of the radiation was 100 W and was demonstrated by the CVLS operating according to the MO–PA scheme, power 259 W. In 1992, the Mitsubishi company launched the CVL with a power of 210 W when operating in the generator mode and 260 watts in the PA mode [20, 206]. By 1993, the CVL with a diameter of the discharge channel of the AE of 80 mm generated 430 W [207]. The task of increasing the radiation power for the LASER-J program was solved in 1995 by Toshiba: the output of one CVL reached 550 W in the generator mode and 615 W in the power amplifier mode [208]. The diameter and length of the discharge tube of the AE were 80 mm and 3500 mm, respectively. According to the LASER-J program, it is planned to demonstrate an industrial uranium enrichment plant 235 U. Therefore, in 1990–1995, research and development of a new technology for separation of isotopes, different from traditional classical methods (centrifugation and gas diffusion separation), received rapid development, and investments amounted to hundreds of millions of dollars. Ideally, AVLIS-technology of uranium enrichment should become a technology of the 21st century, i.e., it is compact, automated and maximally safe [23].

Specialists from Toshiba [209, 210] and [211] published in 1994, independently of one another the data on the use of baffles in large-diameter (60–80 mm) AEs of the CVL discharge tubes, allowing to reduce the radial gradient of temperature and concentration atoms of copper and, accordingly, to increase the specific absorption of radiation power from the active medium. In LLNL a CVL with a diameter of a discharge tube of 80 mm AE in UM mode without partitions generated radiation with an output average power of 255 W with an efficiency of 1.1%, with partitions – 325 W with an efficiency of 1.34% [211].

In 1989, Oxford Lasers (Great Britain) created a small installation based on the CVL and LRK to evaluate the AVLIS program, but in 1997 it was decommissioned [20]. In the same year, the company produced the first commercial CVL with an average output power of 100 watts. It has developed and produces CVL models ACL-25, ACL-35, ACL-45, ACL-100. In 1997, Oxford Lasers created a 10–15 W CVL with air cooling and single-phase power supply from the mains and 100 watts and 120 watts also with air cooling, but three-phase mains power. Another English company (EEV) in 1994 began to put on the CVL market with sealed AEs. By the firm "CI Laser" in 1997, models of CVL with a power from 7–8 W to 25 W, as well as LPBa, LPPb and LPAu were offered. This company specializes in the production of MetVL for forensic science, high-speed photography, on-board systems.

Today, Oxford Lasers advertises the newest CVL models LS20-30, LS20-50 and LS35, designed for high-speed transmission of images and visualization of airborne dust and scientific research [70]. The main parameters of the CVL are the following: the radiation wavelength is 510.6 nm, the CPI is 3.5–20 kHz, the pulse duration is 25 ns, the average radiation power is 20–35 W (depending on the model), the jitter is no more than 2–5 ns and the reliable exploitation. As an option for the laser – fiber delivery of radiation. An important area of application of pulsed CVL is precision microprocessing of materials. To solve the tasks on microprocessing, a laser technological installation of the MP200X model based on the LSPM, operating according to the MO–PA scheme [212], was created. The device provides the possibility of converting the visible yellow-green radiation of LSMW by means of non-linear crystals to ultraviolet with wavelengths of 255 and 271 nm, which expands the field of its practical application. The main parameters of the LMPM of the MP200X device are: radiation wavelengths of 510.6 and 578.2 nm, PRF of 10–20 kHz, a pulse duration of 20 ns, beam divergence of 0.075 mrad, an average power of 60 W with a 10 kHz PIC and a power consumption of about 12 kW. The working field of the horizontal coordinate table XY, where the object to be processed is installed, has a stroke size of 200 × 200 mm and positioning accuracy along the 0.1 μm axes, the length of the vertical table Z travel with a focusing lens of 100 mm and an accuracy of 0.5 μm. A technological plant with such characteristics allows producing efficient processing of metallic materials with a thickness of 0.01–2 mm with a cutting width of more than 5 μm and flashing holes with a diameter of 1–500 μm. The main and significant disadvantage of the MP200X installation is that the AEs in CVLS operate in the neon buffer gas pumping mode, that is, the technology of their manufacture does not allow operating in the sealed mode. It is known that the operating mode of operation reduces the service life of the laser and worsens the stability of the output radiation parameters. In a number of experimental processing operations, Oxford Lasers employees demonstrated unique possibilities of using pulsed nanosecond radiation from CVL for precision microprocessing of various materials [71, 72].

SILVA is the European program on AVLIS technology. The most important components for laser isotope separation equipment were developed in France (Jilas Alcatel) and in England by Oxford Lasers [23]. Great interest in the practical development of laser isotope separation is shown by China, Israel, India, Brazil, Korea and other countries. A partial review of these studies is presented in [23, ref.. 1]. In China, for this purpose, the first powerful CVLV (40 W) was created at the Institute of Electronics of the Chinese Academy of Sciences (Beijing, 1988) [213]. In Israel, works on AVLIS using CVL are being conducted at the Nuclear Research Center in Negev. In 1979, it was reported on the CVL with a discharge tube with a diameter of 40 mm, operating at 4 kHz FPI [214]. In the generator mode, the output power of the CVL radiation was 20 W, in the UM mode 30 W. This CVL was used to process materials: its focused output beam produced holes in a steel sheet 1 mm thick. Then CVL was developed with a power (30 ± 5) W and (100 ± 10) W with a diameter of the discharge tube if the AE (30 ± 2) mm and (80 ± 3) mm, respectively, at PRF (5.5 ± 0.2) kHz. For the fingerprint identification system a soldered LMP with a power of 10 W was constructed, generating a radiation beam with a diameter of 20 mm. Every 1000 hours of operation the AE is replaced with a new one. This CVL employs a solid-state switch with a service life of more than 10 000 h. The AVLIS program with CVL is also being carried out in India – at the Center for Advanced Technology (Indore) and the Bombay Atomic Research Center.

Powerful CVLs used in AVLIS programs usually work in the continuous mode of pumping neon buffer gas through the AE (at a rate of 2–6 liters/hour) and after 300–600 hours of work, a new portion of the active substance (copper) needs to be laid. Bleeding is necessary to remove impurity gases that are continuously generated from materials of AE elements due to high operating temperature (~1500°C).

Based on the information materials available today, more than 20 countries are in the process of researching and obtaining weights of different isotopes by laser isotope separation in Argentina, Australia, Brazil, Britain, China, France, Germany, India, Iraq, Israel, Japan, The Netherlands, Pakistan, Romania, Russia, South Korea, South Africa, Spain, Sweden, Switzerland, USA, Yugoslavia. The following isotopes are of the greatest interest: ^{235}U, ^{130}Ba, ^{176}Yb, ^{43}Ca, ^{48}Ca, ^{70}Zn, ^{150}Nd, ^{25}Mg, ^{203}Tl, ^{87}Rb, ^{155}Cd, ^{157}Cd, etc.

CVL of the German firm Atzevus. Atzevus has advertised copper vapor lasers CVL 175 plus, CVL 275 plus and CVL 375 plus with an average radiation power of 90, 170 and 250 watts respectively, with a CPR of 6.8 kHz (0.7% efficiency) and 70, 125 and 180 W At the CPI of 12.6 kHz (0.5% efficiency). These models work according to the master oscillator–power amplifier (MO–PA) scheme. The diameter of the discharge channel of the AE MO – 25 mm, AE PA – 47 mm. In CVL 175 plus one PA is used, in CVL 275 plus – two PAs, in CVL 375 plus – three PAs. These CVLs are intended for the acquisition of process equipment for high-performance precision processing with micron accuracy of sheet metal materials up to 3–4 mm thick. The most powerful CVL 375 plus is used for cutting and drilling of titanium parts. Today we do not know any new data on the development of this company. The well-known German companies Trumpf and Rofin Sinar Laser develop and produce mainly technological equipment based on their own production lasers and production of other enterprises for productive and high-quality cutting, welding, surfacing, and microprocessing [360]. The applied technological lasers include both continuous and pulsed CO_2 lasers, solid-state lasers on Nd:YAG, in particular discs on Yb:YAG and fiber ytterbium (Yb) lasers.

Development of CVL and GVL in Australia. Most of the theoretical and experimental work on creating and increasing the efficiency of pulsed MetVLs, primarily CVLs, was conducted under the guidance of Professor J. Piper at the Research Center Lasers and Applications at Macquarie University (Sydney) in the period

1988–1996. [20]. J. Piper published in co-authorship more than 260 articles and received 12 patents.

Quentron Optik is the first Australian company that began to develop CVLs and GVLs in 1980. The basic design of the AE, a design developed at Macquarie University was used. Since the firm owned a patent for the use of lasers with radiation corresponding to the red spectrum, the first laser it sold for the treatment of cancer by photodynamic therapy was an GVL with an average radiation power of 5 W (1984) [20, 215]. In 1987 there were already fully automated CVLs with a capacity of up to 40 W. Quentron Optik ceased trading in lasers in 1988, and the company Metalaser (Sydney) was established, which was organized by Australian academic circles and entrepreneurs. In 1989, a new company called Visiray Pty was formed, later called Dynamic Light. This company offered in 1997 a number of CVLs with a power of 8 to 40 W, intended for sale mainly in the medical market (dermatology and photodynamic therapy of oncological diseases). In addition to CVL, BaVL (10 W) and GVL (2.4 and 4 W) were offered. Metalaser, in 1988, after the purchase of Plasma Kinetics, was transformed into Metalaser Technologies [216] and entered the American market with laser equipment for dermatology. Metalaser Technologies continued to trade up to 1993. Before closing (due to intense competition in the market of dermatological equipment), the company sold around 150 systems in the world, 120 of them in USA. In recent years, a group from the University of Macquarie has been studying the work of CVL with additives of hydrogen and hydrogen compounds – so-called CVL with enhanced kinetics [217, 218, 359]. The obtained characteristics for CVL with hydrogen compounds (HCl, HBr) are much better than for CVL with pure hydrogen: the average radiation power was increased by 2.5 times, the duration of the radiation pulses was doubled, and the efficiency was increased by 1.5 times. CVLs with a diameter and length of the AE discharge channel of 25 mm and 1000 mm, respectively, without additives generated 17 W of radiation at a 7 kHz PRF with H_2 additive of 20 W with a 17 kHz PRF, with an addition of H_2 + HCl of 50 W with a PRF of 30 kHz. The CVLs with a diameter and length of the discharge channel of 40 and 1500 mm, without additives, generated 55 W with a 5 kHz PRF, with an addition of H_2 67 W with a 6 kHz PRF, and 101 W with a H_2 + HCl additive at 12 kHz PRF. The possibility of obtaining a higher power in CVL due to the addition of HCl is mainly the result of efficient operation at much higher PRF.

The analysis of information on the development of programs on AVLIS technology on the basis of the CVLS shows that Australia is currently also actively involved in research in the field of laser isotope separation.

CVL and GVL of Bulgarian companies Mashinoexport and Spectronica. The Bulgarian companies Mashinoexport and Spectronica in the past advertised the CVL of the models SCuL 0.5H, SCuL 10H, SCuL 10, SCuL 15, and SCuL 25 with an average output power of 5–25 W, GVL models SAuL 0.5 and SAuL 1.0 power 0.5–1.0 W, a mixed copper and gold vapour laser model SCuAuL 0.3 with a power of 3 W and CuBr – SCuBrL 0.5 and SCuBrL 10 lasers with a power of 5–10 W in the operating frequency range of 5–20 kHz. Lasers are available both in the sealed-off and in the pumped AE. In the AE, quartz tubes are used as a vacuum-tight shell, as a discharge channel – tubes made of single-crystal sapphire (Al_2O_3) with a diameter of 10–40 mm [219–220]. The guaranteed life of soldered AE is 500–1000 h, pumped – up to 2000 h. The operating life of 2000 h is achieved due to the possibility of periodic recharging of buffer gas and active substance (metal) through 300–500 h of operation. Currently, PulseLight offers lasers on copper bromide vapour with an output power of up to 20 W and an efficiency of ~1%.

1.4. The current state and development of the CVL and CVLS in the Istok company

The development of pulsed CVLs at the Istok company began in 1972 in one of the laboratories for gas lasers, where experimental studies were begun on the creation of self-heating metal vapour lasers. At present, the laboratory Lasers and laser technologies is engaged in the development of CVLs and on their basis of technological and medical equipment.

Period 1972–1979. Interest in the development of CVLs in the Istok NPP arose after the P.N. Lebedev Institute of Physics demonstrated the high efficiency of operation of this laser [9, 10, 75–79]. In 1972–1973 the Istok company carried out research and development work Kriostat and Caspian during which the first prototypes of sealed-off self-heating AEs of the CVLs were manufactured, and the issues of the life of AE and the reliability of pulsed high-voltage power supplies with a hydrogen thyratron of type TGI1-2000/35 as a high-voltage switch were studied. The relatively simple design of the self-heating AE of the CVL [222], proposed by the staff of the Istok company and the Institute of Physics of the Russian Academy of Sciences in 1974, was the basis for carrying out a series of research and development works (R & D) in the company. During the implementation of the Kriostat-1 project (1974–1975) the company developed the first in Russia (then in the USSR) and in the world industrial CVL Kriostat with sealed self-heating AE and a thyratron power source. The vacuum-tight shell of the AE is a metal–glass structure, its discharge channel is ceramic of aluminium oxide (Al_2O_3), diameter and length of the channel 12 and 900 mm, respectively. The average radiation power is 3–5 W, the range is 5–20 kHz, the power consumption from the rectifier is 2.2–2.5 kW (efficiency 0.15–0.25%), the guaranteed running time is 200 hours, the beam divergence is 3 mrad and the readiness time is 40–50 min (80% of the steady-state value of the radiation power). This pulsed laser, given the symbol LPMI–75, with the power supply IP–18, was demonstrated in 1975 at the International Exhibition in Munich (Germany). CVL Kriostat was used mainly for pumping LZhI–504 model (λ = 530–900 nm) tunable along the wavelengths of the dye laser.

In the research project Cryogen (1974–1975) and R&D project Cryogen-1 (1976–1977), the amplifying properties of the active medium of CVL were investigated and the first industrial wide-aperture optical quantum amplifier (OQA) of the image brightness was developed with a gain of at least 30 dB, the output average radiation power of 0.5–1 W and a minimum operating time of at least 300 h. The diameter and length of the discharge channel of the AE designed as an OQA are 20 and 400 mm, respectively. The vacuum-tight shell of the OQA is cermet (ceramics 22KhS). OQA on copper vapour was intended for use as an amplifying microprojector in a visual inspection facility for microelectronic products (for example, UVKL–1000). The OQA was used in technological equipment for the operation of evaporation (removal) of matter from microscopic sections of the surfaces of microelectronics products, to enhance the brightness of the image in microfiche readers and other optical systems [24, 78].

In order to increase the efficiency (capacity and efficiency) and reliability of the CVL, the research projects Kriolit (1977–1978) and Kriolit-1 (1978–1979) were carried out. Literature data were systematized and experimental studies of the thermophysical and vacuum properties and chemical composition of various powder and fibrous heat insulators were carried out. The coefficients of heat conductivity of heat insulators as a function of the pressure of the buffer gas, the temperature and density of the heat insulator, the composition of the gases released during the training of the AE were studied. It has been shown that for a combination of positive properties, the most suitable materials for a thermal insulator with an operating temperature of $T_{oper} \leq 1600°C$ in CVL are aluminium oxide and zirconium (Al_2O_3 and ZrO_2) and their combinations with other oxides, for a discharge channel with T_{oper} = 1500–1600°C – pure oxides of Al_2O_3 and BeO_2. General principles of constructing the AE, the operation of annular and tablet tungsten–barium (W–Ba) cold cathodes and with indirect heating up to 1100°C were also considered. The best energy characteristics were obtained with those AEs where the degassing and training were carried out more carefully. With a diameter and length of the discharge channel of the AE (cermet shell) of 20 mm and 900 mm, respectively, the average radiation power reached 18 W at a neon pressure of 20 mm Hg. But at low pressures the durability of the sealed AE was only 100–200 h. In the process of investigation, a strong dependence of the radiation power on the duration of the pump current pulses was observed. The power source was studied on modulating lamps GMI-29A and was made according to the scheme with a partial discharge of the storage capacitance. When using a lamp source, pulses of a current of 50–70 ns duration (PRF 10–25 kHz) were formed, which is 3–4 times less than with the use of the thyratron power source. For example, the average radiation power of the experimental CVL with a thyratron source was 4 W, with a lamp power – up to 14 W.

In the Kriolit research project, in order to significantly reduce the working temperature of the discharge channel of AE in CVL (about 1000°C), experimental studies were carried out using copper halides as the active substance. The best results were obtained with copper monochloride (CuCl). At neon pressures of 10–15 mm Hg the practical efficiency reached 1% (with an average radiation power of 16 W). A quartz tube with external thermal insulation was used as a vacuum-tight shell of the AE. Lightweight fireclay disks with a hole diameter of 20 mm were installed inside the shell to form a discharge channel. The working substance – copper halide – was laid between the washers on the inner surface of the quartz shell. The main deficiencies of this laser were a large consumption of the working substance and relatively high instabilities in the parameters of the output radiation beam.

In the Kriolit-1 research project the main types of AE designs and requirements for its components were considered for long-term operation under high-temperature conditions. Materials used in the AE should have low gas evolution and low heat capacity, heat insulator – low thermal conductivity, and it should be excluded from entering the discharge channel, molten copper should not overlap the aperture of the channel, the materials used must be chemically compatible, etc. Thus, the optimal design of the AE is reduced first of all to the choice of reasonable trade-offs between conflicting demands on its individual elements and nodes. For the first time, a material from fine-dispersed hollow microspheres of Al_2O_3 was used as the heat-insulating material with low thermal conductivity. The maximum operating time of the AE in the Kriolit-1 project at a neon pressure of 100 mm Hg (the diameter and length of the discharge channel, respectively, 20 mm and 900 mm) was about 400 h.

In [223, 224], a CVL was studied at high pressures of the neon buffer gas, which is important for increasing the life of the AE. In the Kriostat project, the lifetime of the sealed self-heating AE was about 3000 hours at a buffer gas (neon) pressure of 300 mm Hg, 10.7 kHz, and the power consumed from the rectifier of the power source was 2.3–2.5 kW [223]. After 2000 hours of operating time, the radiation power was reduced by half (from 4 to 2 W). At the same time, the practical efficiency was very low and amounted to 0.08–0.2%, that is, at the level of the efficiency of an argon laser.

The period 1980–1989. This period begins with the research work Kristall (1979–1980), in which, as a result of extensive research, three types of sealed self-heating AEs on copper vapors – Kulon, Kvant and Kristall with an average radiation power of 1 up to 15 watts. The minimum (guaranteed) operating time of the AE was increased by 2–3 times (up to 500–1000 h), the availability and power consumption decreased significantly. Research work Kristall was the basis for the R & D Kvant, Kristall-1 and Kulon, within the framework of which industrial sealed-off AEs of the new generation with a cermet shell were already developed. When developing AE and creating radiators, lasers, technological and medical devices based on them, the main attention was paid to increasing the efficiency, power, specific characteristics, the quality of radiation, improving the operational parameters and their reproducibility in the process of long operating time.

In the Kvant project (1981–1982), an AE was developed with a minimum operating time of at least 500 h, a readiness time of not more than 50 min and a gain of at least 30 dB for use as an image brightness amplifier in projection microscopes of the CVL-1000 type and technological installations such as Luch-30. The AE Kvant in accordance with the technical specifications (TU) has the symbol UL-102. The ratio of the length of the discharge channel (400 mm) to the diameter of the aperture of the AE (20 mm), which determines the field of view of the microscope, is 20: 1. The radiation power of the device in the generator mode is 5–7 W.

In the Kristall-1 project (1981–1982), the first domestic industrial sealed AE was developed with a relatively high average radiation power (10–15 W) with a PRF of 8–12 kHz, a readiness time of not more than 60 min, a minimum working time of at least 500 h for use as part of a progressive process equipment for the manufacture of electronic products. The Kristall-1 AE in accordance with the technical specification has the symbol GL-201. At the same time, a small-sized Kulon AE (GL-204 according to TU) was created with a radiation power of 2–4 W with a PRF of 10–20 kHz.

Based on two AEs GL-201s in the period from 1983 to 1986, The first domestic CVL Kareliya (LGI-201) with high energy characteristics and high radiation quality, operating according to the MO-PA scheme, was developed and investigated. Pumping of the AE is carried out from a two-channel synchronized thyratron or tube power source. The average radiation power of a two-channel CVL is at least 30 W (pulsed power of 200 kW), it has a divergence of the output beam from several milliradians to 0.1–0.2 mrad (diffraction limit) at a PRF of 8 –12 kHz. With this quality of pulsed radiation, the first experimental studies were carried out in 1984 of the processes of cutting and drilling with a laser beam of various materials with a thickness of 0.3–3 mm (Cu, Al, Mo, Ta, W, D16T, 12Cr18Ni10Ti, U8, VK6, foiled textolite, plexiglas and etc.).

The two-channel CVL Kareliya became the basis for creating a laboratory semi-automated laser technological installation (LTI) Karavella (1986–1987), designed for precision processing of materials used in the manufacture of electronic products. The LTI demonstrated the possibility of precision cutting and drilling of a large group of metals, semiconductor and dielectric materials, many of which were not included in the sphere of laser microprocessing until now. It is shown that Karavella LTI allows to shorten the terms of manufacturing small and medium-sized batches of electronic products by an order of magnitude in comparison with traditional methods of processing, including electric sparking.

At the same time (1986), an experimental laser medical device Yantar'-F with the CVL Kareliya with an average radiation power of at least 10 W was created for localized thermal effects on pathological foci (coagulation, therapy, surgery). The fiber lightguide, transmitting to the object pulsed radiation of CVL, was a flexible quartz monofilament 0.2–1.0 mm in diameter with an outer protective shell. The main advantage of a quartz fiber is its high radial strength (up to 10^{10}–10^{11} W/cm^2). Therefore, in a fiber with small diameters it is possible to transmit 'large' average powers of pulsed radiation (units and tens of watts).

One GL-201 AE was used as a basis in 1989 to develop the Klen radiator (ILGI-202 according to TU), and based on it in 1990 a CVL Kurs (LGI-202) with the upgraded thyratron power source IP-18. The total average output power of the LGI-202 laser is 20–25 W with a 10 kHz PRF and the power consumed from the IP-18 rectifier of approximately 2.7 kW, the guaranteed operating time to failure was 500 h. The laser was designed for pumping wavelength-tunable dye lasers (0.53–0.71 μm) for medical and technological installations and scientific research.

During this period, a large amount of experimental and theoretical work was carried out to increase the power and efficiency of a copper vapour laser, to study the structure and improve the quality of its output radiation [225–233]. It is established that the structure of radiation with an optical resonator is multicellular (usually three to five beams are observed). Each radiation beam has its spatial, temporal and energy characteristics. The use of an unstable resonator of a telescopic type with an magnification factor M = 50–300 leads to the formation of radiation beams with a divergence close to diffraction. In the mode of operation with a single mirror, the radiation structure is of the two-beam type. With one convex mirror the radius of curvature of which is two orders of magnitude smaller than the length of the AE, a beam with a divergence close to the diffraction and high stability of the radiation characteristics is formed [232, 233]. The structure of the output radiation and its characteristics in the CVLS operating according to the MO–PA scheme [227–233] was studied.

The period 1990-2002. This period of development of sealed-off self-heating CVLs is characterized by the search for and creation of new design and technological solutions, effective electrical pumping schemes to increase the guaranteed (minimum) AE production up to 1000 h and above, average radiation power to 50–100 W with a practical efficiency of at least 1%, pulsed radiation power up to 250–500 kW, energy in pulse up to 5–10 mJ. The spatial and temporal characteristics of the output radiation of CVL with such power levels for different optical systems both in the generator mode and in the power amplifier mode are studied. The development of powerful and reliable CVLs with high radiation quality was stimulated by the need to create domestic laser process units for the separation of isotopes for the needs of nuclear power engineering, installations for high-performance precision microprocessing of materials of electronic equipment and modern medical facilities [231, 232, 234–271].

In the Kuban' research project (1989–1990), an experimental CVL with a synchronized three-channel thyratron PS and a modernized radiator Kareliya operating according to the MO–PA scheme with an average radiation power of 100–105 W, a pulsed power of 500 kW, pulse energy 10 mJ and beam divergence close to diffraction (0.3–0.4 mrad) at 10 kHz PRF was constructed. The master oscillator was the Kulon AE (GL-204) with a power of 3–4 W (diameter of the discharge channel 12 mm), the power amplifier – two Kristall-32D AEs, power 50–55 W (channel diameter 32 mm) .

The Kurs CVL was used in the development of the first relatively powerful laser medical devices of the type Yantar'-2F and Yakhroma-2 with flexible quartz light guides for transmitting radiation to a biological object. In the Yakhroma-2 installation, a wavelength-tunable dye laser (λ = 0.58–0.7 μm) is additionally applied to extend the radiation spectrum. The power at the working end of the optical fiber of the Yantar'-2 F installation is 5–10 W (λ = 0.51 and 0.58 μm), in the Yakhroma-2 installations it is 0.5–5 W. The main application of the Yantar'-2 F device is the destruction of atherosclerotic plaques, thrombi and pathological tissues under X-ray control in endoscopic surgery [45, 46], the Yakhroma-2 device – the treatment of oncological diseases using photodynamic therapy [54, 55, 235– 241] using photosensitizers (photohem, photosens, etc.) [44, 54, 55, 234–241]. The Yantar-2 F installation with the application of the Kristall AE on gold vapours with an average radiation power of 4–6 W (λ = 0.628 μm) was also successfully used for photodynamic therapy with photogem. Dermatology and cosmetology is a specific field of medical application of the plants, because effective selective coagulation of pigment and vascular lesions is possible due to a significant difference in their absorption at wavelengths of 0.51 and 0.58 μm [48]. At present, a new generation of small-sized air-cooled medical devices such as Yakhroma-Med and Auran based on copper and gold vapour lasers of the Kulon series is being developed and is being introduced [16, 42, 48, 239].

During the period 1990–2003 The laboratory ALTI Karavella was effectively used for experimental research and fabrication of prototypes of precision parts of electronic equipment, mainly for microwave devices at the Istok enterprise [242–251].

In the SPE "Istok" in the period from 1998 to 2002, the development and production of new models of high-efficiency industrial sealed self-heating AEs on metal vapour Kulon series with an average output power of 1 to 20 W (GL-206 (A, B, C, D, E, F, G, H)) and the series Kristall with a power from 30 to 55 W (GL-205 (A, B, C, D)) was carried out [16, 41, 252–271]. The minimum operating time for the AE on copper vapour is at least 1000 h, on gold vapour 500 h. The practical efficiency for CVL of the Kulon series in the generator mode is 0.3–0.8%, the Kristall series is 1–1.2%. In the power amplifier mode, the practical efficiency for the CVL Kristall increases by 1.3–1.4 times, and the physical efficiency (by the power input into the AE) was about 3%. At the present time, this class of lasers is being improved in order to improve their reliability and the quality of the output radiation, and also to increase the power. Work is under way to increase the efficiency of CVLs due to the addition of hydrogen. New models of AE differ from obsolete models such as TGL-5 (1975), UL-101 (1976), UL-102, GL-201 and GL-202 (1982), GL-204 (1986) by lower power consumption, higher efficiency, longevity and radiation quality. A comparative analysis of the CVL parameters of the production of the Istok company and foreign analogs indicates that the specific power take-off and the minimum operating time of the sealed domestic CVLs are 2–4 times higher [16, 269–271].

Period 2003–2014 During this period of development at the Istok Scientific and Production Enterprise, namely in the laboratory "Lasers and laser technologies", under the guidance of N.A. Lyabin, modern automated laser technological installations (ALTI) of the Karevella type were developed on the basis of industrial pulsed CVL Kulon, and further research was carried out to increase the minimum operating time and quality of the output windows of industrial sealed self-heating AE series Kulon and Kristall and to form a single-beam laser in an CVL with an unstable optical resonator radiation of the diffraction quality, t. e. with a divergence $\theta_{difr} = 2.44 \cdot \lambda/D$ [360–394].

In the Karavella-1 project in 2003 the first domestic industrial ALTI Karavella-1 was developed on the basis of two-channel CVL and a three-coordinate table with PC control intended for productive and high-quality precision microprocessing of metal materials 0.01–0.05 mm thick for electronic equipment components (EEC). The CVL consists of a radiator with two sealed AEs GL-206D from the Kulon series with an average radiation power of 10 W (produced by the Istok company) operating under the MO–PA scheme and a high-voltage two-channel transient power source with a power consumption from a three-phase 5 kW network at PRF of 13–14 kHz and accuracy of channel synchronization ±2 ns (development of Chistye tekhnologii, Izhevsk). The three-axis table is created on the basis of linear synchronous motors and includes a horizontal XY table with a working field of 150 × 150 mm where the material to be processed is mounted and a vertical table Z with a focusing lens and a working stroke of 60 mm with accuracy of positioning for each axis not worse than ± 2 μm (Pretsizionnye tekhnologii, Moscow). The operating modes and processing of ALTI are automated and controlled from the computer through the control unit in accordance with the created basic software. Programs for drawings of manufactured parts are compiled in AutoCad in DXF format. The ALTI Karavella-1 in the period 2004–2005 within the framework of the Kursor project was used for a large volume of experimental studies on microprocessing (cutting and drilling) of various thin-film materials with a thickness of 0.05–0.5 mm with an average radiation power P = 0.4–9.1 W and a focal length of an achromatic lens F = 50, 100 and 150 mm, the cutting speed V = 0.5–5 mm/s and the number of passes N = 1–50. Based on the results of the research, a computer database has been created in the form of Excel spreadsheets and technologies have been developed for the precision manufacture of parts from molybdenum, copper, titanium, stainless steel, silicon and artificial polycrystalline diamond. Annually, from 2005 to 2013, hundreds of thousands of parts of a simple and complex configuration were manufactured at the Istok Research and Production Enterprise in the ALTI Karavella-1, mainly with two-shift operation, in the development and manufacture of microwave devices, with high processing quality and micron accuracy. In 2010, within the framework of the Udar LT project using ALTI Karavella-1, the processing technology for a number of critical parts of atomic-beam tubes on cesium vapour was tested. In 7 years of almost continuous operation of ALTI, a number of design flaws were revealed, which reduced its reliability, quality and manageability, which were taken into account in subsequent developments. With the purpose of further high-quality operation of ALTI Karavella-1 in 2013 the old two-channel power source of the CVL was replaced with a new one, with high reliability and controllability and optimization of the optical system for the formation of a diffraction beam of radiation was carried out.

In 2012, the Istok enterprise developed an industrial ALTI Karavella-1M, which differs from the Karavella installation by higher power and radiation quality, reliability and controllability. The PA of the emitter of the CVL use an AE with a 1.5-fold increase in the power GL-206I (15–20 W), in MO with the AE GL 206D–an unstable optical resonator with two convex mirrors for the formation of single-beam radiation with diffraction divergence. According to preliminary estimations of the developers of the power source (Chistye tekhnologii), the guaranteed operation of the dual-channel upgraded transistor power source, due to the use of new solutions in the nodes of the high-voltage modulator and their placement in liquid oil with high electrical strength and components with high reliability, should increase at least two times – up to 4000–5000 h. Moreover, the new power source should provide not only a packet, as in ALTI Karavella-1, but also a pulse-modulated radiation. The three-coordinate table in this setup is identical to the coordinate table in ALTI Karavella-1. An increase in the mean power of radiation in a beam of diffraction quality to 15–20 W in ALTI Karavella-1M allowed to increase not only the productivity and processing quality, but also the thickness of the processed materials – metal up to 1 mm, silicon and polycrystalline diamond up to 2 mm.

In 2011 ALTI Karavella-2 was developed and put into operation on the basis of industrial single-channel CVL Kulon-15 with packet and pulse impulse modulation at 14–16 kHz PRF manufactured by the VELIT Company. The most powerful sealed-off AE from the series Kulon – GL-205I with power in a high-quality beam of 6–8 W is used in CVL when operating in a mode with a telescopic resonator with $M = 200$. The working field of the horizontal coordinate table XY is 100 × 100 mm, the vertical table travel Z –50 mm. ALTI Karavella-2 provides high-speed and high-quality microprocessing of foil and thin materials: metal with a thickness of 0.01–0.3 mm, nonmetallic to 0.5–0.7 mm. In 2013 ALTI Karavella-2M was launched, which differs from ALTI Karavella-2 only with a large working field of the horizontal XY coordinate table – 200 × 200 mm.

The Udar LT-1 project included a large volume of theoretical and experimental studies carried out to improve the quality of the laser beam in the generator mode in order to improve the quality of precision microprocessing of materials at Karavella-type process units. The basic 'limiting' requirements were formulated for the pulsed radiation of the CVL to ensure the maximum quality of microprocessing. With the use of modern means and measurement techniques, the structure and output characteristics of the laser radiation with an unstable resonator of the telescopic type with a gain of $M = 200$ and with two convex mirrors and using sealed AE GL-206D and GL-206I were studied. In this case, formulas were derived for calculating beam divergence, conditions for the formation of single-beam radiation with diffraction divergence, and conditions for eliminating the instabilities of the position of the axis of the beam pattern and the pulsed energy were determined.

The issues of increasing the service life and the stability in time of the output power of the radiation from industrial sealed self-heating and high-temperature AEs on copper vapour of the Kulon and Kristall series remain valid today. The minimum (guaranteed) operating time of sealed AEs is increased from 1000 to 1500 h, i.e., by a factor of 1.5. In order to further increase the durability and stability of AE parameters, experimental studies are carried out to improve the training technology for degassing and preserving the purity of copper vapour generators, increasing the stability of the pulse-arc discharge from the cold cathode, and the quality of the windows for emission.

With active scientific and technical support and participation of employees of the laboratory Lasers and laser technology in the Design Bureau of the A.N. Lebedev Physics Institute (Troitsk) the medical facility Yakhroma-Med was created for dermatology and cosmetology – treatment of vascular, pigmented and unpainted skin defects in the VELIT company (Istra) – multifunctional medical device Kulon-Med for the treatment of cancer and non-oncologic diseases using the method of photodynamic therapy and wide application in scientific and practical medicine.

1.5. Conclusions and results for chapter 1

1. The main results of the well-known domestic and foreign teams in the study of physical foundations, the state, development and applications of pulsed CVL and CVLS, and the creation of technological equipment and research equipment for various purposes are described.

2. Analysis of a large amount of work (about 250) showed that the CVL and CVLS are the most powerful pulsed sources of coherent radiation in the visible spectral region and possess a unique set of output parameters and therefore have ample opportunities for practical application in science, technology and medicine.

The unique set of CVL output parameters is due to emission wavelengths in the yellow–green region of the visible spectrum – λ = 510.6 and 578.2 nm, a large amplification of the AM – $k = 10^1–10^2$ dB/m and a short pulse duration τ_{pulse} = 20–50 ns, high PRF f = 5–30 kHz and average power pickup from one AE P_{rad} = 1–100 (750) W at 0.5–2% efficiency, but relatively low pulse energy W = 0.1–10 mJ and a divergence close to diffraction and diffraction $\theta = (1–3)\ \theta_{difr}$. With such a combination of parameters, the density of the peak radiation power in the focused spot (d = 5–20 μm), even at relatively small values of the average power (P_{rad} = 1–10 W), reaches very high values $\rho = 10^9 – 10^{12}$ W/cm^2, sufficient for effective microprocessing in the evaporative mode.

3. The most significant contribution to the investigation of physical fundamentals and the development of the CVL and CVLS have been provided by Oxford Lasers (England), Lawrence Livermore National Laboratory (LNL), Macquarie University (Australia), Mashinoexport and Spectronica (Bulgaria), CILAS (France) and a number of firms from Japan, India and Israel, from domestic – P.N. Lebedev Physics Institute of RAS and the General Physics Institute of RAS, The Kurchatov Institute (Moscow), Tomsk Polytechnic University and the Institute of Optics and Atmosphere of the Siberian Branch of the RAS (Tomsk), The Institute of Semiconductor Physics (Novosibirsk), The Istok Company (Fryazino), VELIT (Istra), St. Petersburg State Polytechnic University, N.E. Bauman Moscow State Technical University and OIVT RAS (Moscow).

4. Due to high densities of peak power in the focus spot of pulsed CVL radiation ($10^9–10^{12}$ W/cm^2) and absorption coefficients of emission of the yellow–green spectrum by metals, the CVL is an ideal tool for microprocessing of metallic materials and a large range of non-metallic materials. Such heat-conducting metals as Cu, Al, Au, and Ag are not efficiently treated by irradiation of IR lasers, since the reflection coefficient from metallic materials is very high.

5. CVLs at small radiation power levels (1–20 V) are usually structured in the form of a single unit generator (monoblock) with one low-power AE and a corresponding optical resonator. At medium (20–100 W) and especially high (hundreds and kilowatts) power

levels, an CVLSs is used that operates in an efficient MO–PA scheme with one or more AEs as the PA and often with a preamplifier located before the first LA. In the CVL of the MO–PA type, in comparison with the CVL operating in the single-generator mode, higher efficiency and the quality of the output beam are achieved. Priority for the creation of powerful CVLS abroad belongs to the Lawrence Livermore National Laboratory (LLNL), in Russia – the Kurchatov Institute.

6. An exclusive place is occupied by the use of powerful CVLS in combination with a wavelength-tunable dye laser that also operates in accordance with the MO–PA scheme, as part of the technology for separation of isotopes (ALTI technology), which uses a difference in the absorption spectra of atoms of different isotopic composition. This progressive optical technology makes it possible to produce substances with the necessary level of enrichment and high purity for first of all, in nuclear power engineering and medicine. The most powerful CVLS for AVLIS technology was created in the Lawrence Livermore National Laboratory, 72 kW of average radiation power, and at the Kurchatov Institute 300 W.

7. The Istok Company (Fryazino, Moscow) is a leader in the development and production of highly efficient and durable industrial sealed-off self-heating AEs for the pulsed CVLs. The Istok company produces sealed AE series Kulon of small (1–20 W) and Kristall of the medium (30–100 W) radiation power levels that do not have domestic and foreign counterparts.

The development laboratory Lasers and laser technologies for the first time disclosed the dynamics of the formation and structure of pulsed laser radiation in the regime of a generator with an optical resonator. The structure is multibeam, with each beam having its spatial, temporal and energy characteristics and overlapping beams in space and time. The conditions for the formation of diffraction quality single-beam radiation and high-power beams of this quality in the CVL were determined in the resonator. These studies have made it possible to unambiguously identify the main areas of application of pulses CVL and, first of all, for microprocessing of materials.

8. The Tomsk Polytechnic University and the Institute of Atmospheric Physics of the Siberian Branch of the RAS (Tomsk) made the largest contribution to the study of a CuBr laser, a kind of pulsed CVL. With the addition of hydrogen or its compounds to the CuBr laser, the maximum PRF and power take-off from the volume

unit of the AS are achieved: 200–300 kHz and 1–2 W/cm^3 in the range of 30–60 kHz.

9. The VELIT company (Istra, Moscow) has the priority to develop industrial, efficient and reliable, high-voltage pulsed thyratron power sources (in the near future and transistors) with packet and pulse modulation and control from PC, development and release to their basis on modern industrial CVL Kulon using sealed AE series Kulon of low power (up to 20 W) produced by The Istok company. Chistye Tekhnologii (Izhevsk) has the priority to develop industrial two-channel transistor high-voltage pulse power sources with nanosecond (±2 ns) channel synchronization for pumping and high-speed control of parameters from PC CVLS Kulon-10 and Kulon-20.

CVL and CVLS of this class do no have domestic and foreign analogs and are intended for technological equipment for precision microprocessing of materials.

10. On the basis of the industrial CVL and CVLS, the Istok Company developed and is manufacturing modern ALTI of the Karavella type for productive and high-quality precision microprocessing of metal and non-metallic materials of electronic components, in particular, microwave technology. The spectrum of processed materials includes: heat conducting – Cu, Al, Ag, Au; refractory – W, Mo, Ta, Re and other metals – Ni, Ti, Zr, Fe, etc. and their alloys, as well as a wide range of semiconductors and dielectrics – polycrystalline diamonds, silicon, sapphire, graphite, carbides and nitrides; transparent materials. The ALTI Karavella has no domestic analogues.

11. Medicine is also a promising direction for the development of pulsed CVLs. Multifunctional modern medical devices such as Kulon-Med and Yakhroma-Med for use in oncology, low-intensive therapy, dermatology and cosmetology, microsurgery, etc. have been created on its basis. This class of equipment is the leader of laser non-ablative technologies. Green and yellow wavelengths of pulsed radiation CVL are optimal for removing vascular, pigmented and unpainted skin defects, treating acne, and wrinkle smoothing. A special place is occupied by the treatment of oncological and non-oncological diseases using photodynamic therapy (PDT). Laser pulses at the same time affect the body's defects selectively, without damaging the surrounding tissue and without causing pain (anesthesia is not required).

12. In addition to the use of pulsed CVLs in modern technological and medical equipment, the following areas of application of CVL include: spectroscopic research and analysis of the composition of substances, image intensification amplifiers of micro-objects, nanotechnology, high-speed photography, laser projection systems for imaging on large screens and in open space, lidar installations for sounding the atmosphere and sea depths, navigation systems for wiring and landing airborne vessels, treatment of materials in an aqueous medium, the gas flow visualization, laser acceleration of microparticles, holography, forensics, entertainment industry.

2

Possibilities of pulsed copper vapour lasers and copper vapour laser systems for microprocessing of materials

2.1. The current state of the modern laser processing equipment for the processing of materials and the place in it of pulsed copper vapour lasers

Modern laser processing equipment, developed in many advanced countries of the world (Germany, Japan, England, Italy, the USA, etc.), finds an increasing application in almost all industries. Developments in this direction are currently being conducted in Russia. Laser processing of materials is one of those technologies that determine the current level of production in the industrialized countries. Its distinctive features are the high quality of the products, high process productivity, saving of human and material resources and environmental cleanliness. The development of laser technologies also has a social aspect – jobs are created for highly qualified personnel, which leads to an increase in the educational level of engineers and workers and production culture [16, 20–26].

Today, separate technological lasers are not delivered to the market and the laser technological installations and complexes are the main items. In addition to lasers, the units include external optical delivery systems and beam focusing on the object being processed, controlled tables, manipulators and robots to move the product, software needed to implement a particular technology. Among the laser sources used to process materials, the proportion of CO_2 lasers has traditionally been high. The production of solid-state lasers, especially fiber ones with diode pumping, is growing at a very high rate. Catching up the gas lasers by power radiation, the solid-state lasers are superior in technology, economy and mass and size parameters. Also promising for engineering applications are diode lasers, which have the highest efficiency from all lasers.

The use of laser technologies is extremely diverse: welding, surface processing, alloying and surfacing, cutting and dimensioning, cutting materials, marking and engraving, precision microwelding of electronic components and microprocessing of electronic equipment components (EEC) materials. The lasers help to achieve technical and economic results that can not be realized by other means. The introduction of laser plants, in turn, allows to shorten the cycle of preparation of production, increase the productivity of manufacturing parts, produce microprocessing, improve the quality of products by eliminating the mechanical pressure of the tool or the thermal impact during electrospark machining (ESM), increase the life of the parts produced, etc.

A special place is occupied by laser technological installations for microprocessing, microdrilling, high-resolution marking and engraving of any materials, analysis and healing of defects, etc., the manufacture of precision parts for the EEC. The radiation sources in these systems include efficient and already used short-pulsed, high-frequency lasers of the visible and ultraviolet spectra of the visible and ultraviolet spectra – solid-state, excimer and, in particular, copper vapour lasers (CVL) with an output mean power 5–50 W. The pulsed emission of these lasers, thanks to the high intensity in the focusing spot (up to 10^{10}–10^{12} W/cm^2), allows high-performance and precision microprocessing of metallic and non-metallic materials with a thickness of 0.01 to 2 mm. Experimental studies have shown that in a number of typical operations (cutting through a complex contour, piercing holes, manufacturing fine-meshed nets, etc.), the use of radiation from the CVL practically does not require the time for preliminary preparation (except for tooling, compiling and debugging programs). This makes it possible to shorten by up to 6–10 times the production times for single and small batches of thin sheet metal parts in comparison with traditional methods of shaping, including ESM, chemical etching, microstamping, etc.

2.2. Analysis of the capabilities of pulsed CVL for microprocessing of metallic and non-metallic materials

Features of pulsed radiation of CVL and the spectrum of processed materials. The possibilities of using pulsed CVL for precision microprocessing of materials are determined by the parameters of its output radiation beam: wavelengths in the visible spectral region (0.51 and 0.58 μm), short pulses duration (20–40 ns), high repetition rate (5–30 kHz) and low energy in the pulse (0.1–10 mJ), diffraction quality and high achievable peak power density in the focused radiation spot (10^9–10^{12} W/cm^2). Such a combination of parameters is not possessed by any of the known commercial lasers. Potential possibilities of using CVL radiation for microprocessing were for the first time fairly widely represented in [20]. The CVL is practically unsuitable for relatively 'deep' welding, because of the short duration and energy in the radiation pulse [185, 186].

The CVL beam, focused by an achromatic lens with a focal length F = 50–250 mm, is an ideal non-contact tool with a light spot diameter of 10–30 μm and a length of up to 1 mm (in waist). The method of contour cutting can be used to cut holes of any diameter and profile with high accuracy, both in metallic and non-metallic materials. A unique feature of this tool is the ability to drill holes with a high shape factor (k_s = 50–100). High-speed pulse processing, micron and submicron accuracy of the movement of modern coordinate tables allow the use of the CVLs for the manufacture of precision parts of complex configurations, prototypes and batches, especially small series, as well as in mass production [16, 20, 187–193].

The use of CVL for the microprocessing of materials was reported for the first time in Russia in 1973 in the works of the P.N. Lebedev Physics Institute and later, in 1979 – in the work [214], performed at the Nuclear Research Center in Negev (Israel). A focused beam of radiation with an average power of 30 W [214], formed in a laser system on copper vapour in the master oscillator–power amplifier (MO–PA) scheme, was drilled in a 1 mm thick steel sheet. In the 1980s the possibilities of CVL for microprocessing materials in active optical amplifying systems using a laser projection microscope were intensively studied. The CVLs as micromachining devices began to appear on the market around 1990, when lasers with higher reliability and radiation quality started to appear and demand for them in the industry arose. The first serious studies were carried out in [325], as well as at Oxford Lasers [212], St. Petersburg Technical University [326, 327], and the Lawrence Livermore National Laboratory (LLNL) [328]. The technology of microprocessing with UV radiation, obtained by doubling the frequency of radiation of CVL, was developed in [329].

The radiation energy emanating from the CVL with small pulse portions and with a large peak power at high PRF (pulse repetition frequency) provides a highly controlled and predictable removal of material from the treated area with the formation of minimal heat-affected zones. The short-pulse radiation of the CVL creates a noticeably lower energy threshold for efficient material processing than continuous-wave lasers, which lead to the formation of a shielding plasma [330]. The zone of material removal (processing) is strictly limited by the focused radiation spot, which is smaller in the CVL than in any IR laser. For example, the focusing spot of a beam with a diffraction divergence in a gas CO_2 laser is 20 times larger than for a CVL. In solid-state YAG:Nd lasers, due to the thermal deformations arising in it, the quality of the radiation beam is several times lower than the diffraction limit [331]. Another advantage of the CVL in comparison with the IR lasers is that metals have a lower reflection coefficient in the emission range of the CVL (10–50%) than in the IR band (>95%) [332]. Metals such as Al and Cu can be treated with CO_2 and other infrared lasers only with great difficulties because of the combination of high IR reflection and very high thermal conductivity of metals. Therefore, it is very difficult to obtain a melt with these lasers [333]. The presence of two wavelengths in the emission of CVLs in the yellow–green region of the visible spectrum (0.51 and 0.58 μm) made it possible to easily process both aluminum and copper. Many other materials are also effectively handled by the CVL. For example, the CVL cuts silicon 10 times faster than other lasers similar in purpose [334]. Comparison of the cutting speed performed by a short-pulsed YAG: Nd laser with a modulated Q-factor (τ_{pulse} = 80–140 ns) and the CVL with the same PRF showed that in the second case the cutting speed could be twice as high. And a higher rate of substance removal (more effective interaction with the target, higher peak power and short pulse) allowed to improve the quality of treatment [335].

Because of the high reflection coefficient of metals in the IR wave band, a large amount of thermal energy is required to melt and vaporize them using an IR laser, and therefore a rather large heat-affected zone is formed. The melt should be removed by gas jets, and this makes it impossible to use precision microprocessing. On the other hand, the high density of peak radiation power (10^9–10^{12} W/cm^2) generated by short pulses of the CVL on the surface of the material leads to the removal of the formed vapours and liquid as a result of microexplosions. The heat-affected zone can be an order of magnitude smaller than that of other lasers [325]. Excimer UV lasers can form a smaller heat-affected zone than the CVL, but the CVL processes the material much faster, since its power density and, consequently, the surface temperature of the target are much higher. The use of the CVL is also more effective in cases where it is necessary to make incisions deeper than 0.5 mm [331, 336].

When microprocessing large surfaces of a complex profile with the help of a pulsed CVL, the problem of productivity is solved by increasing the PRF while maintaining the required level of pulsed energy [20]. Therefore, in processing units, CVLs are made according to the MO–PA scheme. In addition, in such systems, increased power levels of the beams with a divergence close to the diffraction are achieved, and by synchronizing the mater oscillator and the power amplifier any programmed sequence of output pulses of radiation with a specific level of pulsed energy can be operatively formed.

The use of pulsed CVL for drilling and cutting. A significant difference between the pulsed CVL and most other technological lasers is that precision microprocessing during drilling and cutting occurs predominantly in the evaporation mode and without blowing gas into the treatment zone [343]. This allows to significantly reduce the heat-affected zone. Since the density of the peak radiation power considerably exceeds the evaporation threshold ($\sim 10^6$ W/cm^2), evaporation has the character of microexplosions and is accompanied by the expansion of vapours and superheated liquid [328, 337, 338]. The latter significantly affects the cutting parameters – efficiency and speed, as well as the roughness of the edge of the cut. With a thickness of material comparable to the width of the cut (10–30 μm), the vapours and drops of metal that fly out from the radiation zone are practically unobstructed and completely removed from this zone. The average radiation power needed to cut such thin materials is about 1 W at a peak power density of $\sim 10^9$ W/cm^2. When cutting 'thick' materials (≥ 50 μm), the expansion of the vapour occurs mainly between the walls of the already formed section on which they condense. In addition, the liquid metal spreads out of the zone of action under the pressure of the vapour. This effect manifests itself the more strongly, the more the thickness of the material exceeds the diameter of the focused spot. The efficiency of removal of a substance depends significantly on the laser power, which determines the initial temperature and energy of the scattered particles [328]. It was established in [328] that for a material thickness of more than 200 μm and an average radiation power of up to 20 W, no complete removal of matter occurs in one pass. It is also noted that with increasing power the width of the cut increases due to the fact that the side 'wings' of the focal spot are already involved in the evaporation. With higher power, it is possible to cut the material in one pass, but the roughness can be too large and does not differ from the roughness in the treatment with a solid-state or CO_2 laser [326, 333]. Therefore, in order to ensure a high quality of cutting, it is necessary to repeatedly pass the beam. Multipass cutting, of course, reduces the processing speed, but it is necessary to minimize the heat-affected zone and edge roughness [336–338].

In [339, 340], calculations were made for aluminium and steel targets in the case of a single-pulse action at a peak power density in the focusing spot of 10^{10} W/cm^2 at $\lambda = 0.51$ μm. The average radiation power of the CVL was 45 W at the PRF of 4.5 kHz, the pulse duration 70 ns, and the spot diameter 50 μm. The calculated maximum surface temperature at these parameters is 12000°C after 5 ns of irradiation, then, after 30 ns, falls to 6500°C. At that time, a layer of steel 0.5 μm thick is removed when heated to a depth of 2 μm. After 3 μs, a 1.1 μm layer is removed and the depth of the melt zone is 8 μm at a surface temperature of 3000°C. These calculations also showed that after 8 ns after the start of irradiation for the next 4 ns the surface heating directly by laser radiation is reduced to 5% as a result of the formation of plasma. In general, during one pulse, 15% of the laser energy reaches the bottom of the target 'directly', 85% is transmitted to the surface by plasma electrons.

For each material, there is a clear threshold for the density of the peak radiation power, after which an absorbing plasma is formed. For carbon steel, this threshold is $1.5 \cdot 10^9$ W/cm^2, and for aluminium it is $2 \cdot 10^9$ W/cm^2. On the basis of the rate of expansion of the vapour it was found that at the end of the pulse for aluminium the density of the vapour atoms in the erosional flare at the surface of the material reaches $3 \cdot 10^{20}$ cm^{-3}, at a depth of 10^{21} cm^{-3} at a duration of 40 ns and a few kHz PRF (the energy in the pulse 4 mJ, peak power density 10^{10} W/cm^2). Measurements have shown that when a carbon target is irradiated with low-power CVL radiation (7 W) at an PRF of 8 kHz and a peak power density of 10^9–10^{11} W/cm^2, 10^{13}–10^{14} atoms are removed per pulse.

The drilling speed by a separate laser pulse depends on its duration and intensity. Four stages of drilling are distinguished: only heating, non-stationary drilling, stationary drilling and plasma-limited drilling. Since the intensity at the beginning of the pulse increases, the drilling speed increases to a certain limit. This limiting speed largely depends on the state of the material in the radiation exposure zone (solid, liquid, gaseous, plasma). Then the stationary drilling mode is set, and at the end of the pulse the speed drops. The depth of a drilled hole in a single-pulse mode for different materials at the same power density level is different. For example, during one pulse in aluminium an aperture with a depth of up to 8 μm is drilled, and in ceramic Si_3N_4 – no more than 1 μm.

Since the energy is supplied to the target and through the absorbing plasma located above it, then with an increase in the plasma volume additional losses of the radiation energy arise. Plasma can cause strong light scattering, heating of the target and the formation of chips in a larger area than is permissible. Due to the creation of this additional plasma barrier, the drilling speed gradually decreases. The arrival of the radiation energy of the CVL on the 'bottom' of the target does not depend to such an extent on the plasma barrier as in other lasers, since the plasma in the visible range has high transparency. The latter is due to shorter radiation waves that cause photoionization. The effect of plasma can be reduced by reducing the duration of the radiation pulse and reducing the size of the focal spot, which causes a faster plasma relaxation [341].

For a laser with specified parameters, the average drilling speed depends on the thickness of the sheet, the material and its structure, its evaporation temperature, and physical properties: thermal conductivity, heat capacity, density and phase state of the ejected matter (plasma, vapour, droplets). Removing a material when drilling a hole presents a serious problem when the form factor is high. The maximum drilling depth is limited to 3–5 mm. It is also clear that there must be a certain critical PRF above which the plasma target screen, due to incomplete plasma relaxation, becomes to some extent a limiting factor for the average drilling speed. In the range of 100 Hz to 6.5 kHz, the drilling efficiency depends little on the frequency. However, even with a 15–20 kHz PRF, the CVL works quite efficiently. The injection of the processing gas into the treatment zone in certain cases results in an increase in the drilling speed. The drilling of foil from copper, iron and titanium in air takes an order of magnitude less time than drilling in an argon atmosphere [332]. Oxygen also increases the speed of metal drilling [339]. The processing speed is influenced by the grain size of the material: the smaller it is, the less the number of radiation pulses is required to start the drilling.

The ability of a pulsed CVL to perform precision microprocessing largely depends on the quality of the beam and the power density of the radiation. The high quality of the radiation beam (minimum divergence, high stability of the position of the axis of the radiation pattern, a clear boundary of the focused spot and, consequently, a uniform distribution of intensity in the far zone) also determines the high quality of the microprocessing.

The choice of the processing method, generally speaking, depends on the thickness of the material and the required form factor. A high form factor can be obtained with direct drilling. In metals up to 1 mm thick, holes of 20–25 μm diameter are obtained by this method. At a power density of 10^9–10^{10} W/cm^2, smaller holes can be made, but these holes at the exit converge on the cone [339]. In direct drilling, the variation in the size of the hole is usually 10% of its diameter. Drilling holes with a diameter above 50–100 μm is produced most often by contour cutting. This method allows one to get deep holes, but, naturally, with a small form factor. The roughness of the processing edge is determined by the intensity distribution in the focusing spot, the degree of stability of the beam axis and the accuracy of the beam moving by the scanning device. With multi-pass scanning, the surface of the cut is levelled and polished. Of course, if it is necessary to make a large number of microholes per unit of time, the first method is more convenient, but it requires higher powers. If high accuracy is not required, optical fibers can be used to feed the CVL radiation onto the workpiece [328]. The quality of the hole in fiber drilling is close to the quality of conventional mechanical processing methods.

Radiation treatment by CVL of ceramics and transparent materials. The rigidity and fragility of ceramics greatly impedes processing by conventional methods. The radiation of CVL can also be used to process ceramics. The CVL like the excimer laser, with relatively short length waves and high peak capacity allows fully remove the ceramics from the processing place without the formation of a glass mass layer (which is observed in the case of YAG:Nd and CO_2 lasers). At then same time, CVL can cut ceramics at a speed more characteristic for the YAG:Nd laser [342, 343]. When processing by CVL radiation Al_2O_3 ceramics it is possible to avoid the formation of microcracks and attain higher speeds of quality cutting. As a rule, the drilling speed of corundum ceramics (0.1–0.2 mm/s at thickness 1 mm) is 1–2 orders of magnitude lower than the drilling speed of the metals. Tungsten carbide with a thickness of 3.7 mm can be cut with speed of 0.2 mm/min. In this the case the incision made by CVL radiation usually has good quality.

The CVL can also be used to drill efficiently micro-holes in optical materials – in glass, quartz, ruby, sapphire [344]. The diameter of these microholes is 10–40 μm, the depth is up to 3 mm, the form factor exceeds 100 [345–350]. Surfaces of holes have good optical quality. The drilling speed is usually about 0.8 mm/s, which is close to the level for metals (0.2–0.9 mm/s). In the sheets of quartz glass the rounded grooves are cut at a speed of 8 mm/min. On the surface of the transparent target there is a clearly defined entrance opening and a long section with a very small taper, the length of which is 100 times the diameter of the inlet. The hole is 'blind' or open ending in the top. Drilling is initiated either with the formation of a colour centre, or after surface breakdown at peak power density levels of $5 \cdot 10^9–10^{10}$ W/cm^2. To do this, even an average radiation power of 2.5 W is sufficient for the diffraction quality of the beam. The drilling speed reaches 0.2 μm per pulse.

After surface breakdown of the target (when an inlet is formed), the laser energy is directed along a smooth hole to the front of the laser drill. The resulting absorbing plasma, associated with the front of the drill, has a bright glow and is clearly visible as the hole deepens. At the end of the pulse, the gas overheating in the hole reaches a maximum, after which a strong expansion begins; hot steam, forming a luminous torch, is pushed out under the action of a shock wave. At the energy of the radiation pulse below ~0.1 mJ, the drilling process of the material ceases. Since 'pulsed' holes in transparent materials are very effectively drilled with pulsed radiation from the CVL, this laser is very promising for the application of volumetric markings and images.

Radiation of the CVL can also be used for the processing of natural and artificial diamonds. The walls of the hole in the diamonds are obtained clean and high quality (as in other materials) which indicates the direct removal of the atoms from the impact zone. At relatively low densities of the peak power (~$2 \cdot 10^9$ W/cm^2) the diamond at first is transformed to graphite (with heating up to

1300°C), which strengthens absorption radiation and accelerates the removal of material due to oxidation and evaporation [351].

Underwater processing by radiation of CVL. The yellow–green region of the visible spectrum of radiation is well transmitted through quartz lightguides and water, so the CVL can be successfully used in underwater operations [352], as well as in repair work at nuclear facilities and offshore platforms. The CVL with an average output power of 110 W can cut under water a 20 mm thick stainless steel at a speed of 5 mm/h. The relatively high transmission coefficient of CVL radiation at a green wavelength (0.51 μm) in the depth of sea water makes it possible to use this laser in submarine–satellite communication systems and for underwater illumination.

Preparation of thin films. The evaporation of the material during its processing by the laser radiation allows this laser to be used to deposit thin films on the selected substrate. The steam flare from the removed CVL material under low pressure (10^{-6} mm Hg) expands and deposits as a film on the target substrate. In [353], films of SbSi, Si, and Ge were obtained. The atoms, whose number reaches 10^{13}–10^{14} in a torch, create up to one monolayer per pulse on a substrate located at a distance of 5–10 mm from the evaporation site. A high deposition rate and good coating properties were obtained on a rotating substrate with radiation defocusing, which created power densities of $4 \cdot 10^8$ W/cm^2. At a substrate velocity of 8 cm/s and a distance of 76 mm to the sprayed material, the deposition rate was 0.0056 nm per pulse. Using this method, the coating can be created in a few minutes, whereas the plasma deposition method requires several hours for this. At a peak power density of $4 \cdot 10^8$ W/cm^3, a film 410 nm thick and 4 cm wide was formed in 23 seconds, and the surface roughness was no more than 10 nm. A 2.5 μm thick film of silicon was deposited in less than 1 min. The deposition rate in the case of CVL application reached 2600 μm · cm^2/h, and with the use of other lasers it was only 10 μm · cm^2/h.

Processing of materials by UV radiation of CVL with frequency doubling. CVL with frequency doubling (due to the use of non-linear crystals) can easily generate pulsed energy in the UV range of hundreds of micro-joules at an average radiation power of about 1 W. This energy is sufficient for microprocessing organic polymers [329, 354–356]. Due to the 'non-thermal' nature of drilling, clean treated edges are obtained, without signs of charring and melt of material on the target. Such radiation can process elements with dimensions of only a few micrometers with submicron accuracy. Acryl, glass, optical fiber, polycarbonate, polyamide, plexiglass, silicone, silicone rubber, etc. are included in a number of materials treated with UV radiation of CVL. Such CVLs with UV radiation can be used in the production of flexible microcircuits, perforation of catheters and nozzles for inkjet printers.

For example, in order to obtain an average UV power of 1 W at a wavelength of 0.2553 μm, the average pump radiation power of the CVL on the green line (λ = 0.5106 μm) was 6.1 W with a PRF of 4.25 kHz. To increase the quality of the beam, the efficiency and the radiation power of the nonlinear crystal, hydrogen was added to the CVL buffer gas. To exclude the effects of diffraction, astigmatism and other distortions, spatial selection of radiation was used. In addition to the fact that CVLs with UV radiation have higher PRF and beam quality, they are much less expensive in terms of investment and maintenance than the base laser installations using ultraviolet radiation – excimer lasers. In addition, the PRF in CVL with UV radiation is more than an order of magnitude higher than that of an excimer laser, which significantly improves the processing capacity.

The processing of thin polymer films with the help of CVL with UV radiation is basically similar to the treatment with an excimer laser, but the speed is 100 times greater [357]. The speed of automatic drilling with such a laser depends mainly on the time required to move the target. To drill a single hole of 20–200 μm in transparent polymer films with a thickness of 75 μm, 10–50 ms was used. The size of the hole can be brought up to 1 μm while the error in the hole area is ~3%.

In the production of semiconductor devices, the contamination of substrates at the micron and submicron levels results in the rejection of half of the products. The qualitative purification of silicon and glass substrates can be effectively carried out with the help of CVL with UV radiation. UV radiation can also be successfully used in high-performance photolithographic operations for the manufacture or modeling of parts.

Thus, the pulsed CVL with respect to the set of parameters is an ideal contactless tool for precision microprocessing a wide range of metallic and non-metallic materials, including transparent ones.

2.3. Equipment MP200X of Oxford Lasers for microprocessing

The processing equipment MP200X is based on the CVLS of the

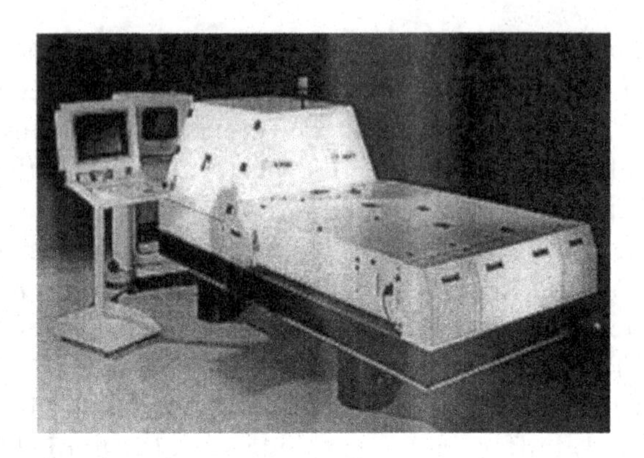

Fig. 2.1. Laser technological unit MP200X of Oxford Lasers for microprocessing materials with a thickness of 0.001–2 mm.

Table 2.1 The main parameters of the laser processing unit MP200X

Parameter	Value
Radiation wavelength, nm	510.6; 578.2
Pulse repetition frequency, kHz	10–12
Duration of radiation pulses (at half-height), ns	20
Average radiation power, W	60
Divergence of radiation beam, mrad	0.075
Power consumption from mains, kW	≤12
The blower of air at zone processing, atm.	7
Consumption of water, l/min	12
Gas: Ne/HCl, l · atm./h	1
Diameter of pierced holes, mm	0.001–0.500
The minimum cut width, mm	≥0.005
Thickness of processed material, mm	0.01–2.00
Working field of the horizontal XY table, mm Vertical Z table, mm	250 100
Accuracy of positioning by X and Y axes, μm, by Z-axis, μm	0.1 0.5
Overall dimensions, mm	1250 × 3250 × 1800
Weight, kg	2400

MO–PA type with the power of 60 W and the precision XYZ three-coordinate table and is the first foreign technological installation with a copper vapour laser designed for productive microprocessing of thin-sheet materials. It was developed by the English firm Oxford Lasers and, judging by the advertising information, in 2002 (Fig. 2.1). The working field of the XY horizontal table, where the object to be processed is installed, has a stroke size of 250 × 250 mm and positioning accuracy along the 0.1 μm axes, the length of the vertical table Z, where the focusing lens is installed, is 100 mm and the accuracy of 0.5 μm. The technological unit with such characteristics allows to productive processing of metallic materials with a thickness of 0.01–2 mm with a cut width of ≥5 μm and piercing holes with a diameter of 1–500 μm. The unit provides an additional module with non-linear crystals to convert the yellow-green radiation of CVLS to UV with wavelengths of 255 and 271 nm, which expands the range of processed materials. UV radiation efficiently processes organic and inorganic polymers. In a number of experimental processing operations, Oxford Lasers employees demonstrated unique possibilities of using pulsed nanosecond laser radiation for precision microprocessing of various materials [71, 72].

The main parameters of the MP200X plant are shown in Table 2.1. The main and significant disadvantage of MP200X is that the AE in the CVLS operate in the continuous pumping mode of the two-component gas mixture (Ne+HCl), i.e., the AE manufacturing technology does not provide a sealed operation mode and requires additional life support elements (a pumping pump, gas cylinders, reducers, control sensors, mixer). It is also known that the operating mode reduces the life of the laser and worsens the stability of the output radiation parameters. Disadvantages of the bleeding regime were confirmed by employees of the Canadian firm Phas Optx Inc. during a visit to the Istok company (Fryazino, Moscow region).

The MP200X unit purchased and maintained by this company did not provide the required level of output power stability, especially in the process of prolonged processing of complex parts. With each new activation extra careful tuning was required to restore the operating power level. For this reason, Phas Optx Inc. purchased for MP200X industrial sealed-off AE Kristall of the model GL-205B with a power of 60 W produced by the Istok company, which have a high stability of the output parameters.

2.4. The main results of the first domestic studies on microprocessing at the Kareliya CVLS and installations EM-5029

In Russia (USSR), the first report on the possibility of using CVL for microprocessing, as mentioned, dates back to 1973 by Lebedev's Physics Institute. However, experimental studies on the application of CVL radiation for microprocessing a wide range of thin-sheet metal and non-metallic materials were carried out in 1983–1986 in the framework of the Kareliya project at the Istok company. In this case, a two-channel synchronized CVLS Kareliya operating according to the effective MO–PFC–PA scheme was used, with an average radiation power of 20–40 W with a PRF of 8–12 kHz and a diffraction quality of the beam [16, 33, 42, 190, 227, 230–233]. In the experiments, the Kareliya CVLS had an average radiation power of 25 W with beam divergence of 0.21 mrad (θ_{diff} = 0.07 mrad) at 8 kHz PRF [16, 190]. The pulse duration was 15 ns at a half-height, and the pulse energy was 3 mJ. The materials used for processing (in the form of plates 50 × 50 mm in size) were fixed on the coordinate X table, which could move at a constant speed within 0.6–1.7 mm/s. To focus pulsed radiation on the target, lenses with a focal length of 85 to 125 mm were used. In this case, the radiation was focused into a spot measuring 20–40 μm and the peak power density was 10^{10}–10^{11} W/cm^2.

As a result of the experiments, it was found that molybdenum (Mo), tungsten (W), tantalum (Ta), 0.6 mm thick, are efficiently cut at speeds of 0.6–0.7 mm/s; stainless steel (12Cr18Ni10Ti), copper (Cu), ceramics from aluminum oxide (Al_2O_3) with thickness up to 0.8 mm; aluminum (Al), and its alloys, textolite, dielectrics and semiconductors, as well as transparent materials. Virtually all of the listed materials were drilled through and through with a thickness of up to 3 mm during a time of 20–50 s. These successfully carried out first experiments testified to the prospects of using the radiation of CVL in technological processes of the production of electronic components.

The two-channel CVLS Kareliya was used to fabricate photomasks for printed circuit boards by evaporating a thin (0.2–0.3 μm) metal coating from a glass substrate with a focused beam of radiation. Such a coating evaporates under the action of one pulse with an energy of 2–3 mJ. The CVLM was used as a basis for the EM-5029 installation for high-speed automated production of large-format photomasks on

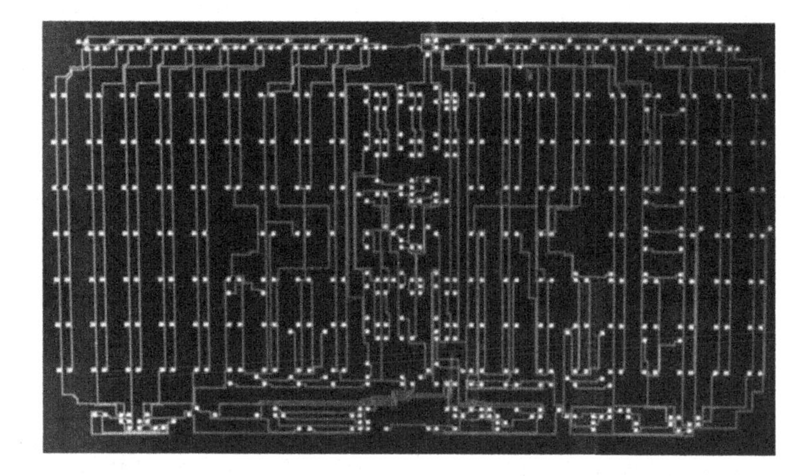

Fig. 2.2. A photomask on a glass substrate made on the EM-5029 installation on the basis of Kareliya CVLS.

a glass with a masking coating (copper, 0.3 μm thickness chromium). The maximum dimensions of the photomasks are 620 × 540 mm, the speed of the processing spot displacement is up to 200–250 mm/s. Even at such speeds, the processing time, depending on the size and density of the photomask, could be from one to tens of hours. The width of the working line varied from several tens of micrometers to 1.5–2 mm/s (the roughness of the edge was ~1 μm). The error in positioning the beam along the working field did not exceed 2–3 μm (provided by laser interferometers). Figure 2.2 shows a photomask produced on the installation EM-5029 with the Kareliya pulsed CVLS .

2.5. The first domestic experimental laser installation (ELI) Karavella

The first domestic ELI Karavella was created on the basis of pulsed CVLS Kareliya in 1987 (Istok company) [16, 33,190]. The ELI had a laboratory design and was intended for experimental studies on precision microprocessing of materials for the IET in the field of vacuum microwave technology [190, 312].

Composition and main parameters of the Karavella. The Karavella ELI (Fig. 2.3) consists of several structurally independent units: a two-channel CVLS radiator Kareliya, two upgraded power supplies IP-18 with a nanosecond synchronization system, two

Fig. 2.3. ELI Karavella for processing materials up to 2 mm thick.

X-axis horizontal tables for moving the processed material, a vertical table *Z* for moving the power focusing lens, the numerical control system 15IPCh-3–001, the television observation system, the suction system of the products of material destruction and blowing gas in the treatment zone and the control panel [16, 33, 190].

Focusing of radiation of the CVLS on the material to be processed, which is installed on a horizontal XY coordinate table, is made with an achromatic lens with a focal length of 100–200 mm. Due to the movement of the *Z* vertical table, the focused radiation spot is directed to the surface of the material. The radiation beam is transported to the working lens by an optical system of three rotating flat mirrors. The main parameters of the Karavella ELI are presented in Table 2.2.

In the radiator of the Kareliya CVLS working according to the scheme MO–SFC–PA, there are two versions of the optical system: with a telescoping unstable resonator (UR) (M = 180) or with one convex mirror (R_3 = 3 or 5 cm) in the master oscillator. When UR is used, as the investigations of the intensity distribution in the far field and pulse oscillograms have shown, the output radiation has a two-beam structure – a central beam with a diffraction divergence (0.07 mrad) and a beam with a divergence of 10 ns ($\Delta t = 2L/c$), 15 mrad (Figs. 2.4, *a* and *b*). When using an optical scheme with a single mirror, the output radiation has a single-beam structure with a divergence of 0.3 mrad, which is 4 times larger than the diffraction limit (Figs. 2.4 *c* and *d*). In the first case, the density of peak power in the focusing plane of the objective lens with F = 100 mm is $0.57–2.6 \cdot 10^{12}$ W/cm^2, in the second $1.4 \cdot 10^{10}$ W/cm^2, i.e. two orders less.

Table 2.2 The main parameters of the Karavella LTI

Parameter	Value
Radiation wavelength, nm	510.6; 578.2
Diameter of radiation beam, mm	20
Average radiation power, W	⊠ 20
Repetition frequency of pulses, kHz	10 ± 1
Energy at pulse, mJ with telescopic UR (M = 180)* total of individual beams with one convex mirror withy $R_z = 3$ (or 5) cm	 2 1 2
Divergence of the radiation beam , mrad with telescopic UR (M =180) with one convex mirror with $R_z = 3$ (or 5) cm	 0,07 (θ_{difr}); 0.15 0.3 (or 0.5)
Duration of radiation pulses (by half-height), ns with telescopic UR (M = 180): total individual beams with one convex mirror with R = 3 (5) cm	 15 10 20
Density of peak radiation power in focal spot, W/cm^2 with telescopic UR (M = 180) with one convex mirror, $R_z = 3$ (or 5) cm	 $(0.57–2.6) \cdot 10^{12}$ $1.4 \cdot 10^{10}$ (or $5 \cdot 10^9$)
Instability of energy at impulse,% with telescopic UR (M = 180) with one convex mirror with $R_z = 3$ cm	 5–10% 1–1.5%
Focal distance of the lens, mm	100
Working field of the horizontal table XY, mm Vertical table Z, mm	140 × 110 200
Accuracy of positioning on each axis, µm	±10
Gain of the observation system, number of times	× 100
Power consumption from network, kW	≤8
Readiness time, min	≥60
Consumption of water, l/min	≤8
Occupied area, m^2	≤12. 5
Weight, kg	≤1400
Working time to failure, h	500

* The structure of the output radiation at $M = 180$ is two-beam.

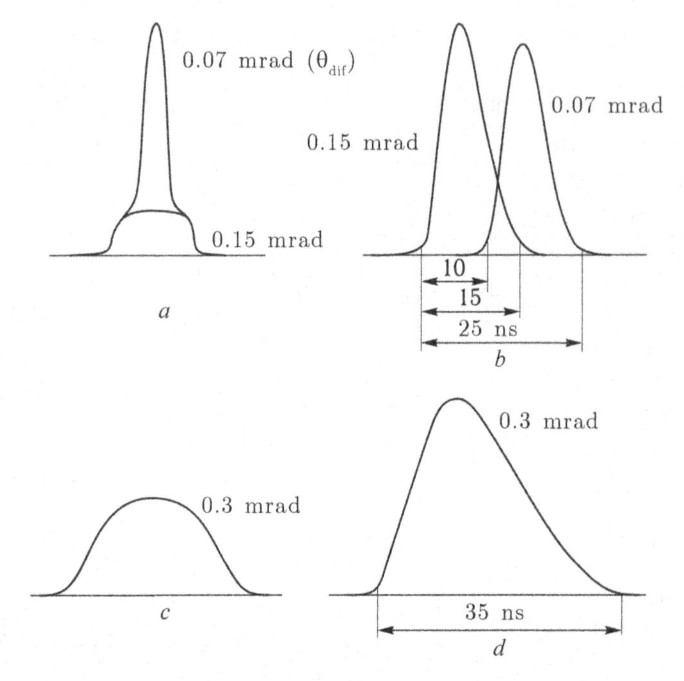

Fig. 2.4. The space–time structure of the output radiation of the Karavella ELI: the intensity distribution in the focusing plane (*a* and *c*) and the oscillograms of the pulses (*b* and *d*) in the modes with telescoping UR with the gain $M = 180$ – *a* and *b* and a single mirror with a radius of curvature of the convex mirror $R = 3$ cm – c and *d*.

Therefore, the speed of drilling and cutting with the telescopic UR will naturally be higher, but the roughness of the cut wall is much greater, since the instabilities of the position of the axis of the beam pattern of the diffraction beam are high and the intensity distribution in the far zone is uneven due to the two-beam radiation structure. The instability of the axis of the radiation pattern, as it was explained, is due to the influence of air–heat fluxes on the resonator field caused by high-temperature AE, and accordingly the design failure. Where a high quality of the cut is required, it is necessary to use a single-mirror mode, since we have single-beam radiation with a high stability of the axis position of its directional pattern [16, 33, 190, 230–233]. But in this mode the width of the cut is large (30–50 μm), which in many cases is unacceptable for microprocessing.

Other significant disadvantages of the Karavella are, first, the fact that the power supplies IP-18 do not provide an automatic mode of operation of the CVLS and require continuous manual adjustment of the power consumption and synchronization of the channels and, secondly, the low positioning accuracy for each axis of the *XYZ* three-axis table, which limits the possibility of manufacturing precision parts. The disadvantage of the ELI also includes a relatively long time to enter the operating mode of the CVLS, which is due to the heating of the discharge channel of the heat-intensive AE Kristall GL-201 to 1600°C. The AE readiness time can be slightly reduced due to the forced warm-up mode. The time to reach the nominal level by the radiation power is always less than the time to establish the position of the axis of the beam pattern of the radiation beam. The latter is associated with a more prolonged establishment of a stationary thermal regime in the entire (massive) installation.

In the Karavella computer system, the characteristics of the output two-beam radiation can be controlled in certain limits by the dissynchronization of the pulses of the emission of the master oscillator relative to the pulses of the power amplifier. In the case of the lead of the master oscillator signal, the diffraction beam receives the predominant gain, and the power in the beam with a greater divergence (0.15 mrad) becomes minimal. But completely to get rid of the influence of the beam with $\theta = 0.15$ mrad due to the dissynchronization is not possible, since the beams partially overlap in time. This problem can be solved either by suppressing the intensity of the beam with greater divergence by the spatial selection method (which turned out to be impossible in the ELI due to the instabilities of the beam axis of the diffraction beam commensurate with its divergence) or by using a new resonator that allows for the formation of single-beam radiation with diffraction divergence and with stable parameters. The ways of solving this problem are shown in chapter 2 of this book.

Main results and conclusions on material processing. During the operation period of the Karavella ELI, a large amount of research was carried out on contour cutting and drilling by direct piercing of various materials. A large number of experimental parts were used, which are used in microwave devices and other electronic devices. But because of the low quality of the laser radiation (as shown above) and the accuracy of the XYZ coordinate table moving, these parts usually did not meet the specified tolerances, roughness and heat-affected zone, so they were not used in specific devices. On the other hand, perspective possibilities of using CVL radiation with a nanosecond pulse duration for promising microprocessing of a large range of both metallic and non-metallic materials were demonstrated.

Table 2.3 Parameters of laser drilling of materials of different thickness in the Karavella installation

Material	Thickness, mm	Average radiant power at the exit lens, W	Focal length of the lens, mm	Drilling time, s
Cu, Al, Ag	0.3	20	110	< 0.1
Cu, Al, Ag	0.3	20	150	< 0.1
Cu, Al, Ag	0.3	22	230	0.15
Cu, Al, Mo	0.6	21	110	0.2–0.3
Cu, Al, Mo	0.6	21	230	
Cu	0.8	20	110	0.8–1.1
Cu	1.0	22	110	1.3–1.6
Cu	1.0	22	230	1.6–1.8
Cu	1.5	22	110	4.2–4.4
Cu	1.5	22	150	4.2–4.8
Cu	1.5	22	230	4.2–4.8
Cu	2.0	20	110	10–12
Cu	2.8	22	110	25–28
Al, D16T	1.8	21	230	5–7
12Cr18Ni10Ti	0.8	23	230	0.6–0.8
U8 A	1.8	23	230	1.0–1.2
W	2.0	20	110	22–25
VK-6	2.3	23	230	45–50

The main results of the investigations were published in [16, 26, 33, 190, 242–251, 271, 312].

Table 2.3 (obtained from experiments) the time of drilling of metallic materials on the Karavella ELI by the direct piercing method.

The average drilling speed of thin ($\leq$0.6 mm) metal plates was 2.5–3.0 mm/s, for thicker (1–2 mm) – 0.2–0.25 mm/s. In this case, the coefficient of the channel shape increases linearly with increasing depth of the hole and reaches values of 30–100. The processing speed is significantly influenced by the peak power density in the focused

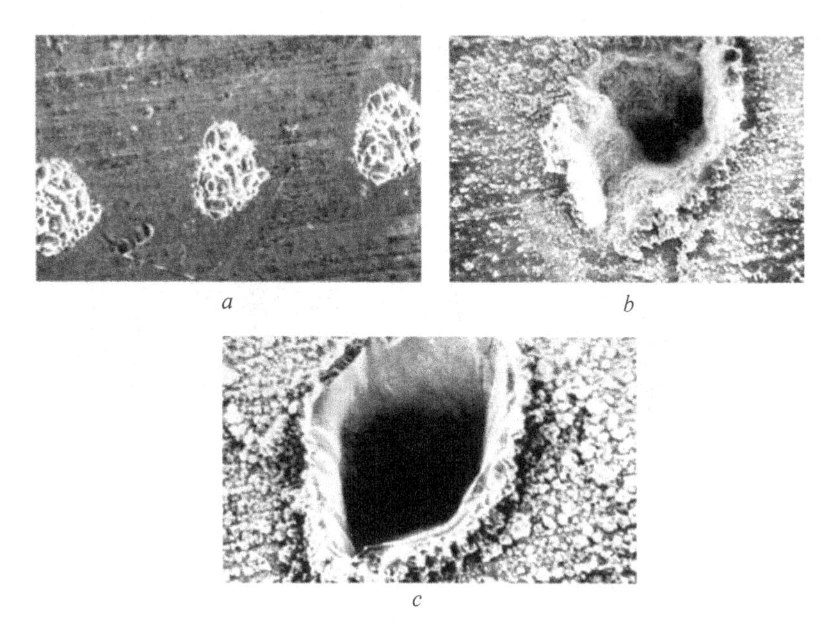

Fig. 2.5. The process of formation of holes in the metal under the action of focused radiation of CVL with a telescopic unstable resonator UP with a gain M = 180 with an average power of 20 W and a pulse duration at half-height τ_{pulse} = 20 ns (f = 10 kHz): under the action of single pulses (*a*) t = 0.2 s – 200 pulses (*b*) and $t \geq 2$ s – 20 000 pulses (*a*).

spot and the PRF. But because of the small amount of energy in the pulse, the dependence on the thickness of the material is the most significant and unambiguous. The decrease in productivity with increasing thickness is associated with the diaphragmatic transition of the beam at the emerging hole, which is confirmed by the similar form of the dependence of the fraction of power transmitted through the hole on the thickness [33, 250].

Figure 2.5 shows the images of metal plates during the formation of holes in them, which indicate a predominantly evaporation mechanism for the removal of material.

From these figures it can be seen that the dimensions of the spots on the surface of the metal plate (*a*) and the holes in it (*b* and *c*) formed under the influence of focused radiation do not correspond to its divergence. With maximum divergence θ = 0.15 mrad (Fig. 2.5 *a*) and a focal length of the objective F = 100 mm, the diameter of the focused spot should be 15–20 μm, and the real dimensions of the spots and holes are 70–100 μm. The spots and holes have uneven edges and a non-circular shape and there is a melt along the perimeter of the holes. Such a pattern at the inlet of the holes is

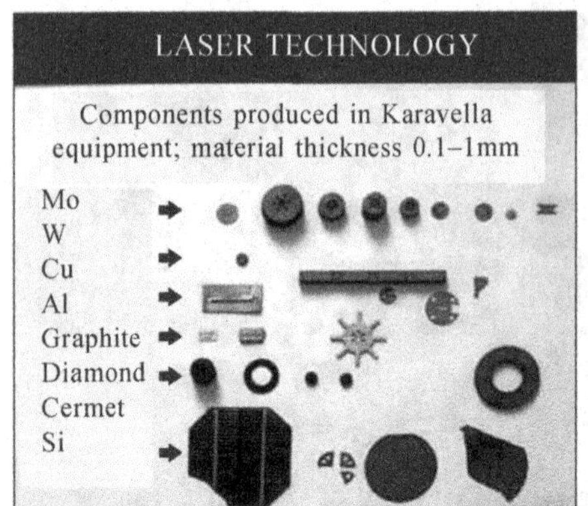

Fig. 2.6. Experimental details made on the Karavella ELI. Left column from top to bottom: Mo, W, Cu, Al, graphite, diamond, cermet, Si.

due, first, to the instability of the position of the axes of the beam pattern, commensurate with their divergence, and, secondly, to the non-uniform intensity distribution in the focused spot (constriction). All experiments were conducted under ordinary conditions in an air atmosphere. The effectiveness of the processing process is also influenced to a certain extent by the blowing of active process gases. The Karavella installation mainly uses oxygen and air, fed into the treatment zone through an axial nozzle with a diameter of 0.8–1.0 mm. The use of active gas intensifying the processes of destruction is most beneficial in the case when the quality of the cut surface of the workpiece at this stage does not have a significant value: for example, if further chemical etching or electric sparking of these parts is envisaged. For example, the drilling speed of a plate made of carbon steel St.10 with a thickness of 0.6 mm was the maximum with air blowing.

When laser processing thin metal plates on the ELI, the average width of the through-cut was usually 30–50 μm throughout the entire length of the sample, which is 5–10 times smaller than the limiting emissivity of the CVL. One of the features of cutting pulse radiation is the appearance of the effect of channeling. This effect is expressed by dragging the diffraction beam (having the maximum power density) into the channel formed by the previous pulses by means of reflection from its wall. The formation of a new channel begins after the entire diffraction beam is shifted beyond the contours of the previous one. This process of channeling and instability of the beam pattern determine the ultimate roughness of the cut wall,

Table 2.4. Comparison of the production capacity of parts by laser microprocessing at the Karavella ELI with traditional processing methods

General view	Name	Material, thickness	Labour content, norm-hours	
			Karavella ETI	Traditional technology
	Spherical mesh EVP	MChVP Molybdenum, 0.1 mm	0,3	5.2–6.5 electroerosion
	Microwave energy absorber	Cermet 0.5 mm	0,2	3.0 with diamond tool
	EVP flat mesh	MChVP Molybdenum, 0.2 mm	0,1	1 drilling
	EVP flat mesh	Pyrographite 0.2 mm	0,1	1.5 drilling
	Plate	Polycrystalline diamond 0.75 mm	0,2	–
	Cooling device part	Copper 0.2 mm	0,2	–

which can be stabilized by multipass processing. The roughness was usually 4–6 μm, which can be considered quite satisfactory for this installation. And it should be expected that when obtaining a diffraction-grade single-beam radiation with stable parameters and using modern precise coordinate tables *XYZ*, the accuracy and roughness of processing can be reduced by an order of magnitude ≤1 μm.

Figure 2.6 shows a number of experimental details of thin sheet metal and non-metallic materials made at the Karavella ELI.

A special problem in the production of electronic components in traditional methods of processing is the shortening of the terms for the preparation and production of single, small and medium batches of molybdenum, copper, tungsten, aluminum, silver, etc. items. The expansion of the range of complex parts leads to the fact that the used for their manufacture of the traditional methods of electroerosion treatment (EET) and microstampings become ineffective, and in some cases – unacceptable, since they require additional time and labour. As demonstrated by the experiments at the Karavella ELI, precision microprocessing of thin metal parts by laser radiation is more rapid and promising, since it does not require the production of special tools and tooling, and the stage of technological preparation consists mainly in compiling and debugging the control program. An example of a quantitative comparison of the characteristic expenditure of time and labor (in the norm-hours) for the production of the same types of parts using the laser processing method and other methods is shown in Table 2.4. Laser processing is more productive than the others (including EET) by more than an order of magnitude [16, 101].

Preliminary studies at the Karavella ELI showed that the pulsed emission of CVL is also promising for the formation of images in the volume of transparent materials [16, 20]. The effect of focused radiation was applied to polished samples of K8 optical glass, fused quartz of KI, KV and KU grades, sapphire, and also artificial polycrystalline diamond. In this case, the destruction of the material occurs both on the surface and in the volume without reaching the surface. When processing inside a volume, the effect of 'accumulating' energy takes place, after which destruction begins. The time of such 'accumulation' is also related to the distance to the surface of the sample, which suggests the possibility of the influence of near-surface defects on the process of destruction. The peak power density required for the destruction of the materials studied increases in the sequence K8–KI–KV–KU–sapphire. The processing of glass by means of the radiation of CVL finds practical application mainly in the decorative and artistic area, but can be used for technological purposes: for volumetric marking of serial samples and creation of fixed defects in reference samples of diagnostic equipment [191, 193].

A separate important practical task is the processing of diamond materials. Its relevance is connected with the appearance in recent years of artificial polycrystalline diamond plates obtained by deposition from the gas phase. This material ($\alpha_T = 1.1 \cdot 10^{-6}$ deg, $T_m = 4000$°C) retains its dimensions and mechanical properties at high temperatures, has a high transparency ($\tau = 98\%$) and a radiation absorption limit of 0.2 μm, electrical strength 10^7 V/cm, the work function is 4.7 eV. Its thermal conductivity is 4–5 times greater than that of copper. Therefore, polycrystalline diamond is a promising material for a variety of applications – as substrates for integrated circuits, windows for the output of high-power microwave energy,

Table 2.5. The results of the thermal impact on artificial polycrystalline diamond in the treatment with radiation of CVL

T °K (not less)	Impact result
1000	Combustion
1700	Graphitization
2300	Fast graphitization
3800	Melting and evaporation

Table 2.6. Threshold values of peak radiation power density for destruction of artificial polycrystalline diamond with different types of pulsed lasers

Laser type	Length wave, μm	Threshold peak power density, n × 10^8 W/cm^2
XeCl	0.31	0.5
CVL	0.51; 0.58	1.2
Nd:YAG	0.53	8
Nd:YAG	1.06	17.5
CO_2	10.6	13.2

waveguides, micromechanical devices, cathodes of electrovacuum devices, etc. The presence of various defects in the crystal lattice of a diamond with a total concentration up to 10^{21} cm^{-3} leads to the appearance of additional absorption levels and photoionization bands, one of which, associated primarily with dislocations and lying in the region 2.0–2.3 eV, coincides yields with the radiation lines of the CVL. It should be noted that of all forms of carbon only diamond is a dielectric. Any structural change in it leads to the appearance of conductivity. When compared with other methods of processing, precision cutting of graphite by radiation of CVL is recognized as the best. The results of the thermal impact on diamond are shown in Table 2.5.

The threshold value of the power density of laser radiation, which causes the destruction of the material, is determined by its nature and concentration of defects. These threshold values for polycrystalline diamond under the influence of nanosecond pulses of different lasers are given in Table 2.6.

The threshold value of the radiation power density of CVL for natural diamond is (1.5–2) · 10^6 W/cm^2, and for a polished plate made of artificial polycrystalline diamond obtained by the method of deposition from the gas phase, $1.2 \cdot 10^8$ W/cm^2 (see Table 2.6). For a noticeable evaporation of this diamond, it is necessary that the power density on the surface exceeds $2.7 \cdot 10^8$ W/cm^2. The average speed of drilling an artificial diamond 1.2 mm thick by CVL radiation with an energy of 1 mJ pulse when focused into a spot with a diameter of 30–40 μm is 8–9 nm per pulse. Processing of thin samples (0.2 mm) is two orders of magnitude more efficient. The increase in power results in a non-linear increase in processing capacity.

A possible prospective direction of technological installations on the basis of CVL is microprocessing of thin-film coatings (0.1–10 microns). Also, there is reason to believe that the use of a fiber-optic cable will allow us to rationally solve the problem of supplying the radiation energy of CVL for vacuum spraying or accelerated dimensional processing of products in chemically active liquids and gas media, which opens the way for the creation of new technological processes and new type installations.

The analysis of foreign and domestic literature conducted in this chapter and the first experimental studies of microprocessing in the Karavella ELI allow us to conclude that the pulsed emission of CVL is a promising precision tool for effectively affecting virtually any metallic materials and a large group of dielectrics and conductors, until recently not included in the sphere of laser microprocessing [16, 26, 33, 242–251, 271, 312].

2.6. Conclusions and results for Chapter 2

1. Laser processing of materials is one of those technologies that determine today the current level of production in the industrialized countries. Its distinctive features are the high quality of the products, high process productivity, saving of human and material resources and environmental cleanliness.

Today, the individual technological lasers are not delivered to the market, but delivered are technological units and complexes using gas CO_2 and excimer lasers, solid-state and fiber diode-pumped lasers and diode lasers for dimensional processing, material cutting, welding, surface treatment, alloying, surfacing, marking, engraving, etc.

2. A special place is occupied by laser systems for precision microprocessing, microdrilling, with high resolution marking and engraving, microwelding of the components for electronics. As quality sources of radiation at these systems may effectively be

used and already are used short pulse (<100 ns), high-frequency (up to 30–50 kHz), with small energy at pulse (0.1–1 mJ) and small reflectivity reflections (<80–90%) and high density peak power (10^9–10^{12} W/cm^2) lasers of the visible and ultraviolet spectrum: solid-state with diode pumping and doubling frequency, gas excimer and nitrogen, and, specifically, CVL with medium power 5–50 W and duration pulses of 20–50 ns.

3. The potential capabilities of the pulsed CVL for processing of materials from use of intrinsic yellow–green radiation and UV-radiation from doubled frequency were first widely examined in the monograph by C.E. Little 'Metal Vapour Lasers' (1999) which included the work of well-known researchers and developers at this region. The monograph presented not only research work, but also concrete examples of the processing of metallic and organic materials, semiconductors and dielectrics. At the same time the Istok company using its own CVL, stimulated and conducted studies of the creation of laser technologies for high-quality micromachining of materials for the new generation products for RF-electronics.

4. The first foreign advertised machine with a copper vapour laser designed for productive microprocessing of thin-walled materials (0.01–2 mm) was the first installation of the MP200X model developed by the British firm Oxford Lasers (2002). The device is based on CVLS of the type master oscillator–power amplifier with an average radiation power of 60 W and a precision XYZ three-axis table with a working field of horizontal XY 250 × 250 mm and the accuracy of positioning along the axes ±0.1 μm.

The main and significant disadvantage of the MP200X installation is the pumping of the operating mode of the active elements of the CVLS, that is, the AE manufacturing technology does not provide a sealed mode and additional elements are required for its support (a pumping pump, gas cylinders, reducers, monitoring sensors, mixer). The pumping mode of operation reduces the life of the laser and worsens the stability of the radiation parameters and, with each new activation, an additional adjustment is required to restore the operating power level.

5. The first domestic experimental studies to assess the possibilities of using pulsed radiation from CVL as a precision tool for microprocessing thin-film materials were carried out at the Istok Co. (1985–1986) on its own two-channel CVLS Kareliya, operating according to the effective MO–PA scheme, with an average radiation

power of 20–30 W, a 10 kHz PRF, a pulsed energy of 2–3 mJ and a peak power density in a focused spot (20–40 μm) of 10^{10}–10^{11} W/cm^2.

The Kareliya CVLS has become the foundation for creating the first domestic experimental technological installations of the type EM-5029 and Karavella LTI.

6. The EM-5029 installation shows the possibility of high-speed automated production of large-format photomasks for printed circuit boards by evaporation with a focused pulsed light beam of thin-film coatings (0.2–0.3 μm) on a glass substrate. The maximum dimensions of the produced photomasks were 620 × 540 mm, the processing speed was up to 200–250 mm/s, the working line width varied from several tens of micrometers to 1.5–2 mm with a roughness of the edge of about 1 μm. The error in positioning the beam along the working field did not exceed 2–3 μm.

7. The Karavella LTI with the working field of the XY horizontal table 0140 × 100 mm and accuracy positioning by the axes of ±15 μm and focusing lens from F = 100 mm on the Z vertical table was used to carry out extensive experimental research by working out technologies of micromachining materials for electrovacuum instruments of microwave technology. It is shown that the radiation of CVL with a peak power density of 10^{10}–10^{11} W/cm^2 (diameter of the processed spots 20–40 μm) can be used in comparison with the traditional methods, including EET, to process with greater productivity (up to 20 mm/s) practically any metal and a large range of non-metallic materials: produce contour cutting (speed to 10–20 mm/s) Mo, W, Cu, Al, Ag, Ta, Ni, D16T, steel, iron, ceramics A-995 and 22KhS, glass, quartz, silicon, germanium, sapphire, artificial diamonds, cermets thick up to 1 mm and direct piercing of holes in these materials thick up to 2–3 mm from with the shape coefficient k_s = 30–100. The average drilling speed of metal plates with a thickness ≤0.5 mm was 2.5–3.0 mm/s, a thickness of 1–2 mm – 0.2–0.25 mm/s, reaching shape coefficients of 30–100.

8. The experimental Karavella laser system has a number of significant deficiencies that do not allow to produce precision components with a small HAZ and roughness for real electronic components. This shortcomings include a high error of positioning at the axes of the XYZ coordinate table (±15 μm), instability of the positions of the axis of the directional pattern of the beam, commensurable with its divergence and caused by the impact of heat flows on the field of unstable resonators, and the uneven distribution intensity at far zone defined by the two-beam structure of radiation.

3

A new generation of highly efficient and long-term industrial sealed-off active elements of pulsed copper vapour lasers of the Kulon series with a radiation power of 1–20 W and Kristall series with a power of 30–100 W

In the active medium of pulsed CVLs, a high degree of population inversion, i.e., maximum generation, is provided at discharge channel temperatures of 1600–1700°C, when the concentration of copper atoms is 10^{15}–10^{16} cm^{-3}. Therefore, when high-temperature active elements (AEs) are created on copper vapours that have high efficiency (power and efficiency), durability, stable and reproducible parameters and retentivity, higher requirements are imposed on its individual elements, nodes and design in general. The choice of materials for the elements of the AE design is limited to a set of stringent requirements: they must have high thermal and chemical resistance, intercompatibility at high temperatures, resistance against the action of molten copper, low gas separation, low thermal conductivity, mechanical strength and vacuum density for prolonged operation under high temperatures, non-toxicity, acceptable cost.

3.1. Analysis of the first designs of self-heating AE pulsed CVLs and the reasons for their low durability and efficiency

The basis of the basic design of industrial sealed-off AEs on copper vapours is the principle of self-heating with an intra-vacuum arrangement of heat-insulating material, proposed by the employees of the P.N. Lebedev Physical Institute and The Istok Company [222]. A schematic representation of self-heating AEs is shown in Fig. 3.1. The AE consists of a high-temperature discharge channel 1, electrode assemblies 2, an active substance – metallic copper 3, a vacuum-tight glass shell 4, a refractory powder heat insulator 5, and optical windows 6 for the output of laser radiation. There is a gap in connections 7, between the ends of the discharge tube 1 and electrode assemblies 2. The gap ensures the free movement of the discharge channel along the axis during the heating and cooling of the AE and the evacuation of the evolved gases from the volume of the heat insulating element 5. To avoid the penetration of the heat-insulating powder into the discharge channel (active volume) the size of the gap must be smaller than the dimensions of the powder particles. In self-heating AE the heating of the discharge channel with the active substance to the operating temperature (1600–1700°C) and maintaining the steady temperature occurs due to the dissipation of the energy of the pulsed gas discharge, which simultaneously excites the copper vapours. In this case, the fraction of the energy converted into laser radiation is 1–3%.

The design of the first generation of industrial sealed-off AE pulsed CVLs of the Kulon and Kristall type with the working

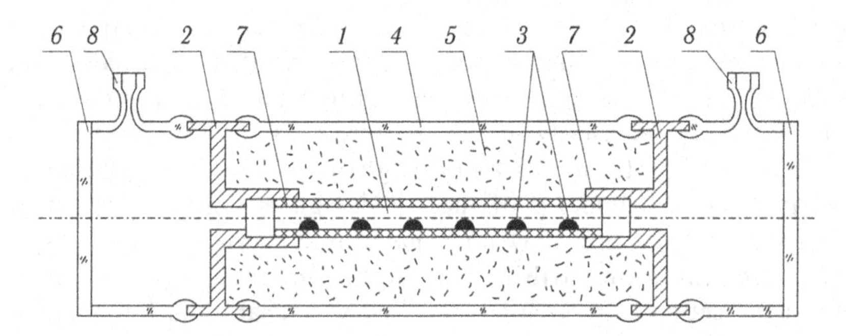

Fig. 3.1. The design of self-heating AE with the intra-vacuum arrangement of the heat insulator: *1* – discharge channel; *2* – electrode assemblies; *3* – active substance - copper; *4* – vacuum-tight shell; *5* – refractory powder heat insulator; *6* – output windows; *7* – connecting units; *8* – tube for filling and pumping gases.

temperature of the discharge channel 1500–1700°C also took the design of a self-heating AE with an intra-vacuum arrangement of the heat insulator [222]. As a result of the analysis of the designs of the first AE models: TLG-5 (1974), UL-101 (1976), GL-201 (1982) and GL-204 (1988), materials used in them, production and training, and excitation conditions of the plant, significant design and technological defects were found, the reasons for the low efficiency of pumping the AM, and the non-compliance of a number of materials with thermophysical and chemical properties with the requirements for high-temperature AE, which are the reasons for low durability, efficiency, and quality of the output radiation beam [16].

The main identified shortcomings of the first industrial sealed-off self-heating AEs are the following:

– the cylindrical layer of the thermal insulator directly surrounding the high-temperature discharge channel (T_{chan} = 1500–1700°C), in the TLG-5 and UL-101 AEs is made of grinding powder of Al_2O_3 oxide with fraction sizes of 125–160 μm with relatively high thermal conductivity (λ = 0.47 W/(m · K)) and specific density (p = 1.9 g/cm^3), the outer shell in the UL-101, GL-201 and GL-204 is a massive cermet. The use of these structural materials in the AE led to high power consumption and availability and low efficiency of the CVL;

– the exit windows of the AE are made of uviolet technical glass of UT-49 brand with a large content of swills and bubbles and are welded to the end glass sections in the flame of the gas burner and at an angle close to 90°, which leads to distortion of the structure and poor quality of the output radiation beam;

– rigid design for fastening the ends of a high-temperature ceramic discharge channel to electrode assemblies, limiting the free movement of the channel along the AE axis during its heating and cooling and leading to cracking of the channel elements;

– the use of a tantalum shell in the copper vapour generator, which causes chemical interaction Ta with ceramic bushings and discharge tubes from Al_2O_3, and the use of thin-walled ceramic tubes in the discharge channel (as in TLG-5), deforming even with insignificant overheating, lead to premature destruction of the discharge channel and, accordingly, a decrease in the life of the AE;

– the applied tungsten–barium cathodes had a 'smooth' surface, which did not provide stable local combustion of a pulsed arc discharge with nanosecond duration and caused instability of the output radiation parameters;

– at the ends of the discharge channel there is an accumulation of condensed copper droplets overlapping its aperture and reducing the power and distorting the radiation structure. The presence of 'large' drops indicates an incorrect location of copper vapour condensers and a poorly designed structure of the near-electrode region of the AE;

– low output power (3–6 W) and efficiency (0.15–0.25%), due to low pumping efficiency and high power consumption of AE (2–2.5 kW). The electrical circuit of the high-voltage pump modulator of the laser power source is made according to the classical (direct) scheme. In this case, the high-voltage pulse commutator (hydrogen thyratron), the storage capacitor and the AE form a single discharge circuit, when the duration of the leading edge and the total duration of the generated pump current pulses are determined mainly by the characteristics of the thyratron and are about 100 and 300 ns, respectively, which is not sufficient for effective conditions excitation of AS;

– reduction of the output radiation power during the operation of the CVL, due to the imperfection of the developed technology of training for degassing the AE and dusting its exit windows with vapours of the working substance (copper), products of erosion of the electrodes and the discharge channel;

– low guaranteed operating time of AE, caused by a number of the above-mentioned shortcomings (criterion according to TU (Technical Instructions) – reduction of the radiation power is not more than 25%): 200–300 h – running of AE of the models TLG-5 and UL-102; 500 h – operating hours of AEs of the models GL-201 and 204.

Despite the relatively low values of longevity, efficiency and quality of radiation of the first industrial sealed-off self-heating AEs of the pulsed CVL models TGL-5, UL-101, GL-201 and GL-204 [16, 26] associated with a number of 'unsuccessful' design and technological solutions, the use of a low-efficiency heat insulator and output windows with low quality, the absence of electrical circuits of a high-voltage modulator of the power source for the formation of short-pulse pump currents with steep fronts, of the structure and the spatiotemporal and energy characteristics of the output radiation from the excitation conditions of the active medium (power consumption, pump pulse parameters, pulse repetition frequency, discharge channel temperature, neon buffer gas pressure, hydrogen additives), the type of optical resonator and its parameters, durability tests, to 1980–1990, the issue of the further development of pulsed CVLs became increasingly urgent.

At that time, the main areas of application of pulsed CVLs in science, engineering and medicine were outlined and the main promising directions were identified, in which the greatest efficiency is achieved with the help of the CVL [12–24]. In such promising areas of pulsed CVL application, it is necessary to include the precision microprocessing of materials for electronic products, the separation of isotopes (for example, ^{235}U uranium enrichment) and the production of high-purity substances for the needs of nuclear energy and medicine, the pumping of wavelength-tunable lasers on dye solutions (DSL), laser projection microscopy (intensification of the image brightness of microobjects), spectral analysis and diagnostics of the composition of substances, sounding of the atmosphere and sea depths, logical and non-neoplastic diseases by photodynamic therapy, dermatology and cosmetology and others. This was the impetus to carry out extensive research aimed at improving the characteristics of the CVL. Already in 1974–1975 in CVLs with an unstable resonator of the telescopic type, beams with a diffraction divergence were obtained with an increase of hundreds of times [91–94], and in 1977 [77] it was reported that an average radiation power of 43.5 W with a practical efficiency of ~1% at the PRF up to 20 kHz. By 1979, the Lawrence Livermore National Laboratory (LLNL) constructed an CVLS, working according to the master oscillator–power amplifier scheme, out of 21 AEs with a total output power of 260 W (within the AVLIS program) [20]. A detailed review of the state and development of pulsed CVLs is given in Ch. 1.

3.2. Investigation of ways to increase the efficiency, power and stability of the output radiation parameters of CVL

To determine the ways of increasing the power, efficiency and stability of the output radiation parameters of sealed-off self-heating AEs, experimental studies were carried out of CVL with the Kristall industrial AE of the GL-201 model and the experimental Kristall of the GL-201D and GL-201D32 type, which are the first most powerful sealed-off self-heating AEs with copper vapour generators with a tantalum sheath, with free copper disposition in the discharge channel, with pseudoalloys of the composition W–Cu and Mo–Cu and generators on a molybdenum substrate at pressures of the buffer gas of neon in the range of 40–760 mm Hg and fixed PRF with different electrical circuits for the execution of a high-voltage pulse modulator of the PID. All research results have been widely presented in co-authorship in the monograph [16, Ch. 3].

The structures of the Kristall AE used for the first time a non-conductive, metal-porous W–Ba cathode with an active substance of the composition $3BaO \cdot Al_2O_3 \cdot 0.5CaO \cdot 0.5SiO_2$ (barium aluminosilicate) of a ring structure and with an annular groove on the inner working surface with which the first positive results on stable local combustion of a pulsed arc discharge in the auto-thermoemission regime, a two-layer high-temperature heat insulator with a low coefficient of thermal conductivity ($\lambda = 0.27$–0.31 W/(m · K)) located between the discharge channel, a vacuum-tight envelope and the electrode assemblies, the two-stage training technique for degassing and cleaning it up to 60 hours, allowed to preserve the purity of the gaseous medium (neon) for over 2000 hours.

A comparative analysis of the experimental data on the pumping efficiency of the CVL AE with different execution of the electrical circuit of the high-voltage pulse modulator of the power source was carried out and it was found that with the electric circuit of the capacitive voltage doubling and the magnetic links of the current pulse compression in comparison with the direct classical high-voltage modulator of the power source and when using a hydrogen thyratron as a high-voltage pulse commutator the duration of discharge current pulses in the AE was halved (from 250–300 ns to 120–150 ns), which leads to an increase in the radiation power and efficiency by about 2 times due to an increase in the optimum concentration of copper vapour in the active medium by about 2–2.5 times (the operating temperature of the discharge channel increases), the service life of the pulsed hydrogen thyratron (up to 1500–2000 h) and the switched power (up to 5–10 kW) are multiplied several times due to a reduction of the power losses in the thyratron, which is a high-voltage commutator. The power source with this scheme works steadily in the PRF range of 3–11 kHz, the amplitudes of the voltage pulses 15–30 kV and the discharge current of 0.2–1.0 kA with a leading edge duration of 40–50 ns with a base duration of 120–150 ns; the advantage of the electrical circuit of the power source modulator with a GI-29A vacuum lamp as a high-voltage pulse commutator is the possibility of forming current pulses with a front duration of 20–30 ns at a total duration of 50–70 ns commensurate with the time of population inversion existence at voltages of 20–30 kV and PRF tens and hundreds of kHz and, accordingly, the possibility of achieving maximum values of radiation power and efficiency, the disadvantage is that the amplitude of the current pulses does not exceed values greater than 300 A because of the limitation of saturation current.

The maximum efficiency and output power of the CVL with the Kristall AE of the models GL-201, GL-201D and GL-201D32 in the sealed-off mode (especially at low neon pressures in the AE – 40–100 mm Hg) is achieved with copper vapour generators on a molybdenum substrate after their reduction by hydrogen at the working temperature of the discharge channel (T_{chan} ~ 1600°C), which is carried out after a full cycle of two-stage degassing of the AE.

At pressures of the buffer gas of neon in the AE in the range 50–250 mm Hg additions of pure hydrogen (with a partial pressure up to 10 mm Hg) lead to an increase in the radiation power by 1.5 times or more, at pressures of neon close to atmospheric and at atmospheric pressure hydrogen has no appreciable effect on the radiation power.

When the neon pressure in the AE varies from 50 to 760 mm Hg, and the optimum power consumption from the power source, the radiation power decreases monotonically. For example, with the industrial AE GL-201 at 10 kHz PRF, the total average radiation power is reduced from 34 W (at a practical efficiency of ~1.0%), to 20 W (efficiency ~0.67%). The decrease in the total radiation power is mainly due to a decrease in power at the green wavelength (λ = 0.51 μm), since the excitation conditions deteriorate with increasing pressure. The power on the yellow line (λ = 0.58 μm) varies little.

With optimum power consumption, the efficiency of the sealed-off AEs pf the CVL of the Kristall series is approximately 2 times greater than the practical efficiency of the laser and is about 2% at relatively low neon pressures and 1.3% at atmospheric pressure.

It has been experimentally established that the highest values of efficiency in CVL can be achieved with power consumption less than optimal values, but with the condition of maintaining a high operating temperature of the discharge channel (~1600 °C). With the active element GL-201 in transient operation modes, it was shown that it is possible to obtain a practical laser efficiency up to 1.8%, an efficiency factor of AE to 3%.

At pressures of the buffer gas of neon in the AE of the CVL close to atmospheric and at atmospheric pressure, to obtain values of the radiation power commensurate with the values at low pressures, the intensity in the gas-discharge gap of the AE should be at least 30 kV/m when current pulses with a front duration of not more than 50 ns with a total duration of ≤150 ns and a rate of increase greater than $4 \cdot 10^9$ A/s are formed. At the neon atmospheric pressure, 9 kHz PRF, and pulsed voltage at the AE GL-201 ~28 kV ($\tau_f \sim 50$ ns and $dI/dt \sim 4.2 \times 10^9$ A/s) the radiation power was 26 W, which is only 1 W less than at $p_{Ne} = 250$ mm Hg.

During the period from 1975 to 1998, more than 600 industrial sealed-off self-heating AEs of the pulsed CVL of the first generation (TLG-5, UL-101, GL-204, GL-201, GL-201D and GL-201D32) were produced.

3.3. Choice of directions for the development of a new generation of industrial sealed-off self-heating AE of the CVLs

The directions for the development of a new generation of industrial sealed-off self-heating AEs of the pulsed CVLs were first identified in the laboratory Lasers and laser technologies of the Istok company. For the period up to the year 2000, a large amount of both experimental and theoretical studies (Chapter 1) was carried out by the staff of this laboratory, in order to improve the longevity, efficiency, quality and stability of the output beam radiation parameters of sealed-off self-heating AEs of the pulsed CVL. At the initial stage, a significant contribution to the development of CVL was made by the work performed within the framework of the R & D projects Kristall (1979–1980) and Kristall-1 (1981–1982), Kvant (1981–1982), and then in the R & D projects Kareliya (1983–1986), Kulon (1987–1988), research works Kuban' (1989–1990) and Kolumbiya (1991–1992). In these works, the physico-chemical properties of the main structural materials and the design of the functional units of the AE, responsible for the longevity, efficiency and reproducibility of the parameters in the process of long-time production (Chapter 1), were analyzed and studied at a high scientific and technical level. Within the framework of these activities, three main types of sealed-off self-heating AEs were created (Fig. 3.2): the Kulon of the GL-204 model, the Kristall GL-201 and the Kvant UL-102. They were the basis for choosing the directions of development of modern industrial sealed-off self-heating AEs of the pulsed CVL.

Instead of the old Kvant device of the UL-101 model, the AE of the new model – UL-102, designed as an image brightness intensifier and visual control of microelectronic products [24], was created, with twice the efficiency and minimum operating time. Due to the decline of domestic microelectronics, the demand for laser image

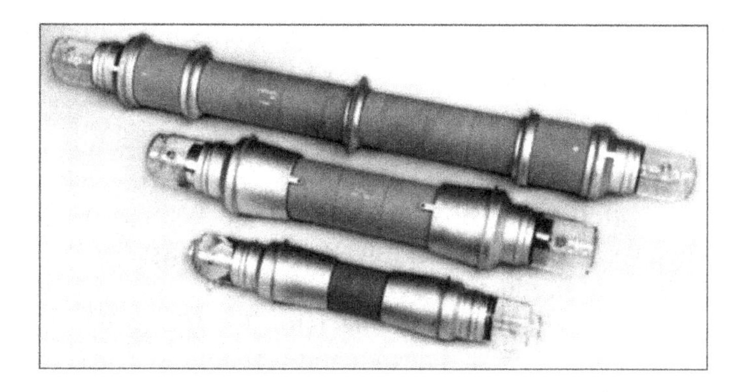

Fig. 3.2. The first industrial sealed-off self-heating laser AEs on copper vapours: Kristall (GL-201), Kvant (UL-102), Kulon (GL-204) (from top to bottom).

amplifiers of microobjects has fallen, and this class of devices has not been further developed.

The Kulon AE of the model GL-204 with an average radiation power of 3–4 W became the basis for the development of the direction of low-power AEs of the CVL. A new generation of small-sized industrial sealed self-heating AE series Kulon and based on them the CVL with air or water cooling, designed for the technological and medical research of the composition of substances, spectroscopic studies, nanotechnology, etc. has been developed. Today the AEs Kulon (GL-206 according to the Technical Instructions (TU) on copper vapour with an average radiation power of 1–20 W and on gold vapour with a power of 1–2 W [16, 41, 42, 268–271] are produced.

The second direction, which has been successfully developed, is the development of a new generation of industrial sealed-off self-heating AE series Kristall (GL-205 according to TU) on the basis of the first sealed-off self-heating AE of the Kristall models GL-201, GL-201D and GL-201D32 with the average power of radiation of 30–100 W. This series of industrial AEs is mainly used in pulsed CVLS of the MO–PA type, intended for modern technological equipment for precision microprocessing of 'thick-plate' materials (1–2 mm), isotope separation and production of highly pure substances for nuclear power and medicine, etc. [16, 41, 42, 227–233, 252–264, 269–271].

3.4. Appearance and weight and dimensions of industrial sealed-off AEs of the pulsed CVL of the Kulon and Kristall series

A new generation of industrial sealed-off self-heating AEs of the pulsed CVL Kulon series of low power (1–20 W) and the Kristall series of the average power level (30–100 W) was first developed and prepared for serial production over the last decade in the laboratory Lasers and laser technologies of the Istok Co (Fryazino, Moscow region). The work to improve design and technology of AE training, in order to further improve their durability, efficiency, quality and stability of output parameters, continue to this day.

Appearance of the new generation of industrial AE series Kulon and Kristall pulsed CVLs is shown in Fig. 3.3. Conventional designation of the AE series Kulon in terms of technical specifications (TU) – GL-206A, GL-206B, GL-206C, GL-206G, GL-206D, GL-206E, GL-206ZH and GL-206I, Kristall series – GL-205A, GL-205B, GL-205V, GL-205ZH and GL-205G [16, 41,42, 227–233,252–264, 268–271]. The operating position of the AE is horizontal and fixed usually behind the connecting points in the electrode assemblies (in Figs. 3.4 and 3.5 – the size *a*). In Fig. 3.4 is a schematic representation, in Table 3.1 – overall and connecting dimensions and weight of new industrial sealed AE of the Kulon series. Figure 3.5 is a schematic representation, in Table 3.2 – overall

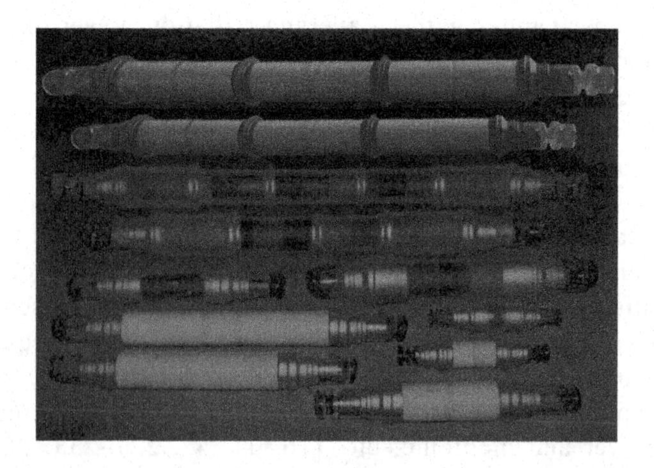

Fig. 3.3. Appearance of industrial self-heating sealed-off laser active elements on copper vapour of the Kulon series with low power (1–20 W) and Kristall of medium power (30–100 W).

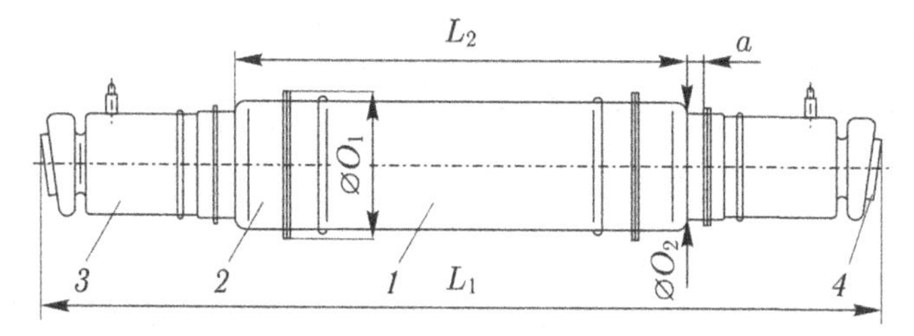

Fig. 3.4. Schematic representation of industrial sealed-off laser AE of the Kulon series: *1* – vacuum-tight shell; *2* – electrode assemblies; *3* – end sections; *4* – output windows; L_1 is the length of the AE; L_2 is the length of the vacuum-tight shell with electrode assemblies; D_1 is the diameter of the AE; D_2 is the diameter of the connecting place; a – width of the connecting place.

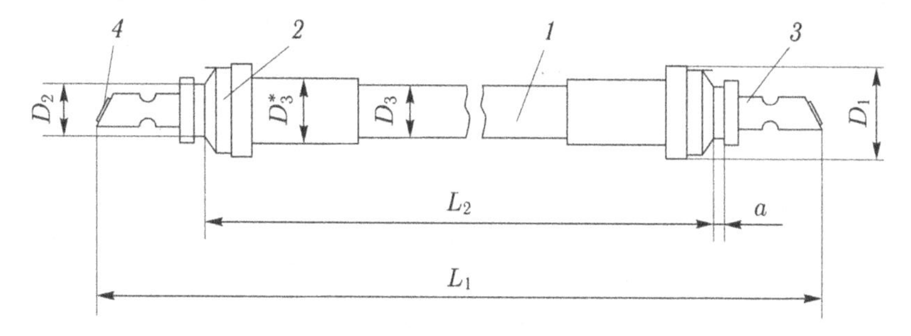

Fig. 3.5. Schematic representation of industrial sealed-off laser AE of the Kristall series: *1* – vacuum-tight shell; *2* – electrode assemblies; *3* – end sections; *4* – output windows; L_1 is the length of the AE; L_2 is length of the vacuum-tight shell with electrode assemblies; D_1 is the diameter of the AE; D_2 is the diameter of the connecting place; a is the width of the connecting place; D_3 is the diameter of vacuum-tight shell; D_3^* is the diameter of the broadened end of the shell.

Table 3.1. Overall and connecting dimensions and mass of industrial sealed-off AEs of the Kulon series

Model	L_1, mm	L_2, mm	D_1, mm	D_2, mm	a, mm	Weight, kg
GL-206A	330	157	84	55	14	1
GL-206B	360	192	84	55	14	1
GL-206C	600	328	100	73	14	2.5
GL-206D	600	328	100	73	14	2.6
GL-206F	600	328	100	73	14	2.6
GL-206E	770	500	100	73	14	3.3
GL-206G	770	500	100	73	14	3.3
GL-206J	900	626	100	73	14	4.0

Table 3.2. Overall and connecting dimensions and mass of industrial sealed-off AEs of the Kristall series

Model	L_1, mm	L_2, mm	D_1, mm	D_2, mm	D_3 (D_3^*), mm	*a*, mm	Weight, kg
GL-205A	1315	916	134	90.5	96 (114)	19	11
GL-205B	1621	1222	134	90.5	96 (114)	19	13,5
GL-205C	1627	1227	134	90.5	114	19	15
GL-205D	1315	916	134	90.5	96 (114)	19	11
LT-50Cu-D	1930	1530	134	90.5	114	19	19
LT-75Cu	1627	1227	134	90.5	114	19	18.5
LT-100Cu	1930	1530	134	90.5	114	19	24

and connecting dimensions and mass of new industrial sealed self-heating AE series Kristall.

From the AEs of the Kulon series, the minimal dimensions and weight are those of the GL-206A model (L_1 = 330 mm, D_1 = 84 mm, D_2 = 55 mm, M = 1 kg), maximum – model GL-206J (L_1 = 900 mm, D_1 = 100 mm, D_2 = 73 mm, M = 4 kg). From the AEs of the Kristall series, the minimum dimensions and weights are those of the GL-205A model (L_1 = 1315 mm, D_1 = 134 mm, D_2 = 90.5 mm, M = 11 kg), maximum – model LT-100Cu (L_1 = 1930 mm, D_1 = 134 mm, D_2 = 90.5 mm, M = 24 kg).

3.5. Construction, manufacturing and training technology, basic parameters and characteristics of industrial sealed-off AEs of the Kulon and Kristall CVL series

The AEs of the pulsed CVLs are high-voltage gas discharge devices operating in the pulsed arc discharge mode and at high temperatures. As the basic structural materials in industrial sealed-off AEs of the new generation are the materials and technologies widely used in the electrovacuum applications, in particular, in the electrovacuum microwave equipment produced by the Istok company were used.

In the design of the sections, in the technology of manufacturing and training of self-heating AEs of new industrial models, all the positive results that were achieved during the 30 years of experimental and theoretical studies of CVL were used. The results of these studies are generalized and presented in [16, 26]. The developed sealed-off AE of the Kulon and Kristall series, despite the relatively large difference in power, have identical design, manufacturing technology and training in degassing and cleaning. The AEs and their main functional units are practically different only in overall dimensions, mass and duration of the process for degassing and cleaning. New industrial models of the AEs of the Kulon and Kristall differ from the first generation by lower power consumption, higher power output and efficiency, service life, quality, stability and reproducibility of the output beam parameters.

Maximum efficiency, power and durability, high stability of power and axis of the radiation pattern and high quality of radiation in industrial sealed-off self-heating AEs of the Kulon (1–20 W) and Kristall (30–100 W) series of the pulsed CVLs are achieved due to:

– using a ceramic discharge channel with blind grooves, in each of which a copper vapour generator is installed in the form of a molybdenum substrate with holes wetted by the active substance – molten copper and perforated end tubes;

– the developed technology for surface purification of copper vapour generators in a hydrogen atmosphere with neon at T_{work} = 1600°C after complete degassing of AE at T = 1700°C with a duration of 30–60 h (depending on the AE model);

– creation of a non-spark auto-thermionic metal-porous tungsten-barium (W–Ba) cathode of a ring structure with an annular groove on the inner surface;

– application of output enlightened windows with an angle of inclination to the optical axis of the AE not exceeding the values

$$\propto = \mathrm{arctg}\frac{(2-ab)}{(2a+b)}, \tag{3.1}$$

where $a = D_{ch}/l_{ch}$ (D_{ch} is the diameter and l_{ch} is the length of the discharge channel), $b = D_{ch}/l_{w}$ (l_{w} is the distance from the end of the discharge channel to the window along the optical axis).

The design of industrial AEs is protected by 4 patents of the Russian Federation: No. 2 191 452, No. 35177, No. 30468 and No. 20617.

The design, manufacturing and training technology of AEs of the Kulon series with a power of 1–20 W. The design of industrial

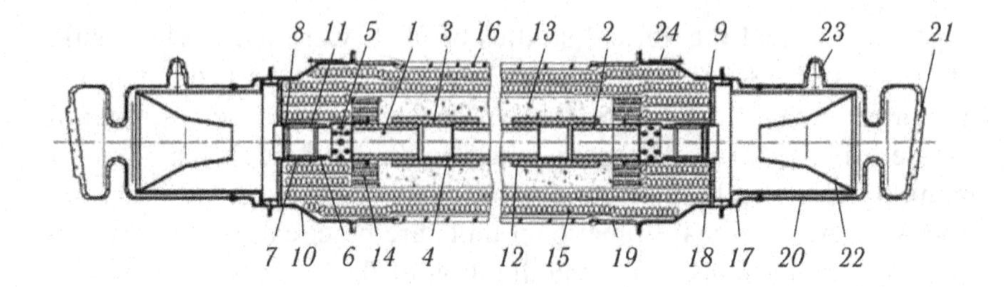

Fig. 3.6. Design of industrial sealed-off self-heating laser AE of the Kulon series of the pulsed CVL: *1* – sectioned discharge channel; *2* – discharge channel tubes; *3* – connecting bushings; *4* – copper vapour generators; *5* – copper vapour condensers; *6* – near-electrode (end) tubes; *7* – cathode holder; *8* – cathode; *9* – anode; *10* – electrode cups; *11* – aluminophosphate cement (APC); *12* – high-temperature sealing cement; *13, 14, 15* – three-layer heat insulator; *16* – vacuum-tight shell; *17, 18, 19* – metal cups; *20* – end sections; *21* – output windows; *22* – screen-traps; *23* – glass tubes; *24* – metal lobes.

sealed-off self-heating AE of the Kulon series of the pulsed CVL is shown in Fig. 3.6.

The outer part of the AE structure is a cylindrical metal–glass casing (item 16) with optical windows (item 21) on the ends for laser output and metal lobes (item 24) on the electrodes for connection of high-voltage pulse from the power source. The envelope of the AE is vacuum-tight and is the carrier of the entire structure, installed horizontally on the metal cups (item 19). On the AE axis there is a sectioned discharge channel (position 1, 2, 3) with generators (item 4) of copper vapour in the blind grooves (joints of tubes with bushings) and condensers (position 5), electrode sections at its ends 10) with an annular cathode (item 8) and an anode (item 9), in the space between the metal–glass sheath (item 16), the discharge channel (item 1) and electrode assemblies (item 10) – a three-layer heat insulator (positions 13, 14 and 15).

The partitioned discharge channel (position 1) of the AE, as can be seen from Fig. 3.6, consists of ceramic tubes (item 2), interconnected by ceramic bushings (item 3), two or three copper vapour generators (item 4), two copper molybdenum vapour condensers (item 5) and two near-electrode ceramic tubes (item 6). The gaps in the joints of ceramic pipes (item 2) with bushings (item 3) are sealed with high-temperature cement (item 12) from fine powder 98–99% Al_2O_3 + 2–1% TiO_2. When training the AEs for degassing, when the temperature of the discharge channel rises to 1600–1700°C, cement is caked. The design of the discharge channel becomes integral and mechanically rigid and removal of copper vapour from the active volume through the gaps in the joints is prevented. (The design of a sectioned ceramic discharge channel with copper vapour generators is protected by patent No. 2 191 452 RF 'Discharge tube of a metal vapour laser' [317].) At the same time depletion of the active substance from generators (item 4) is determined by the diffusion loss of copper vapour along the discharge channel to its relatively cold ends, where the perforated condensers are located (position 5). The service life of the AE is determined by the formula

$$t = \frac{m}{D(N / l)S\gamma}, \tag{3.2}$$

where m is the mass of copper in the generators (g), D is the diffusion coefficient of copper (cm^2/s) [297], N is the concentration of copper vapour in the generator region (cm^{-3}), l is the distance from the generators to the copper vapour condensers (cm), S is the area of the aperture of the discharge channel (cm^2), and γ is the mass of the copper atom (g).

As a material of the tubes (items 2 and 6) and bushings (item 3) of the discharge channel (item 1) with an operating temperature of up to 1600°C, by means of a compromise analysis of the properties of high-temperature oxides, was A-995 ceramics of the composition 99.8% Al_2O_3+ 0.2% MgO with a melting point of 2050°C produced

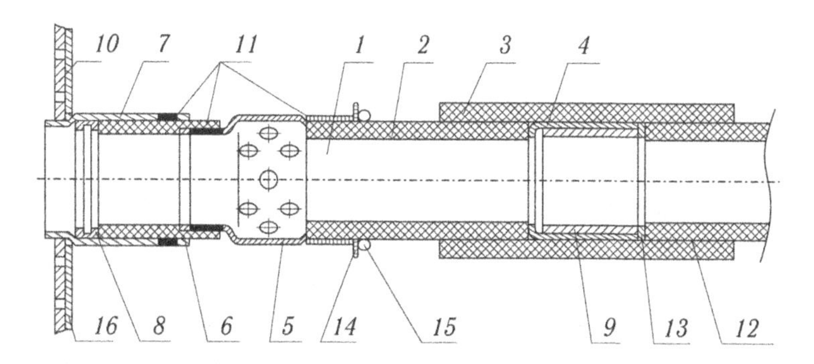

Fig. 3.7. Fragment of the discharge channel with cathode node of laser AE of the Kulon series: *1* – sectioned discharge channel; *2* – a tube of the discharge channel; *3* – connecting sleeve; *4* – molybdenum substrate; *5* – condenser of copper vapour; *6* – near-electrode (terminal) tube; *7* – cathode holder; *8* – cathode; *9* – copper ring; *10* – base of the electrode cup; *11* – aluminophosphate cement (APC); *12* – high-temperature sealing cement; *13*, *14* and *15* – restrictive molybdenum rings and wire; *16* – a foil molybdenum disk.

by the Istok Co. [16, Ch. 2]. Although as regards the physico-chemical properties the preferred material for tubes and bushings of the discharge channel is ceramics from beryllium oxide (BeO) [13]. The melting point of ceramics from BeO is above 2500°C, the vapour pressure of the compound at operating temperature is less than 10^{-5} mm Hg (Al_2O_3–10^{-3} mm Hg). The BeO oxide in comparison with Al_2O_3 has 2.5 times higher thermal conductivity (14.5 and 5.8 W/(m·deg)) and 1.4 times less specific gravity (2.7–2.8 and 3, 7–3.8 g/cm^3) and, accordingly, better withstands thermal shock during heating. But the beryllium oxide has a significant drawback – high toxicity, which creates serious technological problems in the manufacture of devices. In addition, in Russia the production of ceramic tubes from BeO does not exist anymore and it is expensive.

For a more precise representation of the design of the partitioned discharge channel Fig. 3.7 shows its enlarged fragment with the cathode unit adjacent to its end. The aperture of the discharge channel (item 1) and the active medium (AM), respectively, is determined by the internal diameter of its tubes (item 2), the length of the discharge channel is the distance between the cathode (item 8) and the anode (item 9), the length of the active medium by the distance between the condensers of copper vapour (item 5).

In the course of this work, six AE models on copper vapours and two models on gold vapours with different diameters and the length of the discharge channel were developed and investigated. In the low-power AE models GL-206A and GL-206B, the aperture of the discharge channel was 7 mm, in GL-206C – 12 mm, GL-206E and GL-206I – 14 mm. The length of the 'cold' discharge channel of these AEs and its active medium was 140 and 110 mm, 175 and 150 mm, 340 and 260 mm, 490 and 420 mm, 625 and 550 mm, respectively. The volume of the active medium, defined as the product of the channel aperture by the length of the active medium, for AE GL-206A was 4.2 cm^3, for GL-206B = 5.7 cm^3, GL-206C = 28 cm^3, GL-206E = 65 cm^3 and GL-206J = 85 cm^3.

In the operating mode, the relative elongation of the discharge channel for the AE of GL-206A with a temperature T_{chan} ~ 1700°C is about 1.5 mm, (coefficient of thermal expansion $\alpha_{TAl_2O_3} = 78 \cdot 10^{-7}$ K^{-1}) GL-206B – 1.7 mm, GL–206C with T_{chan} ~ 1650°C 4.2 mm, GL-206E with T_{chan} ~ 1600°C – 6 mm and GL-206J – 7.8 mm. Thin-wall perforated base plates of electrode assemblies (item 10) play the role of membranes that provide relatively free movement of the discharge channel when it is heated and cooled along the axis of the AE.

Copper vapour generators (item 4 in Fig. 3.6) are placed on the connecting ceramic sleeves (item 3), in the space between the tubes (position 2) of the discharge channel, without overlapping the channel aperture. They consist of a cylindrical molybdenum substrate (item 4 in Fig. 3.6) that is directly adjacent to the inner surface of the connecting sleeve (item 3), the side restricting disk of molybdenum (item 13) and the copper ring (item 9) as the active substance. Molybdenum MChVP is used, copper grade MV or Mob. The fact that molybdenum is chosen as the substrate material is explained by its two positive properties: firstly, high wettability by molten copper and, secondly, it does not interact with ceramics from alumina [16, Ch. 2], [26, Ch. 1]. The copper vapour generators of this design and such composition, in terms of their physico-chemical properties and efficiency, fully comply with the discharge requirements channel with operating temperature T_{work} = 1500–1700°C. At this temperature molybdenum is well wetted by molten copper and spreads over the surface of the molybdenum substrate (item 4 in Fig. 3.7). But usually, after the end of the many hours of training of the AEs for its degassing in a discharge with the pumping of the buffer gas, the molten copper (item 9 in Fig. 3.7) does not spread over the surface of the substrate (item 4), but accumulates in its lower part in the form of a cusp with a 'ragged' and uneven shape, significantly overlapping the aperture of the discharge channel. In the process of training, there is a strong gas separation, mainly from the heat insulator and the formation of metal vapours and, naturally, contamination occurs, including refractory oxides, of the surfaces of both the molybdenum substrate and molten copper. Even the heating of the channel to 1700°C did not lead to the desired effect. In order to restore the surface of metals from oxides to the working volume of the AE, high-purity molecular hydrogen with a partial pressure of up to 50 mm Hg is introduced (until the generation disappears completely). After holding for a certain period of time high-purity neon is pumped through the AE in the operating mode until the 'contaminated' gas mixture is completely purified. If the molten copper does not spread over the molybdenum surface or even spreads, but no increase in the radiation power is observed, then the technological process is repeated. The process of hydrogen purification is repeated until full wetting occurs, and the radiation power does not reach its maximum value, which in turn attests to the achievement of a high degree of purity of the surface of the molybdenum substrate and molten copper and, accordingly, the evaporation rate of copper vapour. The efficiency of the AEs with pure copper vapour generators increases significantly with a decrease in the working pressure of neon, for example, from 300–400 mm Hg up to 50–100 mm Hg. It was also observed that the addition of several mm Hg of pure hydrogen in the AE after its complete purification leads to an additional increase in efficiency and radiation power.

Condensers of copper vapour (item 5) are molybdenum bushings with a perforated section (with holes) located in the condensation zone of copper vapour, between the electrode tubes (item 6) and the discharge channel pipes (item 2) (item 1). The holes are designed for free passage of copper vapours from the condensation zone to the fibrous heat insulator VKV-1, which prevents the formation of copper droplets at the ends of the discharge channel and, as a consequence, the overlapping of its aperture. Overlapping the aperture with droplets of copper degrades the quality of the radiation beam and reduces its power. (The design is protected by the author's certificate No. 1 572 367 of the USSR 'Active element of a laser on vapours of chemical substances' [321].) The design works as follows. The active substance (copper) vapour under the action of longitudinally directed temperature and concentration gradients diffuses along the discharge channel from its central high-temperature part to the colder ends. Reaching the perforation zone, the copper vapours under the action of a radially directed temperature gradient deviate to the walls of the tube and through the holes diffuse into the heat insulator, where they condense, i.e., beyond the aperture of the AE. Finely dispersed aluminophosphate cement (APC, item 11) based on aluminum oxide is applied for rigid fixing of the condenser sleeves in the near-electrode ceramic pipes (item 6) on the surface of their mutual overlap. APC was developed in the ceramic department of The Istok Co. specially for fastening together the metal and ceramic internal parts of electrovacuum devices with an interval of operating temperatures from –60°C to +1200°C.

The cathode material (item 8) was a metal-porous tungsten-barium (W–Ba) cathode with an active substance of the composition $3BaO \cdot Al_2O_3 \cdot 0.5CaO \cdot 0.5SiO_2$ (barium aluminosilicate) with a minimum work function (2 eV) developed at the cathode department of The Istok Co. [16, Ch. 2]. The W–Ba cathode is unheated and operates in the regime of auto-thermoemission. The cathode is structurally a ring (position 13 in Figs. 3.6 and 3.7) mounted in a molybdenum cylindrical holder (item 12) and pressed tightly from the side of the discharge channel by a ceramic tube (item 6). An enlarged image of the new cathode and a fragment of its working surface is shown in Fig. 3.8 a. In this case, the position of the tube (item 6) relative to the holder (item 12) is fixed with the APC (item 10) applied to its surface and into the technological holes in the holder. The cathode holder (item 12), to ensure a reliable electrical contact, is soldered at the end to the base of a cup of alloy 47HD (item 15 in Figs. 3.6 and 3.7) of the electrode assembly. At the base of the cup, annular grooves are made to improve the radial thermal isolation of the cathode assembly. The dimensions of the cathode ring for low-power AE models of the GL 206A and GL 206B models is 15 × 12 × 2.5 mm, the mass is 3.5 g, for the GL 206B, GL 206D and GL-206I 20 × 16 × 4 mm and 2 g. Of this mass, about 7–8% is the active substance – barium aluminosilicate, which is impregnated with a porous tungsten cathode substrate. The melting point of tungsten is 3400°C. The work function of the active substance is about 2 eV. The activity of the cathode during operation is ensured by the continuous formation of free barium due to the reduction of the active substance by tungsten ($6BaO + W = Ba_3WO_6 + 3Ba$). On the inner surface of the cathode there is an annular groove with a width of 0.5–0.8 mm, which ensures local and stable burning of a pulsed arc discharge in the regime of auto-thermoemission and, accordingly, stable parameters of the output radiation beam. A localized cathode spot of about 1 mm in size progressively moves along the perimeter of the groove as the barium is depleted. The moving speed of the spot, as shown by the long tests of the AE of the GL-206E is no more than 0.014 mm/h, which corresponds to a service life of the cathode of at least 3500 h (guaranteed operating time at 1500°C on AE). Traces of erosion and penetration of tungsten from intense ion bombardment are distinctly observed on the working surface of the cathode (Fig. 3.8 *b* and *c*).

In the operating mode, the cathode at the discharge localization locations is brightly glowed and heated to a temperature close to the melting point of tungsten (T_m = 3400°C), which indicates its effective operation in the auto-thermoemission regime. The anode is identical in design to the cathode, but is made of pure molybdenum. The anode is sprayed to a much lesser extent. The erosion of the anode is caused by the vibrational (damped) character of the development of the discharge and, accordingly, the anode work partially in the cathode mode, but at lower pulsed current values. At present, tungsten is also used as the anode material, the melting point of which is 1.3 times greater than that of molybdenum.

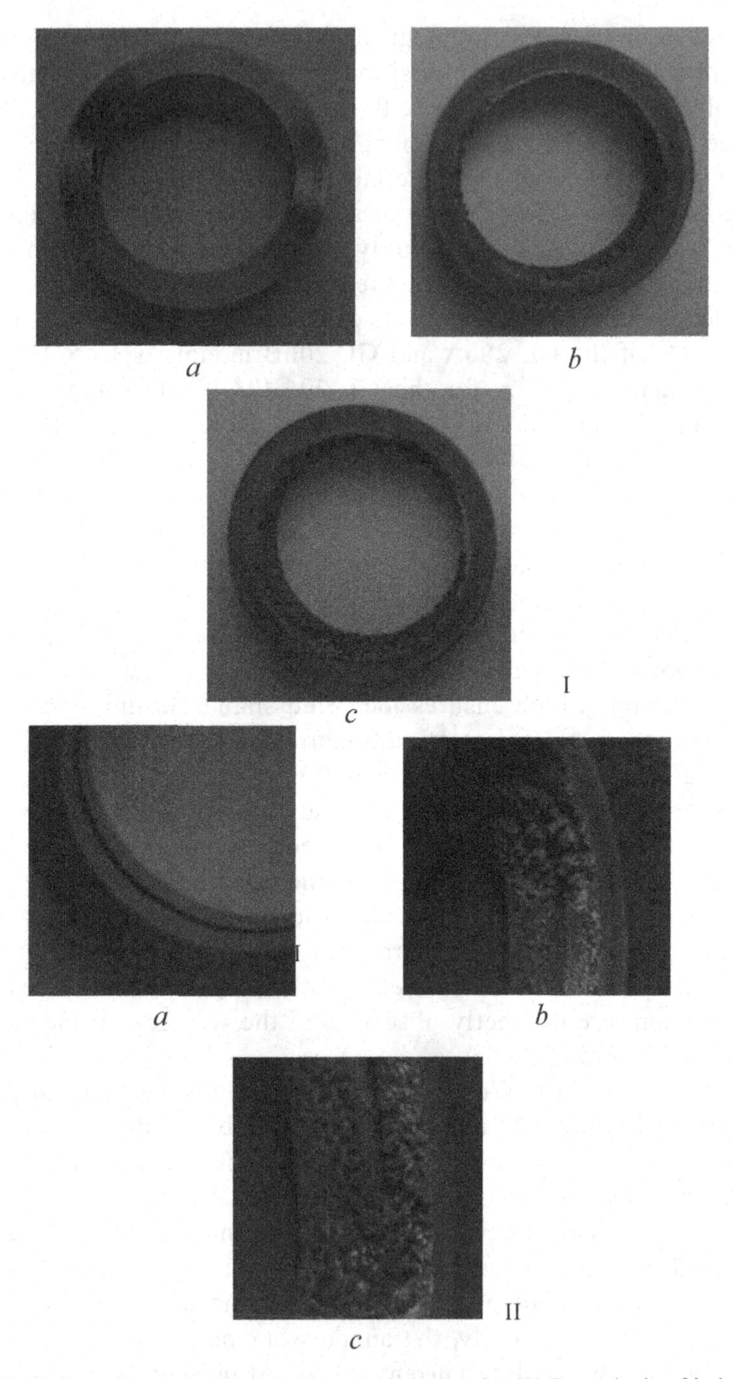

Fig. 3.8. Unheated auto-thermoemission metal porous ring W–Ba cathode of industrial laser AEs of the Kulon series (GL-206): I – external appearance; II – fragment of the working surface; *a*) new, *b*) after working for 2000 h; *c*) after 3200 h.

The heat insulator used in self-heating AE is a three-layer one – hollow microspheres of T brand (item 13 in Fig. 3.6) of 97% Al_2O_3 +3% SiO_2 composition with operating temperature (T_{oper}) 1600 °C and kaolin fiber VKV-1 (position 15) of 55–45% Al_2O_3 + 45–55% SiO_2 with T_{oper} = 1100°C produced by NPK Term (Moscow-Zelenograd) and a fibrous material of the type Pyrofiber 1600 (item 14) of 95% Al_2O_3 + + 5% SiO_2 with T_{oper} = 1600°C from Didier (Germany). The advantage of a three-layer thermal insulator is a low thermal conductivity (λ) at high operating temperatures and a small specific weight (ρ), which leads to a decrease in the power consumption and heat capacity of the AE and, as a result, to an increase in efficiency and a decrease in the laser readiness time. The coefficient of thermal conductivity and specific gravity of the heat-insulating layer made of the T material with sizes of the fractions of 20–200 μm (item 13), immediately adjacent to the 'hot' discharge channel (position 1), are $\lambda = 0.31$ W/(m·K) and $\rho = 0.32$ g/cm^2, the heat of the insulator from the material of grade VKV-1 (item 15), adjacent to the vacuum-tight shell (item 16), $\lambda = 0.27$ W/(m·K) and $\rho = 0.5$ g/cm^2 and Pyrofiber 1600 insulator (item 14), which keeps the microspheres moving along the outer surface of the discharge channel to its ends and their spilling through the holes in the condensate Orach (item 6 in Fig. 3.7) in the working volume of the channel, – $\lambda = 0{,}27$ W/(m·K), and $\rho = 0{,}9$ g/cm^2. These heat insulators, prior to laying in the AE, for preliminary cleaning, are annealed in a high-temperature furnace (up to operating temperatures). The diameter of the separation boundary between the layers of thermal insulators of the T brand (item 13) and the VKV-1 (item 15) in low-power AE models of the GL-206 A and GL-206B models is 39 mm, the models GL-206 V, GL-206D and GL-206I – 50 mm, the temperature in this zone is not higher than 1100°C. The thermal insulator VKV-1 is sintered at higher temperatures and its thermophysical properties deteriorate. The near-electrode regions of the AE, where the temperature is less than 1100°C, are filled with a fibrous heat insulator VKV-1, which along the discharge channel is separated from the loose hollow microcircuit of the heat insulator T by the layer of the Pyrofiber 1600 fibrous heat insulator with the operating temperature up to 1600°C. Thermal insulator Pyrofiber 1600 was not used in the first industrial self-heating AEs. The new heat insulator is located in the zone where the temperature of the discharge channel varies from about 1400 to 1100°C. Therefore, it does not cake and, consequently, a gap along the discharge channel is not formed and the heat insulator from the microspheres does not pour through the condenser holes into the channel volume, does not overlap its aperture and does not dust out the output windows. In the first (old) AE models the condenser zone contained the fibrous material BKV-1, which, due to the relatively low operating temperature (no more than 1100°C), was sintered over time, which facilitated the entry of the heat insulator from the microspheres into the channel and the exit windows, respectively, to a decrease in the power and quality of the radiation. Thus, the three-component combination of the heat insulator in terms of its physico-chemical properties and the thermal protection of the discharge channel fully meets the requirements for self-heating AE.

The vacuum-tight shell of the AE, which also determines its appearance, includes three assemblies interconnected through metal cups by argon-arc welding. These include the metal–glass sheath (item 16), the electrode assemblies with a cup (item 10) and a glass (item 18) and end sections (item 20). In the first industrial nuclear power plants of the type GL-204, UL-101, UL-102 and GL-201, as well as in the new models GL-206A and GL-206G, the casing (item 16) had a cermet structure with increased mechanical strength. It consists of one or three ceramic cylinders made of 22KhS material (94.4% Al_2O_3) and two metal cups at the ends of alloy 47ND, joined together by copper brazing. Soldering the parts into a single unit is done in a hydrogen furnace at a temperature of about 1100°C. The disadvantages of the cermet structure include high weight, high cost, the emergence during the cyclical operation of microleaks in the places of soldering the parts and unfit for repeated use. The metal–glass construction does not have these drawbacks and, as experience shows, can be re-used up to three times. The metal–glass casing (item 16) is a glass cylinder of the C52-1 grade with a diameter of 90 mm (for GL-206C, D and I) or 70 mm (for GL-206B) to the ends of which cups of alloy 29NK (Kovar) are soldered by means of a high-frequency glass generator (item 19). To relieve stresses in the soldering zone, the unit is annealed in a muffle furnace at T = (550 ± 10)°C. A cup (item 10) and a glass (item 18) from a non-deficient 47ND alloy are welded into a single electrode assembly with a copper solder form a vacuum-tight section connecting the sheath (item 16) to the end section (item 20).

The end section (item 20) consists of a glass cylinder C52-1 with a diameter of 70 mm, one end of which is soldered by a high-frequency generator with a metal beaker made of alloy 29NK (item 17), and to the other – by the flame of a gas burner, the optical window (item

21) from A-151 glass (C49-2-OM) [295] with parallelism of planes no more than 20″. The finished unit is annealed in the furnace at $T = (550\pm10)$°C until the stresses arising at its manufacture are completely removed. The glass cylinder has a corrugated structure (see item 20) with an internal diameter equal to the diameter of the output radiation beam. This design allows to reduce the dusting of optical windows. (This technical solution is protected by patent No. 2 023 334 RF 'Active element of a chemical vapour laser' [319].) The corrugated structure is one or several constrictions separated by reservoirs and operates as follows. Hot convective flows of a buffer gas of neon contaminated with vapours of the working substance and the products of erosion of the cathode and anode emerging from the discharge channel on the path to the optical window pass through a series of reservoirs. As a result of convection, gas circulation occurs in each of them, which causes a predominant deviation of the contaminated flow from the axial direction vertically upwards. The circulation of gas in a relatively narrow gap along the relatively cold walls of the reservoir leads to the condensation of particles on their surfaces. The lower the wall temperature, the more actively the condensation process proceeds. Moreover, the high activity of natural cooling of the walls of the reservoir is ensured by the development of their cooling surface. The relatively small thickness of the glass cylinder walls (1.5–2 mm) promotes good heat exchange between the heated gas and the surrounding medium, as well as an increase in thermal resistance along the frame of the corrugated element. The latter allows reservoirs to reduce the flow of heat from heated electrode assemblies and thereby reduce their temperature for a given area of the cooled surface.

To eliminate in the AE an inverse parasitic connection with the active medium (AM), which arises from the reflected radiation from the output windows (item 21), the windows are located at an angle to the optical axis of the AE. The effect of incident reflected radiation in the AM is the distortion of the spatial–time structure of the output radiation beam, that is, the deterioration of its quality. The maximum allowable angle of the window is determined by the edge beams of the beam of superradiance formed by the aperture of the discharge channel. From the consideration of the geometric variation of aperture rays propagating at an angle D_k/l_k, it was established that in order to completely eliminate the parasitic coupling, the angle of inclination of the window to the optical axis of the AE should not exceed the value determined from formula (2.1).

This technical solution is protected by the certificate No. 20617 RF 'Pulsed metal vapour laser' [316]. The calculated values of the slope of the window for the AE series Kulon (GL-206) are in the range α = 85–86°. In real devices, in order to create a constructive reserve, the angle of the window is reduced by several degrees.

Inside the end sections (item 20), metal blackened screen–traps (item 22) of an ink-well construction with a continuous aperture close to the diameter of the radiation beam are coaxially installed. The trap screens, like the corrugated structure of glass sections, are designed to protect the inner surface of optical windows from dusting with erosion products of electrodes, copper vapour and other substances formed during training and operating AE. The dusting of the windows leads to a decrease in power and a decrease in the degree of spatial coherence of the radiation.

In order to increase the service life, efficiency, power and stability of the output parameters of the sealed-off CVL, a three-stage training technology for degassing and cleaning AE with a total duration of 40–50 h was developed. This technology allows preserving the purity of the gaseous medium in the AE until the end of its operation (more than 3000 h). At the first stage of the AE training, a 10-hour degassing operation is performed at the pumping station when the temperature is raised to 450°C (the output windows are welded in the flame of the gas burner) and a 2 h holding at this temperature point. The AE is pumped through the glass exhaust tubes (item 23), soldered to the end sections (item 20), to a pressure of 10^5–10^6 mm Hg. At the second stage of training, the AE is degassed in its own pulsed arc discharge with continuous pumping of buffer gas through it (at first cheap argon, and at the end the working neon) at the training and AE testing stands. To increase the temperature of the outer vacuum-tight shell of the AE there is a cylindrical aluminum screen around it, which also performs the function of the reverse current conductor. The latter leads to a decrease in the inductance of the discharge circuit of the power source and, correspondingly, to an improvement in the conditions for the excitation of the AM and to an increase in the optimum operating temperature of the discharge channel of the AE. At this stage, due to a gradual increase in the power input from the power source to the AE, the temperature of the discharge channel rises from room temperature to 1700°C, the vacuum-tight shell to 420°C, exceeding the operating values by about 100°C. Buffer gases argon and neon are of high purity and are stored in 40 liter metal cylinders at a pressure of 150 kg/cm^2

(atm.). The purity of gaseous argon is not less than 99.998% of the volume fraction (TU 6-21-12-94), neon – not less than 99.994% (TU 2114-00 153 318-03). A pure gas is introduced into the AE through a glass exhaust tube from the side of the cathode (item 8), and the evacuation of the 'dirty' gas is via the exhaust tube (position 23) from the side of the anode (item 9). The process of degassing the AE is completed when the evacuated neon gas at the outlet (from the side of the anode) when excited by a high-frequency device of the Teslo type has the same bright red glow as the pure neon at the input. Depending on the AE model, the degassing time in the second stage of the training, i.e., in the discharge with the pumping of the buffer gas, is within 30–40 h. In the third stage of the training, in order to achieve maximum AE efficiency, the surface of the copper vapour generators (item 4) and other elements from the two-stage is cleaned from the refractory oxides formed during the process. For this purpose, gaseous hydrogen of spectral purity (H_2–99.9999% by volume, TU 2118-06-18 136 415–06), to a partial pressure of 50 mm Hg is introduced into the AE with neon at the operating temperature of the discharge channel of about 1600°C. After not less than half an hour, the neon is pumped through the AE until the 'contaminated' gas mixture is completely purified. The process of hydrogen purification is repeated until the generators are completely wetted (spreading) by molten copper over the surface of the molybdenum substrate, and the radiation power does not reach its maximum value and, accordingly, evaporation of copper vapour – maximum rate. After this technological operation, the device is filled with pure neon to the required operating pressure and soldered through the glass exhaust tubes (item 23).

Preservation of the purity of the gaseous medium in sealed self-heating AEs and their relatively high power, as shown by long-term tests and operation of sealed-off AEs of the GL-206C and GL-206D (2000–3000 h) types in technological and medical equipment, attest to the high efficiency of the developed three-stage technology training on degassing and cleaning instruments. In studies with gaseous hydrogen, it was found that the addition of 2–5 mm Hg pure hydrogen in the AE, after the purification step, leads to an additional increase in efficiency and radiation power. Therefore, in order to maintain the radiation power of sealed self-heating AEs at a high level, hydrogen is added to a certain partial pressure before they are sealed-off before the gas is decanted into the buffer gas. Increasing the efficiency of CVL, which occurs when hydrogen is added, many authors explain by the intensive decrease in the electron temperature in the afterglow period due to elastic and inelastic collisions with hydrogen atoms and molecules [20].

The design, technology of manufacturing and training of AE of the Kristall series with a power of 30-100 W. Development of industrial sealed-off self-heating AEs of the Kristall series of the pulsed CVL (GL-205 according to TU) was carried out in parallel with the development of the AE series Kulon (GL-206 according to TU) [16, 41, 42]. The basic materials used in these AEs, the design and manufacturing technology of the main units, and the technology for degassing the instruments are largely identical. The AE of he Kristall series has several times the length – 1.3–1.6 m, weight – 11–15 kg (see Tables 3.1 and 3.2) and the power consumption – 3–10 kW (see Tables 3.3 and 3.4), which imposes additional requirements on their design, manufacturing technology and training. As a part of the technological equipment the AEs of the Kristall series are usually operated in cylindrical water-cooled metal heat sinks with a flow rate of at least 5–10 l/min. The heat sink also performs the function of a reverse current lead, which leads to a decrease in the inductance of the discharge circuit and, accordingly, to an increase in the steepness of the front of the pump current pulses.

AEs of the Kristall models GL-205A (30 W) and GL-205B (40W). Designs of industrial sealed-off self-heating AEs Kristall on copper vapour model GL-205A with an average output power of 30 W (open name Kristall LT-30Cu) and GL-205B with a power of 40 W (Kristall LT-40Cu) are almost identical. The diameter of the discharge channel of these AEs is 20 mm (Table 3.2). The AE GL-205B is 30 cm longer than the GL-205A AE (1230 and 930 mm, respectively) and has six copper vapour generators, which is two more generators. The AE GL-205A is a modernized variant of AE GL-201 [16], GL-205B – modernized variant of AE GL-201D [16], but the length of the discharge channel and overall dimensions did not change (diameter 134 × 1315 mm for AE GL-205A and diameter 134 × 1621 mm for the AE GL-205B). The outer part of the AE design is a cylindrical cermet shell (item 16) with electrode assemblies (item 10), which are soldered with metal lobes (item 24) for connection to a high-voltage impulse power source, and end glass sections (item 20) with optical windows (item 21) for the output of laser radiation (Fig, 3.9).

The discharge channel (item 1) of the AE of the Kristall series, like the AE of the Kulon series, is partitioned, consists of ceramic

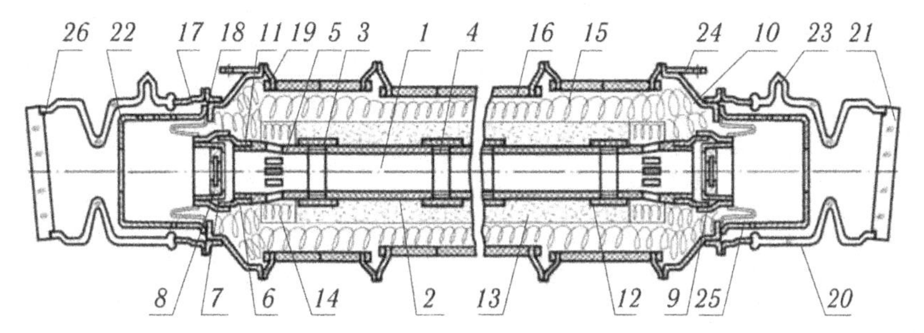

Fig. 3.9. The design of industrial sealed-ff self-heating laser AE Kristall models GL-205A and GL-205B: *1* – sectioned discharge channel; *2* – a tube of the discharge channel; *3* – connecting sleeve; *4* – copper vapour generator; *5* – near-electrode (end) slotted tube; *6* – collar; *7* – cathode holder; *8* – cathode; *9* – anode; *10* – electrode cup; *11* – aluminophosphate cement (APC); *12* – high-temperature sealing cement; *13, 14, 15* – heat insulators; *16* – vacuum-tight cermet shell; *17, 18, 19* – metal cups; *20* – end section; *21* – the output window; *22* – screen-trap; *23* – glass exhaust tube; *24* – metal lobe; *25* – contact loop; *26* – adhesive grade MSP-1.

tubes (item 2), connected by ceramic bushings (item 3). Ceramics A-995 with the Al_2O_3 oxide content of 99.8% was used. In the junction points of the channel sections, that is, in the blind grooves, copper vapour generators are located (item 4). The design of the sectionalized discharge channel is protected by two RF patents: for invention No. 2 191 452 [317] and a utility model No. 35177 [318]. A tungsten–barium cathode (item 4) provides a high degree of localization and stability of burning of a pulsed arc discharge. Inside the end sections (item 20), trap screens (item 22) are installed to protect the exit windows from copper vapour and other substances and sputtering products of the cathode and anode. The three-layer heat insulator (positions 13, 14 and 15), located between the discharge channel (item 1) and the vacuum-tight shell (item 16), provides the required operating temperature of the discharge channel (up to 1600°C) at relatively low power consumptions. High-purity neon is used – not less than 99.994% as a buffer gas in which a pulsed discharge burns, .

The discharge channel of the AE GL-205A (item 1 in Fig. 3.9.) consists of three central ceramic tubes with a length of 186 mm (item 2) and two near-electrode slotted ceramic tubes of 123 mm length (position 5) with an internal diameter of 20 mm and a wall thickness of 2.85 mm. The discharge channel of the elongated AE GL-205B uses five central ceramic tubes (item 2), each of which has a length of 165.5 mm. The central tubes of the channel are connected with each other by ceramic bushings (item 4) with a length of 50 mm,

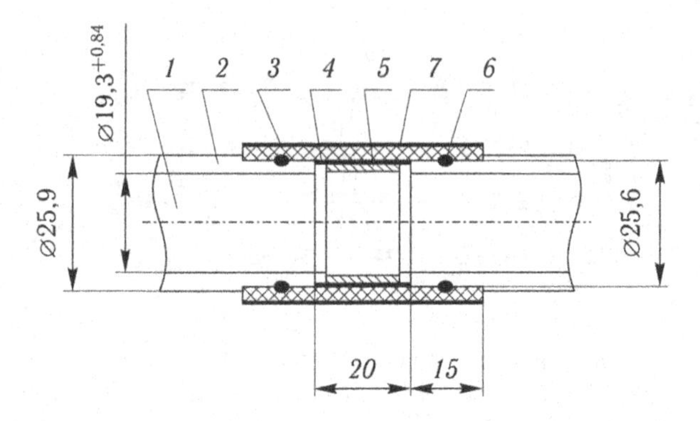

Fig. 3.10. Fragment of a partitioned ceramic discharge channel: *1* – partitioned discharge channel; *2* – ceramic tube; *3* – connecting ceramic bushing; *4* – molybdenum substrate; *5* – copper ring, (*4* and *5* – copper vapour generator); *6* – high-temperature sealing cement; *7* – molybdenum foil.

with electrode tubes 60 mm long. The inner surface of the connecting bushings is ground and has a fitting diameter 25.7 mm. The thickness of the walls of the sleeves is 2.65 mm, the length of their mutual overlapping with the tubes is 15 mm. In the old (basic) AE models GL-201 and GL-201D, to connect the tubes to the sleeves, the ends of the ceramic tubes were ground over the outer surface by a length of 15 mm and had annular projections for fixing the bushes, which turned out to be concentrators of mechanical stresses. The latter often led to the destruction of the discharge channel. Therefore, in the new models of the AE GL-205A and GL-205B, the discharge channel tubes were made either with minimal fixing protrusions (Fig. 3.10) or smooth along the entire length (Fig. 3.9).

Assembling the discharge channel with a smooth tube surface (Fig. 3.9) differs from the assembly of the old design and includes the following technological operations. First, a high-temperature sealing fine cement of the composition 98–99% Al_2O_3 + 1–2% TiO_2 (item 12) is applied to the length of 15 mm from the ends of the tubes along the external surface.

The working temperature of the cement is 1600°C (item 12). Then these tubes (item 2) are serially connected by ceramic bushings (item 3), with copper vapour generators installed in them (item 4). The sleeves relative to the tubes are successively fixed with special mandrels. After the cement binder has dried, the fixing mandrels from the discharge channel are removed. In the process of training AE before T_{chan} = 1650–1700°C, the cement is sintered and the channel design becomes sealed and rigid. The lifetime of an AE

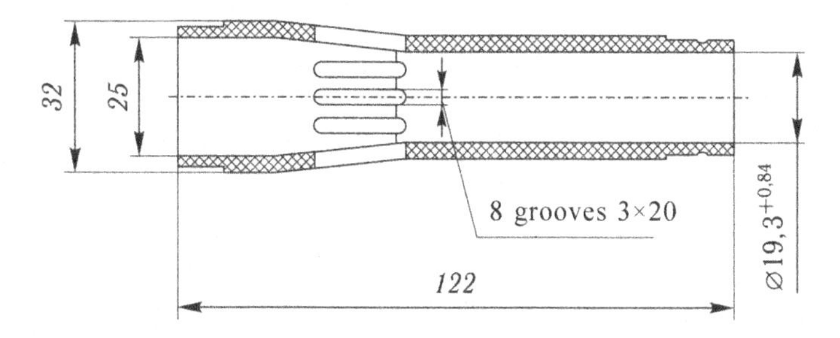

Fig. 3.11. The near-electrode slotted tube of the laser AE discharge channel "Kristall" models GL-205A and GL-205B.

with a sealed discharge channel is determined practically only by the diffusion of copper vapour along the discharge channel (position 1) to its relatively cold ends. In the end electrode tubes (position 5) of the discharge channel, there are longitudinal slotted holes of size 20 × 3 mm (Fig. 3.11), designed to discharge copper vapour into heat insulators (items 13, 14 and 15), where they condense. (The design is protected by copyright certificate No. 1 572 367 of the USSR 'Active element of a chemical vapour laser' [321].)

In the old design of the discharge channel (AE GL-201 and GL-201D), the function of each slit ceramic tube was performed by three parts: an electrode ceramic sleeve, an end tube of the channel and a cylindrical molybdenum tube with slots connecting a ceramic tube and a sleeve. The molybdenum tube limited the life of the AE, since when it is used for a long time in high-temperature conditions, it becomes brittle and collapses in a cyclic mode of operation.

The active substance vapour generators (item 4 in Fig. 3.9) with a cylindrical molybdenum substrate (item 4 in Fig. 3.10), which are the most efficient sources of copper vapour, are placed between ceramic tubes on the inner surface of the connecting sleeves (item 3), i.e., in the blind grooves of the discharge channel. In order not to deform the molybdenum substrate, the substrate has openings for free passage of molten copper from the surface of the substrate to the gap with the ceramic sleeve and vice versa (Fig. 3.12). (This generator design is protected by a patent for utility model No. 35177 [318].)

Through the holes, the excess metal will be forced out, not mainly into the ceramic tubes of the discharge channel, but to the surface of the molybdenum substrate, where it is spread out without forming convex droplets and not overlapping the aperture of the AE channel. At the same time, the dimensions of the gaps should

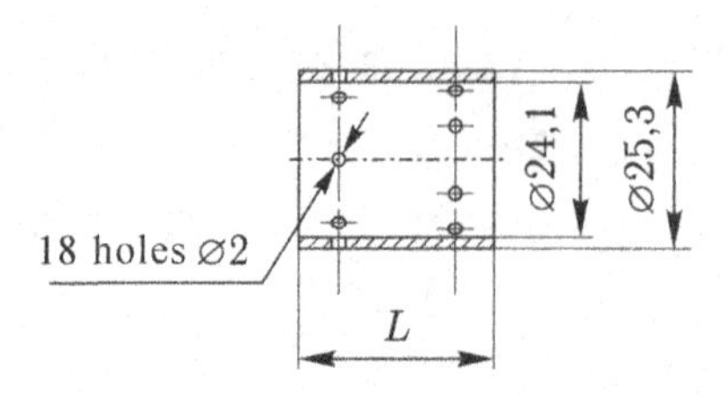

Fig. 3.12. Molybdenum substrate of the copper vapour generator of laser AE Kristall models GL-205A and GL-205B; L = 19.6 mm in the centre and 29.5 mm in the two extreme generators.

be such that the elements of the unit structure do not collapse when the channel is heated. This requirement is ensured by the fulfillment of two conditions. The first condition is that the radial dimension of the molybdenum cylindrical substrate is smaller than the groove size (inner radius of the ceramic sleeve) by the value $\Delta R = R \cdot T \times (k_2 - K_1)/(1 + k_2 \cdot T)$, the second is the axial dimension of the substrate less than the axial dimension of the groove $\Delta H = H \cdot T \cdot (k_2 - k_1)/(1 + k_2 \cdot T)$, where R and H are the radius and width of the groove where the copper vapour generators [m] are located, T is the operating temperature of the discharge channel [°C], k_1 and k_2 are the coefficients of thermal expansion of the substrate and the material of the elements forming the grooves [deg^{-1}]. The perforation of the substrate and the presence of gaps in the grooves have made it possible to increase the overall reliability of copper vapour generators and adjacent nodes.

In the design of the discharge channel, the width of the groove (distance) between the central ceramic tubes of the channel is 20 mm, between the outer central and near-electrode slit tubes – 30 mm. According to the above formulas, at the working temperature of the discharge channel T = 1600°C and the coefficients of temperature expansion of the materials, $k_1 = 6.65 \cdot 10^{-6}$ deg^{-1} and $k_2 = 7.8 \cdot 10^{-6}$ deg^{-1} radial and the axial clearances should not be less than ΔR = 54 µm, ΔH_{20} = 43 µm and ΔH_{30} = 64 µm. These gaps with a margin were taken into account in the allowances for responsible details. The mass of copper in the central generators is 11.4 g, in the extreme generators – 16.2 g. Since the extreme generators are located closer to the gap openings of the electrode tubes, the consumption of copper in them is larger and it is necessary to lay more copper in them (see formula 3.2)).

In the operating mode, the relative thermal elongation of the discharge channel of the AE GL-205A is about 12 mm, AE

GL-205B – 16 mm ($T \sim 1600°C$, thermal expansion coefficient $\alpha_{TAl_2O_3} = 78 \cdot 10^{-7}\ K^{-1}$). Therefore, the structural elements of the AE electrode assemblies should not interfere with the free movement of the discharge channel along the axis during its thermal elongation and cooling. In the old AE GL-201 under the thermal elongation of the channel, the thin-walled bases of the cups of electrode assemblies with radial grooves played the membrane function; in the elongated AE GL-201E, a stainless steel bellows installed in the anode assembly. In the new models of AE GL-205A and GL-205B, the electrode assemblies are structurally made without membranes and bellows. This construction is simple, and no stress arises when the discharge channel is expanded in it. In this case, the cathode (item 8 in Fig. 3.9) and the anode (itcm 9) units are installed directly on the ends of the electrode ceramic tubes (item 5) and connected to the electrode cup (10) by flexible current-carrying strips in the form of a loop (item 25). The material of the loop is nickel, aluminium or copper strip 0.1–0.2 mm thick. The design of the new electrode assembly is protected by a patent for utility model No. 30468 RF Active element of a metal vapour laser [320].

The unheated metal-porous tungsten–barium cathode (item 8 in Fig. 3.9) is structurally a ring that has been rolled into a molybdenum cylindrical holder. The latter, to ensure a reliable electrical contact, is welded along the end to the molybdenum cylindrical cuff of the electrode assembly. The size of the cathode ring is 32 × 24 × 3.5 mm, and the mass is 14 g. Of this mass, approximately 7% (1 g) is made up of an active substance of the composition $3BaO \cdot Al_2O_3 \times 0.5CaO \cdot 0.5SiO_2$ (barium aluminosilicate), which impregnated the porous tungsten ring cathode substrate. The work function of the active substance is about 2 eV. On the inner surface of the cathode there is an annular groove with a width of 0.5–1 mm and a depth of up to 3 mm, which ensures local and stable burning of the pulsed arc discharge. The annular W–Ba cathode with an annular groove on the inner working surface, rolled into a cylindrical molybdenum holder, is shown in Fig. 3.13.

As the barium is depleted the localized cathode spot about 1–1.5 mm in size gradually moves along the perimeter of the groove as the barium is depleted. The moving speed of the spot, as evidenced by the prolonged tests of the AE, is ~0.013 mm/h, that is, the life of the cathode in the Kristall AE is about 6000 hours. Traces of erosion from intense ion bombardment are clearly observed on

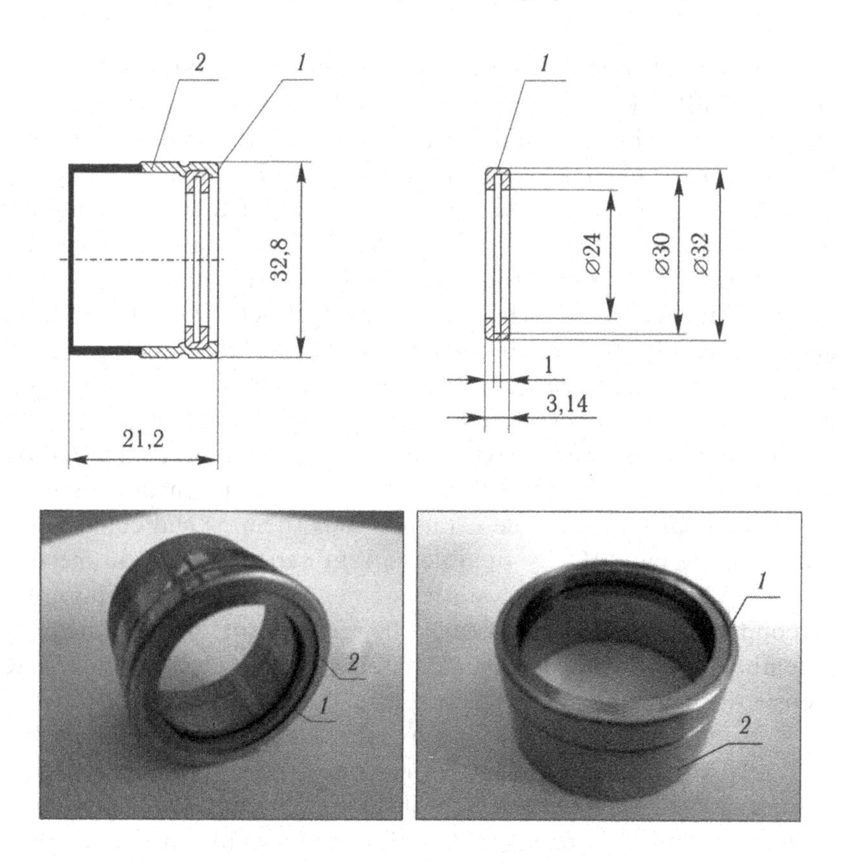

Fig. 3.13. Unheated ring tungsten–barium cathode (new) in a molybdenum holder in industrial laser AE GL-205A and GL-205B; *1* – cathode; *2* – cathode holder.

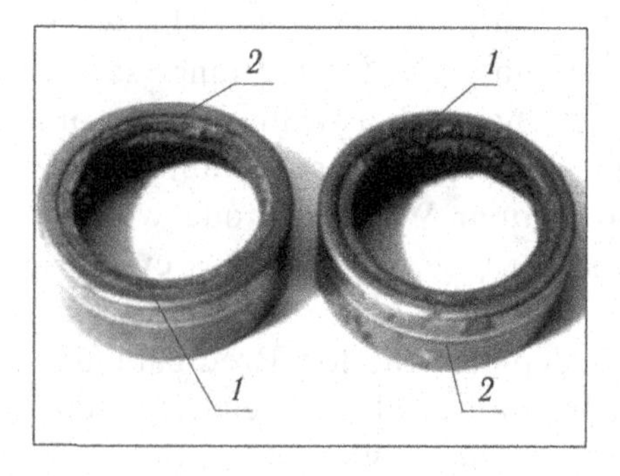

Fig. 3.14. Unheated ring tungsten–barium cathodes in the molybdenum holder of industrial laser AE Kristall of models GL-205A and GL-205B, used for 1500 hours (left) and 2000 hours (right); *1* – cathode; *2* – cathode holder.

the cathode working surface. In this case, the tungsten substrate is partially sprayed.

Figure 3.14 shows W–Ba cathodes in molybdenum holders, spent about 1500 hours (left) and 2000 hours (right).

At the place of discharge localization, the cathode is heated up to a bright glow, i.e., the unheated W–Ba cathode, as in the Kulon AE series, operates in an auto-thermoemission regime. The anode of this construction is identical to the cathode and was made of molybdenum. It turned out that the molybdenum anode is partially atomized, but much less than the cathode. The erosion of the anode is caused by the vibrational (damped) character of the development of the discharge and, accordingly, by the anode work partially in the cathode mode, but at lower pulsed current values. At present, pressed tungsten, the melting point of which is 1.3 times greater than that of molybdenum, is used as an anode material. The tungsten anode is almost unaffected in the modes of pulsed CVL erosion.

The three-layer effective thermal insulator of the AE (items 13 and 15 in Fig. 3.9), located between the discharge channel (item 1), and the vacuum jacket (item 16) provides thermal protection of the channel with an operating temperature of about 1600°C. The inner layer (position 13) adjacent directly to the discharge channel is formed of a fine powder based on T-shaped hollow microspheres (95% Al_2O_3 + 5% SiO_2) at an operating temperature of 1600°C, the top layer (item 15) (55–45% Al_2O_3 + 45–55% TiO_2) with the working temperature of 1100°C (see Table 3.2). The diameter of the boundary separating the layers of the heat insulator is 55 mm, the temperature in this zone is not more than 1100°C. The near-electrode regions of the AE are filled with the heat-insulator VKV-1, which is separated from the main two-layer heat insulator by a layer of a new fibrous heat insulator (item 14), such as Pyrofiber 1600 by Didier company (95% Al_2O_3 + 5% SiO_2) at an operating temperature of 1600°C or Altramat 80 – with the same operating temperature. In previous models, heat insulators of this quality were not used. A new heat insulator is located in the zone where the temperature of the channel varies from about 1400 to 1100°C. Therefore, it does not cake and, consequently, a gap along the discharge channel is not formed and the powder heat insulator do not escape through the slotted holes of the electrode tubes into the channel volume, does not block it and does not pollute the exit windows. In the old models of AE this zone contained the fibrous material VKV-1, which eventually sintered, which contributed to the ingress of powder into the channel

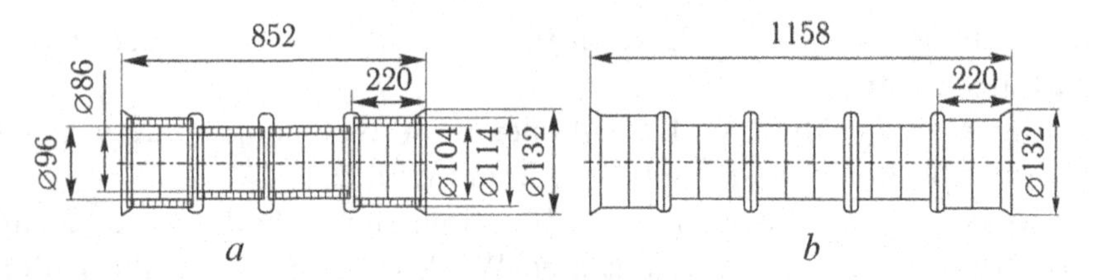

Fig. 3.15. Cermet vacuum-tight shells of laser AE of the Kristall series: GL-205A–*a*, and GL-205B–*b*.

through the cracks in the tubes. The sintering of the material also led to an increase in the non-uniformity of the temperature distribution along the axis of the discharge channel, an increase in the power consumption and, accordingly, a decrease in the radiation power. New models of AE have practically no such drawbacks. They are also increased by approximately 1.25 times, both the density of backfilling of hollow microspheres, and the packing density of fibrous material VKV-1, which led to a decrease in the power consumption by about 10–15%. If the mass of the microspheres and fibrous material in the GL-201 AE was 0.6 and 2.3 kg, then in the ALE GL-205A – 0.7 and 3 kg, in GL-201E – 0.8 and 3.2 kg and in GL-205B – 1.0 and 4 kg.

In the AE GL-205A and GL-205B, the broadened ends of the cermet vacuum casing (item 16) were elongated in comparison with the old AEs GL-201 and GL-201E from 125 to 220 mm (Fig. 3.15 *a* and *b*). This led to an increase in the temperature of the ends of the discharge channel and, correspondingly, to an increase in the length of its active zone. The inner diameter of the ceramic cylinders 22KhS (94% Al_2O_3) in the shell extensions is 104 mm, the thickness is 5 mm, in the central part 96 and 5 mm, respectively. Structurally, the GL-205A shell (Fig. 3.15, *a*) consists of four welded cermetal assemblies, with metal 'trays' 1–1.2 mm thick of alloy 47ND or 29NK with compensating ceramic rings 6 mm in height.

The nodes are interconnected by means of argon-arc welding of the end plates. The length of the envelope of the AL GL-205B is 306 mm larger than that of the GL-205A, due to the addition of a single cermet assembly (Fig. 3.15 *b*). Today the industrial AE of the Kristall series is produced mainly with a metal-glass vacuum-tight shell (Fig. 3.16).

The metal-glazed shell of the AE GL-205A (Fig. 3.16 a) consists of two identical sections connected by argon-arc welding through the end trays (diam. 120 mm) from the alloy 29NK (Kovar). The C52-1 glass cylinders used in the sections have a diameter of 100

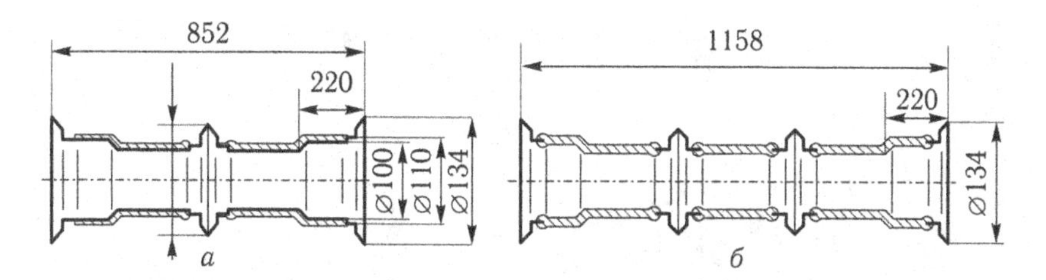

Fig. 3.16. Metal–glass vacuum-tight shells of laser AE of the Kristall series: GL-205A (*a*) and GL-205B (*b*).

and 110 mm. In the elongated AE GL-205B the shell (Fig. 3.16 *b*) consists of three sections: the central and two identical end. When assembling the AE, the shell, also by means of argon-arc welding, is connected to the cups (Fig. 3.9, item 10) of the electrode assemblies through the end plates, the latter with the kovar cups (pos.17) of the end glass sections. The metal-glass casing in the manufacture is more technologically and the yield coefficient of suitable products is almost 100%. At the same time the cost of AE is 20–25% lower than with the cermet shell. The most important for sealed AEs with metal–glass construction is the fact that in the course of their long-term operation there were practically no failures associated with violation of the vacuum density of the shell. In the case of a cermet structure, especially when operating in a cyclic mode, non-renewable failures occurred due to the depressurization of the casing, preferably at the metal–ceramic junction. Usually, after the end of operation, AEs are dismantled by nodes for the purpose of their evaluation for subsequent use in new devices. For this purpose, the outer shell of the EA is cut at the seams of argon-arc welding of metal 'trays' to the following sites: a vacuum-tight shell (item 16), two electrode assemblies (item 10) and two end sections (item 20). When mechanically sawing, the cermet shell (Figure 3.15) usually becomes not sealed, and first of all, the metal–ceramic junctions and the recovery for reuse are not subject. The vacuum density of the shell of the metal–glass structure when disassembling the AE is not violated. It is suitable for two-three times the repeated use, depending on the technological margin of the diameter of the welded end plates (diam. 134 mm in Fig. 3.17). This is another significant advantage for industrial AEs with the metal–glass sheath.

A metal–glass or metal–ceramic shell (item 16 in Fig. 3.9), electrode assemblies (item 10) and end sections (item 20), interconnected at the ends by argon-arc welding, form a single

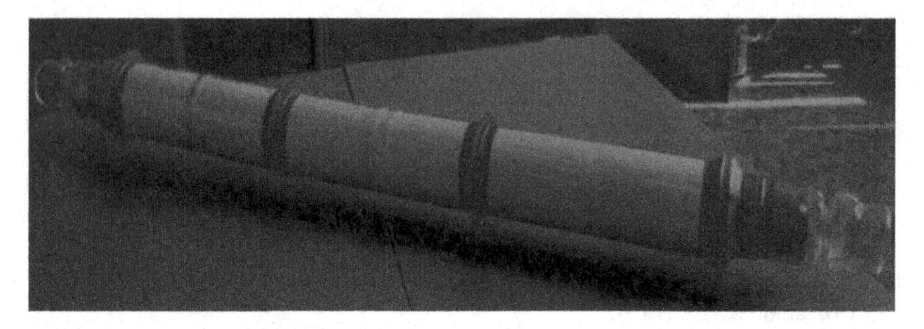

Fig. 3.17. Appearance of the industrial sealed-off self-heated laser AE GL-205C (Kristall LT-50Cu) with a metal–glass shell.

outer vacuum-tight shell. The main element of the end sections is the optical window (item 21) for the output of laser radiation. At the first stage of the manufacturing process of the section, a glass cylinder with a diameter of 85 mm of C52-1 grade (item 20) is soldered to a glass of 29NK alloy (item 17) using a high-frequency generator. To completely remove stresses in the soldering zone, the finished assembly is annealed at T = (550 ± 10)°C for 30–40 min, for example, in a muffle furnace. After this operation, the flame of a gas burner is used to produce a constriction (corrugation) with an internal diameter close to the diameter of the radiation beam in the glass cylinder, the end of the cylinder narrows to the fitting size of the exit window, and a glass exhaust tube (position 23) is soldered for pumping and evacuation of gases during AE training (see items 20, 21 and 23). Then the unit is annealed again at the same temperature and holding time and the end of the cylinder is ground at an angle to the optical axis determined by formula (3.1). The exit windows are made of A-151 (C49-2-OM) optical glass [295] and have a diameter of 72 mm, a thickness of 7 mm and a deviation from the parallelism of their planes of not more than 10–20, the windows (item 21), using structural high-strength adhesive grade MSP-1 with an operating temperature of 200°C developed by the Istok Company (TU 2252-003-07 622 667–99) are glued to the end glass sections (item 20). The maximum temperature of the output windows during the operation of the AE in a closed cylindrical heat sink is 130–150°C. At the AE tests for more than 2000 hours (about 200 cycles of switching on and off) the depressurization of the glued nodes did not take place. This way of connecting the output windows to the end sections does not introduce deformations into the window material and leads to an improvement in the quality of the output radiation beam. In the old models GL-201 and GL-201D, optical windows were welded to the glass sections in the flame of the gas burner, which is why the quality of the output radiation has been distorted to some extent. At the same time, the rejects associated with the deterioration of the quality of the window when welding, are in the range of 20–50% and depend on the individual qualities and qualifications of the glass blower. Promising is the use as a material of the output windows of KV quartz instead of glass, which absorbs less heat from the heated discharge channel and does not lead to the occurrence of internal stresses. But this requires a new design of the end section, since the coefficient of thermal expansion of glass and quartz differ by an order of magnitude.

For the Kristall AE of the model of the GL-205A model, the calculated value of the slope angle of the output window to the optical axis of the AE, in accordance with the formula (3.1) [316], is $\alpha = 85.8°$, GL-205B – $\alpha = 86.1°$. In real Kristall devices, as in the previously reviewed Kulon systems, in order to create a constructive reserve, the angle of the window with respect to the calculated data was reduced to 78°. At this angle, the inverse parasitic connection with the active medium is completely excluded and there is no distortion of the spatiotemporal structure of the output radiation beam. To reduce the radiation power losses at the exit windows of AE associated with Fresnel reflection ($\rho = 8\%$), the window surfaces in the new models are enlightened. The enlightened film is resistant to ultraviolet radiation from the discharge plasma and the transmission coefficient remains at 98% for the entire life of the AE.

In order to effectively protect the inner surface of the exit windows from sputtering with the products of electrode erosion, active substance vapours and other particles during the training and operation of the AE, trap screens (item 22 in Fig. 3.9) and constrictions (corrugations) are used in the end glass sections (position 20) [316]. The internal diameter of the trap screens and corrugations is equal to or close to the diameter of the radiation beam. In the upper part of the trap screens (item 22), on the side of the glass exhaust tubes (position 23), there is a series of holes intended for free passage of the buffer gas in the pumping mode during the training of the AE.

The technology of training and assembly of high-temperature self-heating AE on copper vapour has its own specific features and know-how, which are not considered here. We note only a few points. Training on the degassing of the AE of the Kristall model

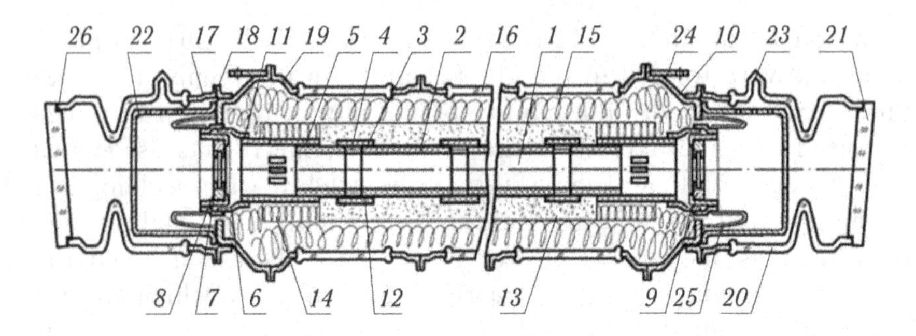

Fig. 3.18. The design of the industrial sealed-off self-heating laser AE Kristall, model GL-205V: *1* – sectioned ceramic discharge channel; *2* – a tube of the discharge channel; *3* – connecting sleeve; *4* – copper vapour generator; *5* – near-electrode slotted tube; *6* – collar; *7* – cathode holder; *8* – cathode; *9* – anode; *10* – electrode cup; *11* – aluminophosphate cement (APC); *12* – high-temperature sealing cement; *13, 14, 15* – heat insulators; *16* – metal–glass vacuum-tight shell; *17, 18, 19* – metal cups; *20* – glass end section; *21* – the output window; *22* – screen-trap; *23* – glass exhaust tube; *24* – metal lobe; *25* – contact loop; *26* – adhesive grade MSP-1

GL-205A is performed by analogy with the three-stage training regime of the AE Kulon and the two-stage basic AE GL-201. The purpose of AE training is to achieve high and stable parameters of output radiation and to preserve them in the course of long-term operation by removing from the AE the impurity gases and vapours released from the elements of its design. At the first stage of training the AE GL-205A for degassing, its 10 h evacuation is carried out with simultaneous heating in the furnace to a temperature of 180°C (limited by the operating temperature of the glue MSP-1). At the second stage of training, when AE is placed inside a thermal aluminum cylindrical screen, degassing is carried out in the burning mode of its own arc pulsed discharge with pumping argon and neon buffer gas from the cathode to the anode. The screen allows to raise the temperature of the vacuum-tight shell by 100–150°C above the operating temperature. In the third stage, the device is cleaned in a reducing environment – in an atmosphere of molecular hydrogen. To date, the technology of AE training without a pumping phase with heating in the furnace is available. Without this stage, the duration of the AE GL-205A training is approximately 45 hours, which is 1.4 times more than for the base GL-201. The latter is due to 1.2 times higher density of packing of the heat insulator and somewhat larger volume. The training mode of the AE GL-205B is similar to the training mode of GL-205A and takes about 60 hours, which is 1.33 times more.

Design and manufacturing technology for industrial sealed-off self-heating AE Kristall model GL-205C (55 W). The industrial sealed-off self-heating AE Kristall model GL-205V (open name – Kristall LT-50Cu) with an output power of 50–55 W is shown in Figs. 3.3 and 2.17. In Fig. 3.3 shows the appearance of AE with a cermet shell, in Fig. 3.17 (color insert) – with a metal-glass sheath, and in Fig. 3.18 – its construction.

AE GL-205V is a modernized version of the AE GL-201D32 and has the same overall dimensions and length of the discharge channel as the GL-205B. The discharge channel has a diameter of 32 mm and a length of 1230 mm, the volume of the active medium is about 900 cm^3. The mass of the AE is 15 kg (see Table 2.2).

The discharge channel (item 1) AE GL-205V consists of five ceramic tubes (item 2), each 151 mm long and two 65 mm long, having an internal diameter of 32 mm and a wall thickness of 2.65 mm, and two near-electrode end) slotted tubes (item 5) with a length of 115 mm with an internal diameter of 37.3 mm and a wall thickness of 3 mm. Eight longitudinal slots of size 30 × 3 mm, available in the electrode tubes (item 5), are intended for the care of copper vapour into the heat insulator. The central tubes (item 2) of the discharge channel are connected with each other by four ceramic bushings (item 4) with a length of 50 mm, and the end (5) by two lengths of 60 mm with an internal diameter of 37.3 mm and a thickness of 3 mm. The outer surface of the used ceramic tubes (item 2) of the channel is smooth, i.e. without annular steps at the ends for fixing the connecting sleeves (item 4). Steps, as shown by the experience of AE operation in a cyclic mode, lead to the appearance of ring cracks in thick-walled tubes and, as a consequence, to the destruction of the discharge channel due to the occurrence of mechanical stresses. Therefore, when assembling (practically gluing) the discharge channel, the fixation of the tubes and connecting sleeves with respect to each other is carried out using special mandrels. The design of the sectionalized discharge channel is protected by two RF patents: for invention No. 2 191 452 [317] and utility model No. 35177 [318].

The construction and technology of manufacturing the discharge channel (item 1, figure 3.18), copper vapour generators (item 4), cathode (item 8) and anode (item 9), electrode assemblies (item 10), electrode tubes (5) heat insulators (item 12, 13, 14), vacuum-tight shell (item 16), output optical windows (item 21), end sections (item 20) and trap screens (item 22) AE GL -205V are similar to those used

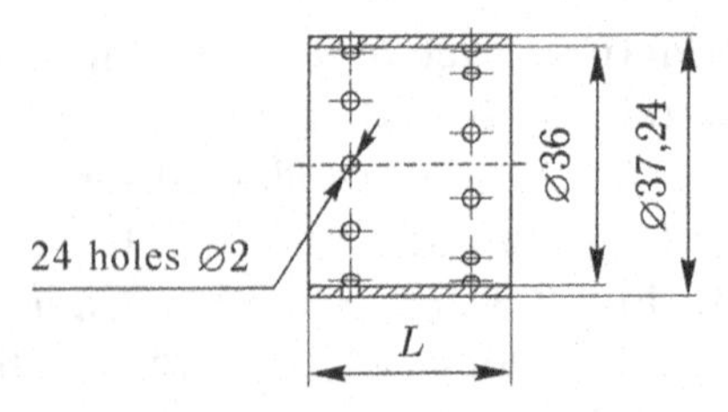

Fig. 3.19. Molybdenum substrate of the copper vapour generator of industrial laser AE GL-205V; L = 19.6 mm in the center and 29.5 mm in the two end generators

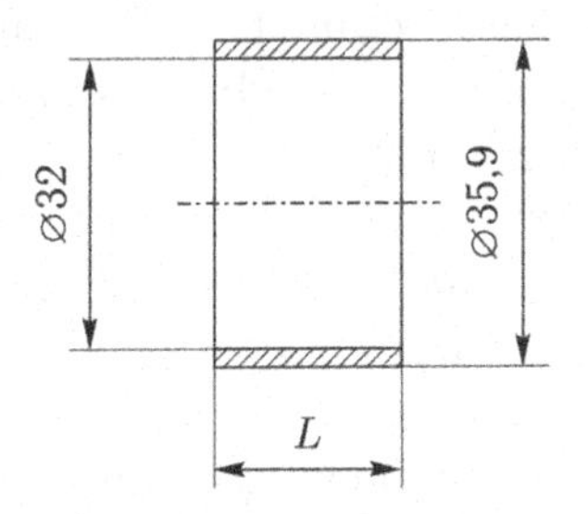

L, mm	Weight, g
13	22,8
21	36,8

Fig. 3.20. Copper ring of the copper vapour generator of industrial laser AE GL-205C.

in GL-205A and GL-205B. Elements of the copper vapour generator (item 4) are separately shown in Fig. 3.19 and 3.20.

The beznakalny tungsten-barium ring cathode AE GL-205V size 40 × 33.5 × 3.7 mm with an inner annular groove 1 mm wide (item 8 in Figure 3.9), containing about 7% of the mass of the active substance of barium aluminosilicate, as well as in other models of AE Kristall, provides a stable local combustion of the pulsed arc discharge during the entire service life. In Fig. 3.21 (color insert) shows new and partially used cathodes of AE GL-205V (see also Figs. 3.8, 3.13 and 3.14). A vacuum-tight cermet shell with an outside diameter of the ceramic cylinder of 114 mm and a wall thickness of 5 mm (Fig. 3.22 *a*) has no broadening at the ends, as ceramic cylinders made of 22KhS material with a large diameter are not produced by the Istok company. At the same time, the length of the core of the GL-205C AE relative to the GL-205B, which has the same length of the discharge channel, decreases somewhat due to a smaller layer of thermal insulation at the ends of the channel. Therefore, in order to increase the temperature at the ends of the discharge channel, the packing density of the fibrous heat insulator of grade VKV-1 along the ends of the shell was increased in comparison with the central part. Currently, industrial soldered AEs GL-205C are produced only

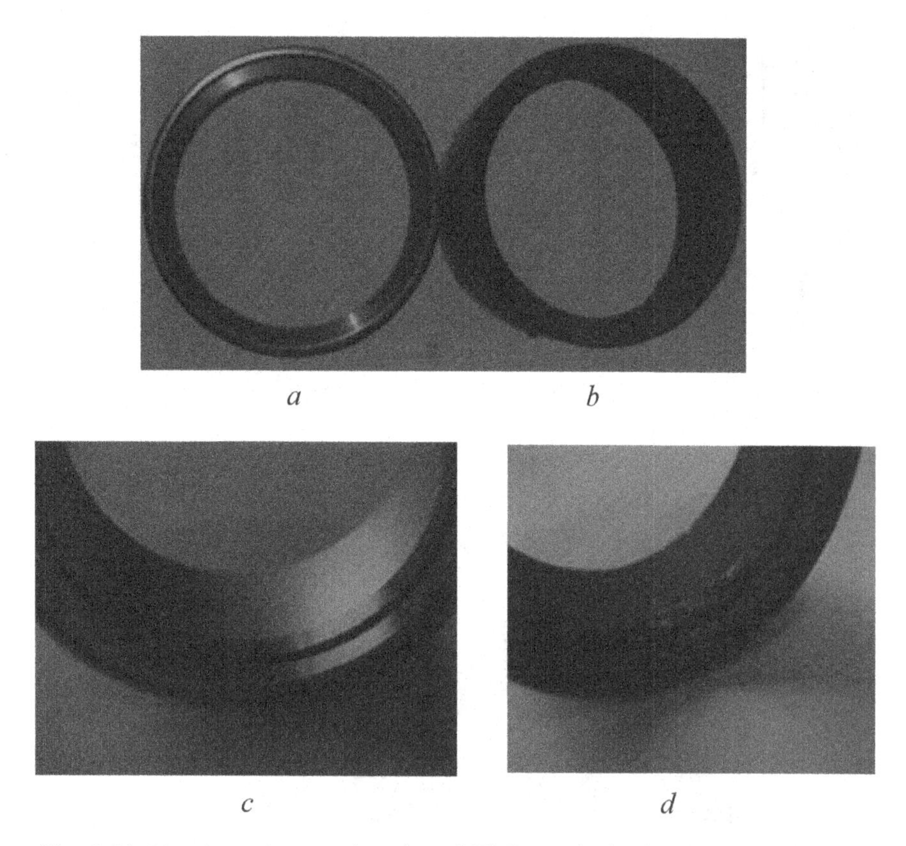

Fig. 3.21. Non-incandescent ring-shaped W–Ba cathodes in a Mo holder of the industrial Kristall laser AE, model GL-205C: *a* – new cathode; *b* – cathode after operation for 1500 h; *c* – fragment of the working surface of the new cathode; *d* – fragment of the working surface o the cathode after 500 h operation.

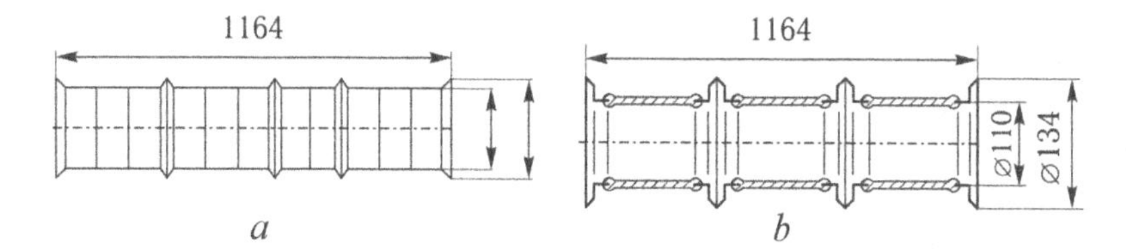

Fig. 3.22. The design of the cermet (*a*) and metal–glass (*b*) vacuum-tight shells of the laser AE Kristall model GL-205C.

with a metal–glass sheath (Figs. 3.22 *b* and 3.23), since it has a significant advantage over the ceramic series.

The metal–glass envelope consists of three sections, interconnected by argon-arc welding. Each section is a glass cylinder of the C52-1

Fig. 3.23. Appearance of the metal–glass vacuum-tight shell of the laser AE Kristall, model GL-205C.

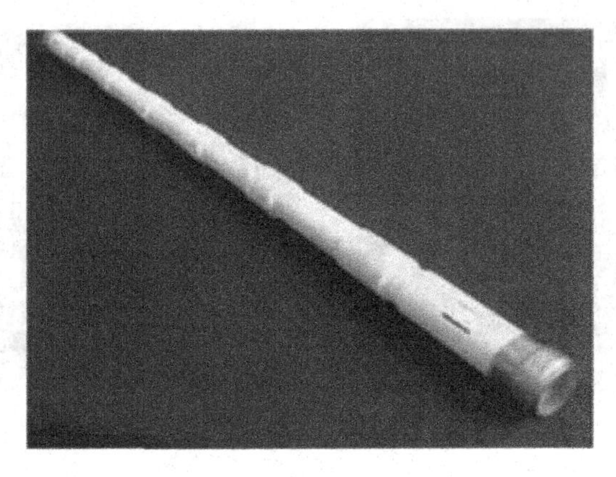

Fig. 3.24. Sectioned ceramic discharge channel with electrode assemblies at the ends of the laser AE Kristall, model GL-205C.

brand with a diameter of 110 mm and a wall thickness of 3 mm, to the ends of which the glasses are soldered from 29NK alloy (kovar). The training mode for industrial sealed self-heating AE GL-205V for degassing and its purification is similar to the three-stage training regime for industrial AE GL-205B. The total duration of training AE GL-205C is, like the GL-205B, about 60 hours, as the masses of their thermal insulators are approximately equal. The analysis of 2000–3000 h of operaton of the sealed-off AE GL-205C, when the operating temperature of its discharge channel under the conditions of the usual effective pumping of the AM is 1550–1600°C, shows that the design of the AE and the functionality of the main units are preserved. For example, Fig. 3.24 is a sectioned ceramic discharge channel of a device that has worked more than 2000 hours. And, as can be seen, the integrity of the discharge channel design with electrode assemblies at the ends has been preserved.

The results of analysis of long-term tests and operation of pulsed CVLs in continuous and cyclic regimes attest to the high reliability of the designed structures and technology of training of industrial sealed-off self-heating AEs of both the Kulon (GL-206) and the Kristall (GL-205) series. It should also be emphasized that all materials used in the construction of sealed AE CVL (except for heat insulators) are widely used in domestic and foreign electrovacuum technology.

In all new models of industrial sealed-off AEs on copper vapour of the Kristall series after the completion of the full cycle of training, high and pure hydrogen gas with partial pressure up to 5–10 mm Hg is added to increase and maintain efficiency with prolonged operating time.

A powerful series of AE Kristall. The first experimental studies to further increase the radiation power of sealed-off AEs were carried out on a sample whose construction is similar to the industrial AE GL-205C (LT-50Si Kristall), but has an enlarged discharge channel length by 30 cm (interelectrode distance 1530 mm, V_{AC} = 1200 cm^3). The average radiation power of this sample, designated as LT-50Si-E Kristall, in the case of using an effective thyratron power supply and a plane-spherical resonator with a radius of curvature of the 'blind' mirror of 3.5 m at a neon pressure of 100 mm Hg and the 10 kHz PRF was about 60 W, in the power amplifier (PA) mode 80 W; when using a two-lamp power supply and 12 kHz PRF in the PA mode 85–90 W.

Among the most promising of the Kristall series of powerful AEs are the experimental AEs Kristall LT-70Cu and "Kristall LT-100Cu". The design of the LT-70Cu Kristall is completely identical to the construction of the Kristall LT-50Cu [323, 324], with the exception of the diameter of the discharge channel: in the AE Kristall LT-70Cu it is 45 mm, i.e. by 13 mm more. The exit windows are enlightened and inclined to the optical axis at an angle of 76–78°. The AE Kristall LT-70Cu was processed using the standard technology of Kristall devices and filled with neon of high purity with a small addition of hydrogen (2–3%). All measurements of its output parameters were carried out in the generator mode using a plane-spherical resonator with a radius of curvature of the 'blind' mirror R_3 = 4 m. The AE was excited from a power source with a high-voltage pulse modulator based on a powerful hydrogen thyratron TGI1-2500/50, with voltage

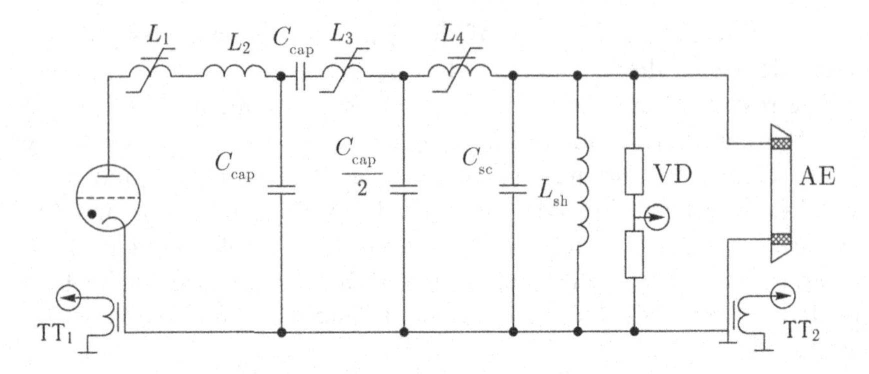

Fig. 3.25. The circuit of a high-voltage pulse modulator with capacitive voltage doubling and two links of magnetic compression of current pulses: C_{cap} = 1600 pF, C_{sc} = 500 pF; TT_1 and TT_2 – current transformers; VD – voltage divider.

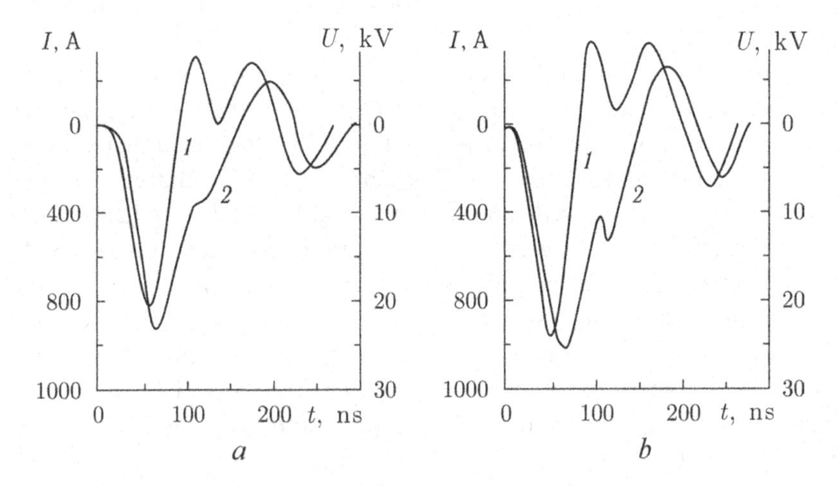

Fig. 3.26. The oscillograms of the voltage pulses (*1*) and the current (*2*) of the AE Kristall LT-70Cu (*a*) and LT-100 Cu Kristall (*b*) of the CVL.

doubling and two optimized links of magnetic compression of pump current pulses (Fig. 3.25).

With two links of magnetic compression in the modulator, compared to one link, power losses in the thyratron are minimal and its service life becomes more than 2000 h. The capacitance of the storage capacitor C_{cap} during the testing varied from 1600 pF to 2200 pF, which resulted in a slight power change from the radiation. The maximum value of the average radiation power was obtained for C_{cap} = 1600 pF and PRF 9.6 kHz.

When the pressure of the buffer gas of neon in the AE Kristall LT-70Cu varies from 50 to 120 mm Hg, the power of the radiation gradually decreased, by about 10%. The amplitude of the discharge current pulse was approximately 850 A with a duration of 160 ns at the base and a leading front of 50 ns (Fig. 3.26 *a*). The amplitude of the voltage pulse was ~20 kV with a duration of 75 ns on the base. In this mode, an output power of about 74 W was achieved with a power consumption from the rectifier of 6.6 kW, (the discharge channel temperature was 1500°C), which corresponds to a practical efficiency of 1.12%. The use of a two-link magnetic line for the compression of current pulses made it possible to reduce both the loss in the thyratron of the TGI1-2500/50 PS modulator and to increase the pumping efficiency of the AE.

The design of the LT-70Cu Kristall LT has become the basis for the development of a more powerful LT-100Cu Kristall model [322]. Its difference from the basic AE lies in the longer length of the discharge channel (1520 mm instead of 1230 mm) and more evaporators of copper (seven instead of five). The gas content of the AE was similar to that described above. Measurement of the output characteristics of the AE was also carried out in the mode of the generator, with a radius of curvature of the 'blind' mirror of the resonator R_3 = 5 m. To excite the AE, an PS with the same high-voltage modulator was used (see Fig. 3.25). The pulses of the voltage and current of the LT-100Cu Kristall (Fig. 3.26 *b*) had the same duration as the pulses of the LT-70Cu Kristall, but were slightly larger in amplitude. The maximum output power of the LT-100Cu Kristall reached 90 W at 10 kHz. The duration of the generation pulse along the base was 25–30 ns at a neon pressure of 90 mm Hg. The radiation power on the yellow and green lines is about 46% and 54% of the total power, respectively. At the same time, the power consumed from the rectifier was 9.0 kW with losses in the thyratron ~2 kW (losses were measured by the calorimetric method), and practical efficiency, respectively, 1.0%.

The achieved values of the average radiation power in a pulsed CVL with experimental AEs Kristall LT-70Cu and LT-100Cu Kristall – 74 and 90 W – were obtained in the mode of the generator, i.e. with an optical resonator. However, from a practical point of view, it is more preferable to use these long Kristall AEs in high-power laser systems on copper vapour (CVLS) of the MO–PA type as a PA when the output power increases by more than 30% (see Table 3.2) . Therefore, it can be argued that AE Kristall LT-70Cu and Kristall LT-100Cu when used as amplifiers will give an increase in the radiation power to 97 and 117 W, respectively. These values correspond to

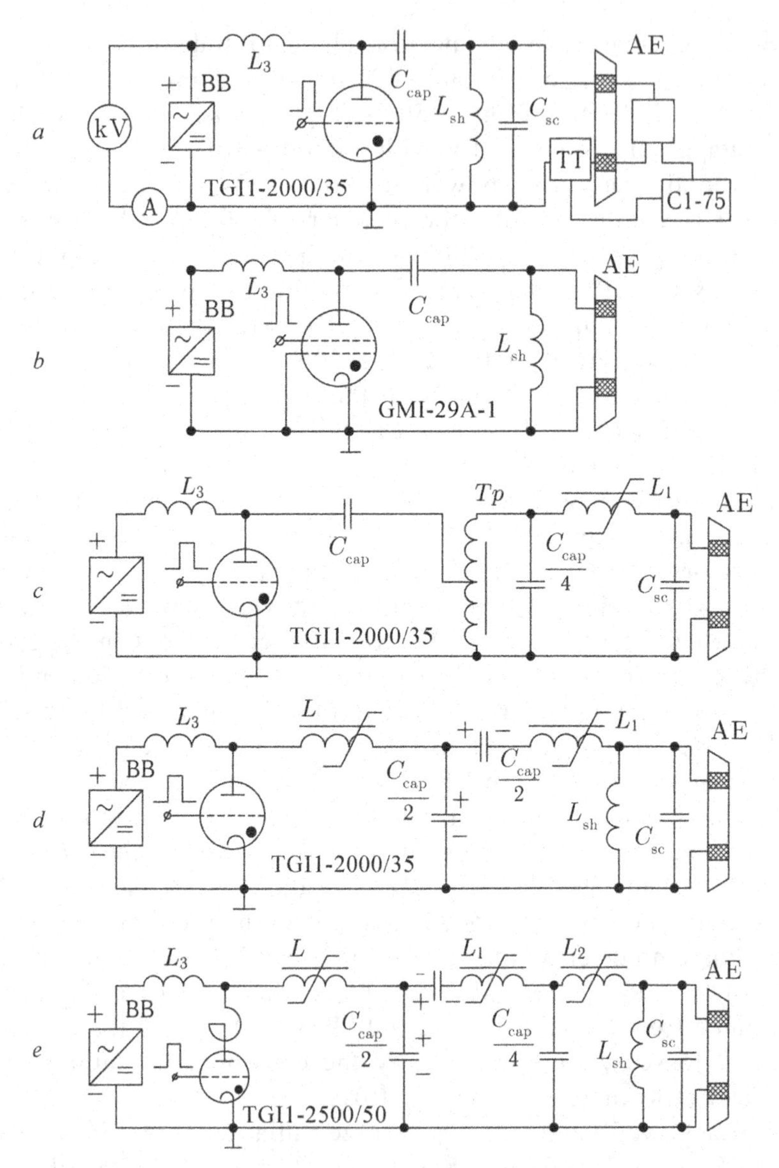

Fig. 3.27. Basic electrical schemes for the execution of high-voltage pulse modulator of the power source of the CVL: *a* and *b* are direct circuits with thyratron and lamp switches; *c* and *d* – schemes of transformer and capacitive voltage doubling with a link of magnetic compression of current pulses; *e* – scheme of capacitive voltage doubling with two links of magnetic compression and anode reactor.

a specific power takeoff of about 0.054 W/cm^3, which is 1.5 times less than that of the analogue – LT-50Cu Kristall (GL-205C). The decrease in the power take-off is due, first of all, to the increase in the diameter of the discharge channel, since it leads to a decrease in the degree of restoration of the AM in the interpulse period.

As it was mentioned above, the important parameters for the high-power AE Kristall class are those when operating in the PA mode, which include the radiation power, specific power take-off, practical efficiency and AE efficiency. High values of radiation power (80-120 W), practical efficiency of CVL (1.3–1.5%) and AE efficiency (2.6–3%) testify to their effective use in high-power laser systems on copper vapour (CVLS) of the type MO–PA, used in technological equipment for the selective separation of isotopes.

Unfortunately, in Russia, for example, in the Kurchatov Institute, in recent years, the development of a progressive technology for the laser separation of isotopes and the production of highly pure substances has been suspended for 'incomprehensible' reasons, which in turn inhibits the implementation of new developments to create a powerful class of industrial sealed-off AE of the pulsed CVL.

Studies of the efficiency of pumping CVL with different execution of the electrical circuit of the high-voltage pulse modulator of the power source. In addition to design and technological solutions, the conditions for pumping (excitation) of the active medium of the AE provided by high-voltage power sources of nanosecond pulses have a significant effect on the power and efficiency of the CVL. Our experimental studies and investigations by many authors show that effective excitation conditions are achieved primarily by reducing the duration of the leading edge of discharge current pulses, i.e., increasing the rate of its increase (steepness) and total duration. But the maximum radiation power and efficiency are reached at current durations close to the time of existence of the population inversion, i.e., practically the duration of the radiation pulses [13–20]. These characteristics of the excitation pulses essentially depend on the execution of the electrical circuit of the high-voltage modulator of the power source and the inductance of the discharge circuit.

Figure 3.27 shows schematic circuit diagrams of high-voltage research with industrial sealed-off AEs of the Kulon and Kristall series [16, 26, 41, 42, 147–152, 227–233, 253–271, 313–322].

In the schemes *a*, *c*, *d* and *e*, water-cooled hydrogen thyratrons of the type TGI1–1000/25, TGI1-2000/35 and TGI1-2500/50 [146] were used as a high-voltage pulse commutator. The durability of thyratrons with circuit *a*) is usually 100–200 h and at a PRF above 11–12 kHz they operate unstably, which is manifested in transitions from the pulsed regime to the regime of continuous arc discharge. To ensure operation at high PRF (up to 21 kHz) water-cooled hydrogen tacitrons of the type TGU2-1000/25 were used. In the electrical circuit *b*) a water-cooled vacuum lamp GMI-29 A-1 is used as a pulse switch, which makes it possible to generate short pump pulses with a 10–10^2 kHz PRF [307].

The exciting current pulses in self-heating AEs also provide heating of the discharge tube with an active substance (copper) up to 1500–1600°C, therefore it is necessary to observe certain relationships between the duration of the current pulses, their amplitude and PRF. The electrical circuit of the nanosecond pulse modulator, shown in Fig. 3.27 *a*, is the simplest scheme for excitation of the active medium of the CVL. This scheme is called direct, in some works – classical or traditional. The name 'direct scheme' largely reflects its physical essence. The operation of the direct scheme was considered in many works, but in more detail in [79]. The circuit operates in the full discharge mode of the storage capacitor through a thyratron. In this scheme, resonance charging of a storage capacitor with a capacitance Cn via a nonlinear charge choke with an inductance of $L_{ch} \approx 0.5$ H occurs and a shunt inductor with an inductance of $L_{sh} \approx 60$ μH from a high-voltage rectifier (HR). Using the choke L_{ch} as a charging line allows increasing the voltage on the capacitor C_{cap} to values that are twice the voltage of the explosive. In real working conditions, the voltage on the C_{cap} as a result of recharging increases by another 20% or more. The greater the mismatch between the modulator and the AE, the greater the voltage will be on the C_{cap}. For example, as the pressure of the buffer gas increases, the resistance of the AE increases and the degree of charge exchange of the storage capacitor decreases, which causes a decrease in the power consumed from the rectifier. The inductance of the choke L_{sh} is usually 40–80 μH – this value is sufficient to not shunt the AE discharge during commutation. In addition, the inductor with inductance L_{sh} short-circuits the discharge gap of the AE during the interpulse period. A sharpening capacitor with a capacitance C_{sc}, connected in parallel with the AE, performs the function of exacerbating the front of the current pulses, i.e., an increase in the rate of current rise. One terminal of the sharpening capacitor is connected directly to the cathode (high-voltage electrode) of the AE, and the other through a coaxial metallic (copper or aluminum) current lead to its anode. The inductance of such a circuit is insignificant and amounts to 0.5–2 μH. Usually the optimum capacitance of C_{sc} is

3–5 times less than the capacitance C_{cap}. The sharpening capacitance in combination with the return current leads to an increase in the radiation power by 10–20%. Despite the simplicity, a direct circuit can not always be used, since the pumping of the AE is ineffective in its application. This is due to the fact that during the discharge of the current, the thyratron, the storage capacitor, and the AE are turned on in series, that is, they form a single discharge circuit, and the parameters of the pulses directly depend on the rate of discharge development in the thyratron [146].

In the first industrial CVL Kriostat, the modulator of the power supply IP-18 was made in a direct scheme. With a storage capacitance C_{cap} = 2200 pF and a 10 kHz PRF, the average radiation power is 3–5 W, and the practical efficiency is 0.15–0.2%. One of the main reasons for the low efficiency is the long duration of the discharge current pulses (~250 ns). A large volume of experimental studies using a direct modulator execution scheme is given in Refs. [16, 26] for the sealed-off AE GL-201. In the range of 3–18 kHz PRF, capacitances C_{cap} = 1320–6800 pF and buffer gas pressures of 50–760 mm Hg the duration of the front and the total duration of the current pulses were 70–150 and 200–400 ns, respectively, at amplitudes of 150–350 A. The current pulses through the AE were detected by a Rogowski coil operating in the current transformer (CT) mode, the voltage pulses by the compensated voltage divider (VD) and the oscilloscope C1-75 (see Fig. 3.27 *a*). Since the duration of the current pulses for a direct pumping scheme is 5–6 times, and in some cases and by an order of magnitude greater than the duration of the radiation pulses, it was impossible to expect high values of the radiation power and efficiency. The efficiency was about 0.5% at radiation powers of 12–15 W. In this scheme, the service life of thyratrons is too short (due to large initial losses) – 100–400 hours. To ensure stable operation of the thyratron it is necessary to stabilize the heating voltage of its hydrogen generator and cathode. Nevertheless, this scheme was used in CVLs for about 15 years, until the early 1980s.

To obtain the same high radiation power and efficiency, it is necessary to reduce the total duration of the current pulses to 50–100 ns with the front duration up to 10–30 ns, i.e., to a value close to the time of existence of the population inversion. Such characteristics are almost impossible to obtain in the direct modulator design described above, since the thyratron can not commute large average powers when operating with short current pulses following with a high frequency.

The first work in Russia (USSR), in which the conditions of excitation of CVL were significantly improved, was published in 1983 [225]. The principal electrical circuit of a pulsed pump modulator, applied in [225], is shown in Fig. 3.27 *c*. This circuit allows one to double the amplitude of the voltage and reduce the duration of the current pulses by half. The scheme uses a pulsed autotransformer and a link of magnetic compression on ferrites (non-linear saturable choke). The pulsed autotransformer with the 1:2 transformation ratio is made on the basis of six 400NN ferrite rings with dimensions K 125 × 80 × 12 mm and gaps between them (1 mm) to improve cooling. The winding is made with a cable with fluoroplastic insulation PVTFE, the number of turns is 23. The braid and the central core are connected in series. The turns are isolated from each other and from the ferrite core by means of fluoroplastic rings with grooves. The transformer operated with forced air cooling. The non-linear choke with inductance L_2 consisted of 70 ferrite rings 3000NMA with dimensions K 16 × 8 × 6 mm with copper wire passed through them. The choke was placed in a glass tube 30 mm in diameter and cooled by running water. The capacitance of the storage capacitor C_{cap} was 2200 pF, the working capacitor $C_{cap}/4 \approx$ 600 pF, the sharpening capacitor $C_{sc} \approx 165$ pF.

In [225, 231–233, 253], the circuit with transformer voltage doubling and the link of magnetic compression was used to increase the excitation efficiency of the Kristall-type AEs of the GL-201 and GL-201D type. In [253], the pulsed autotransformer is made on three ferrite rings of the 200NN brand with the dimensions K 180 × 110 × 20 mm. Between the rings, an air gap (2–3 mm) is formed with the help of cardboard pads. The winding is made with a wire with a copper conductor cross-section of 1.5 mm^2, the number of turns is 16. To isolate the wire from the core and the turns, discs from sheet plexiglass with holes for stretching the winding wire were used from each other. This transformer is more reliable in electric strength than the one used in [225]. The design of the non-linear choke (L_1) given in [253] is simpler: a water-cooled copper tube with M1000NM ferrite rings strung on it with sizes K 20 × 12 × 6 mm, the number of which reaches 150.

The circuit in Fig. 3.27 works as follows. Resonant charging of the capacitor C_{cap} is realized from the high-voltage rectifier through the throttle L_{ch} and the input winding of the autotransformer T_p. After opening the thyratron, the capacitor C_{cap} is recharged through the input winding of the autotransformer T_p to the capacitor

$C_{cap}/4$. The parameters of the choke L_1 were selected so that it would become saturated only after the capacitor $C_{cap}/4$ was fully charged. After saturation of the throttle L_1, the capacitor $C_{cap}/4$ is rapidly discharged to C_{sc} and through the AE. Due to the fact that the working capacitance for AE is the capacitance $C_{cap}/4$, the total duration of the current pulse through the AE is half as much as in the direct circuit, where the working capacitor C_{cap} discharges directly to the AE through the thyratron. The thyratron thus works in the facilitated mode on the speed of increase of a current as the load is not AE, and an input winding of the transformer T_p. The service life of thyratrons increases to 1000 hours or more. This circuit (*c*) worked steadily on the PRF from 3 to 13 kHz with an average switching power of up to 5 kW. The duration of the front of the pulses of the exciting current, depending on the parameters of the circuit, could vary from 25 to 100 ns, the amplitude from 0.2 to 1.0 kA with a change in the voltage across the AE from 15 to 30 kV. For the first time in [225], the radiation power of the AE TLG-5 (Kriostat) with circuit b at the 8 kHz FPI was increased from 5 to 10 W (twice), GL-201 from 10 to 18 W.

The circuit *d* in Fig. 3.27 [299], by the principle of operation and efficiency of excitation, practically does not differ from scheme *c*. But from the point of view of design it is simpler and leads to less power losses. This is due to the fact that the doubling of the voltage in the case of using the scheme *d* (according to the Blomlein scheme [299]) is carried out on high-frequency capacitors with low losses. In scheme *b* about 10% of the thyratron-switched power is dissipated in a ferrite transformer T_p, which requires additional (forced) air cooling. In the circuit, resonant charging of the working capacitors with a capacitance $C_{cap}/2$ from the high-voltage rectifier HR is carried out through a charge choke (L_{ch}), a non-linear choke (L) and an air reactor (L_0). One of the working capacitors with a capacitance of $C_{cap}/2$ (the upper one in the diagram) is connected to the 'ground' through the chokes L_1 and L_{sh}, and the other (bottom) – directly. After opening the thyratron, the bottom capacitor $C_{cap}/2$ is recharged (inverted) and, as a result, the voltage doubles on the capacitors connected in series. After saturation of the reactor L_1 (as well as in circuit *c*), there is a rapid discharge of working capacitors $C_{cap}/4$ to C_{sc} and via AE. The non-linear choke L (thyractor), designed to reduce the starting losses in the thyratron [146], is a water-cooled copper tube with ferrite rings threaded on it, for example, M1000NM with dimensions K 20 × 12 × 6 mm, the number of which reaches

20–30. Such a thyractor retards the development of the anode current of the thyratron with respect to the voltage by 50–70 ns. With a longer delay, with one compression link, the characteristics of the pump pulses of the AE deteriorate. The L_0 choke is necessary to match the recharge time of the lower capacitor $C_{cap}/2$ with the delay in the development of the discharge in the AE and is units of microhenry.

The circuit *d* with capacitive voltage doubling has practically no fundamental limitations on switched power, which is important for pumping AE with a large power consumption. To do this, firstly, to increase the duration of the pulses of the anodic current of the thyratron by two times or more (up to 300–400 ns) and in the same way to its amplitude, which leads to a decrease in the initial losses to a minimum (due to an increase in inductance *L*) and, secondly, to increase the number of magnetic compression links for the subsequent reduction of the duration of the current pulses to the minimum value [157–164].

Figure 3.27 *e* shows a circuit with a capacitive voltage doubling with two links of magnetic compression and an anode reactor. The most powerful domestic thyratron TGI1-2500/50 in this scheme allows switching medium power to 8–10 kW. In this scheme, thyratrons operate in a lightweight mode (in the minimum loss mode) and their service life is 2000–3000 h.

The indisputable advantages of vacuum lamps as controllable devices are the possibility of forming the fronts of current pulses with a duration of 10–20 ns with a total duration of 40–70 ns and operation at a PRF of 10–10^2 kHz. The main disadvantage of vacuum lamps is the relatively small values of the peak amplitude of the anode current pulses due to the saturation current limitation. In the case of using the GMI-29 A-1 lamp, the amplitude of the current pulses does not exceed 300–350 A, and the values of the switched average power are 4.5–5 kW. In circuit *b*, the GMI-29 A-1 lamp operates in the partial discharge mode of the storage capacitance C_{cap}.

At present, work is underway to replace water-cooled hydrogen thyratrons and vacuum lamps with more compact solid-state switches. [308] considered a power source using thyristor switches with a power consumption of 1 kW, providing a passport mode of operation for the AE Kulon LT-1.5Cu [16, 26, 42]. The authors of [308] suggest the development of such a power source for the Kulon LT-5Cu AE [16, 26, 42, 389, 393]. A power source with a transistor wrench for pumping a CuBr laser was created [309]. The switched capacity was 0.66 kW, the radiation power was 0.5 W.

Together with a group of specialists of the Materialy mikroelektroniki, a two-channel transistor power supply for the CVLS Kulon-15 and Kulon-20 with commercial sealed-off AE Kulon LT-10Cu (GL-205D) and Kulon LT-15Cu ГЛ-205I) [16, 34], working according to the scheme MO–PA was constructed In each channel with two pulse transformers and three magnetic compression links, voltage pulses with an amplitude of ~17 kV and current pulses with an amplitude of ~250 A and a duration of about 100 ns are generated at a PRF of 12–16 kHz. The power of each channel is 2.5 kW. Study [310] reported on the creation of a pulsed power supply on transistor switches with a switched power of up to 25 kW for pumping a 280-watt hybrid copper vapour laser (Cu–Ne–HBr).

In addition, together with the VELIT company the work on the creation of modern industrial CVL Kulon-10 with the AE model GL-205D and Kulon-15 with the AE model GL-205I with the use of transistor switches is being completed with the high reliability and efficiency. These commercial lasers have no domestic and foreign analogs.

At present, the Kulon and Kristall CVLs are mainly pumped with power sources with capacitive voltage doubling, magnetic compression links and anode reactor (circuit *e*), which, due to a 2-fold decrease in the duration of the current pulses (up to 100–120 ns), in comparison with the classical version of the power source, enable doubling the radiation power (Fig. 3.28) [16, 26, 41, 42].

However, the working temperature of the discharge channel of the AE increases by about 100°C (from 1550°C to 1650°C). With a lamp power source, the temperature rises even more – up to 1700°C. Therefore, the temperature of the discharge channel of the AE CVL during the training on degassing is adjusted to 1700°C.

Basic parameters and characteristics of industrial sealed-off AE of the Kulon series. The main parameters of industrial sealed-off self-heating AEs of the Kulon series (GL-206 according to TU) pulsed CVL and GVL are presented in Table 3.3. The AE of the Kulon series have relatively small dimensions – length 0.33–0.9 m (see Table 3.1), power consumption in the range of 0.65–2.2 kW (see Table 3.3), a metal–glass version of a vacuum-tight shell and are operated mainly in the air cooling mode. The most powerful of this series of AE GL-206D and GL-206I are used in technological installations such as Karavella, intended for microprocessing of electronic

Table 3.3 The main parameters of a new generation of industrial sealed self-heating AE of the Kulon series

Parameter	Model								
	GL-206A	GL-206B	GL-206C	GL-206D	UL-102	GL-206E	GL-206I	GL-206F	GL-206G
Laser environment	Copper vapours							Gold vapours	
Radiation wavelength, nm	510.6; 578.2							627.8	
The pressure of the buffer gas Ne, mm Hg	600		450	300	300	300	220	300	300
Diameter of the discharge channel, mm	7	7	12	14	20	14	14	14	14
Length of the discharge channel, mm	140	175	340	340	470	490	625	340	470
Active medium volume, cm^3	4.2	5.7	28	38	116	65	85	38	65
Pulse repetition frequency (PRF): optimum operating range, kHz	15–18 8–25	15–18 8–25	14–17 8–17	14–17 8–17	10–12 8–14	14–17 8–17	14–17 8–17	14–17 10–17	14–17 10–17
Average radiation power in the generator mode (for the optimum PRF *), W	1.0–1.4	1.5–2.0	3–4	5–6	7–8	15	20	1.2–1.4	1.8–2.0
Radiation power ratio **	3 : 2	3 : 2	3.5 : 2	3.5 : 2	5 : 3	3 : 2	3 : 2	–	–
The duration of the radiation pulses (at half-height), ns	10–15	10–15	17–20	17–20	20–25	17–20	17–20	15–18	15–18
Divergence of radiation with a plane resonator, mrad	6	5.5	6	6.5	7	4.5	4.3	6	4
Divergence of radiation with a telescopic unstable resonator, mrad	0.1–0.2 mrad								

Table 3.3

Power consumed from the rectifier of the power source, kW	0.65	0.75	1.15	1.2–1.5	1.7–2.1	1.5–1.8	2.1	1.4	1.9
Readiness time (at optimum power consumption), min	25	25	45	50	60	60	60	60	60
Minimum operating time, h	≥1000	≥1000	≥1500	≥1500	≥1500	≥1500	≥1500	≥500	≥500
Service life not less than, h	2000				3000			1000	

* High-voltage pulse modulator power source is executed under the scheme of capacitive voltage doubling with two links of magnetic compression of nanosecond current pulses. The capacity of the storage capacitor is $2C = 2000$ pF, the sharpening capacitor is $C_{ob} = 110$ pF. ** P_{isla} ($\lambda = 510.6$ nm)/P_{is} ($\lambda = 578.2$ nm).

materials. Therefore, to ensure the stability of the position of the beam axis, the heat-loaded AE is installed inside the water-cooled cylindrical metal heat sink, which simultaneously performs the function of the reverse current conductor.

As can be seen from Table 3.3 the working pressure of neon buffer gas in the AE of GL-206A with a radiation power of 1 W and GL-206B with a power of 1.5 W is 600 mm Hg, GL-206C power of 5 W – 450 mm Hg, GL-206D with a power of 10 W – 300 mm Hg. st and GL-206I power of 15 W – 220 mm Hg. Art. The buffer gas pressure in sealed AEs is selected by a compromise solution between the longevity and the radiation power. The durability of the sealed AE is higher, the more the mass of copper in the copper vapour generators, the distance from the generators to the copper vapour condensers, and the neon pressure. The high pressure of neon in malaborite AE is the main and necessary condition for ensuring their high longevity, since large amounts of active substance can not be constructively made in short discharge channels, and the length of the path for diffusion of copper vapour from extreme generators to condensers is always limited. In the AE model GL-206A only two copper

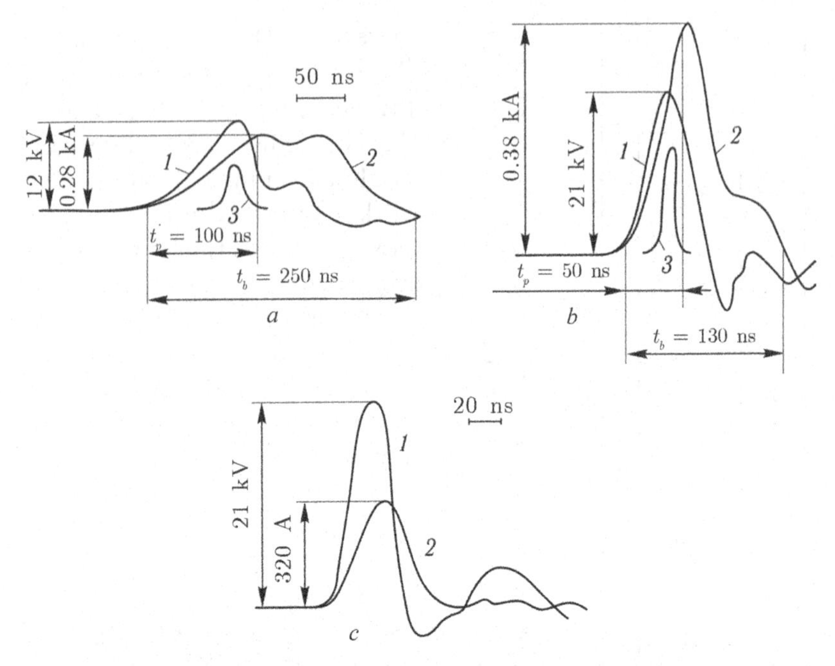

Fig. 3.28. Oscillograms of voltage pulses (*1*), current (*2*) and radiation (*3*) of the CVL with the Kristall GL-205A AE with different execution of high-voltage modulator of the power source.

vapour generators are placed, 4.5 g each, GL-206B – three and also 4.5 grams, in GL-206B – three 9 g each, GL-206D three, GL -206I – four for 11 g. In practically all known works on pulsed CVLs both in the sealed-off mode and in the pump gas pumping regime, the maximum radiation powers are attained at relatively low neon pressures (10–100 mm Hg). Together with the employees of VELIT Co, a high-voltage pulse thyratron power source was developed and optimized, the modulator of which was designed in accordance with the capacitive voltage doubling scheme with two links of magnetic compression of nanosecond current pulses and an anode reactor (Fig. 3.27 *d*) [160, 162, 163], which makes it possible to produce highly effective pumping of the AE of the Kulon series even at neon pressures close to atmospheric. The power source was cooled with forced air. With this circuit, with amplitude voltages of 15–20 kV, current pulses with a total duration of 100–140 ns and an amplitude of 250–350 A are generated with a front duration of no more than 40 ns, which corresponds to the time of existence of population inversion in the active medium, i.e., practically the duration of radiation pulses. The the average radiation power levels

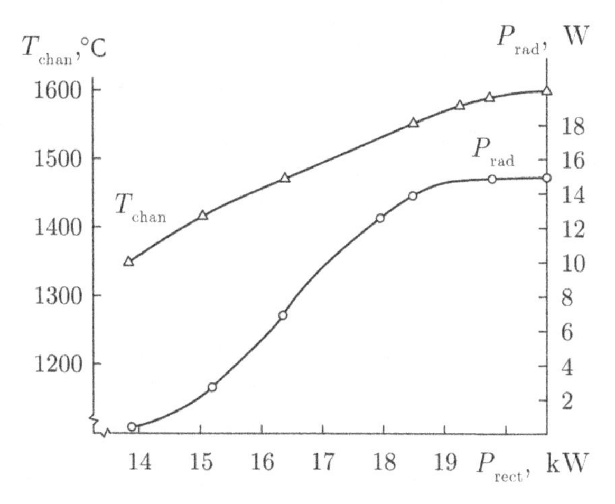

Fig. 3.29. Dependence of the average radiation power (P_{rad}) and temperature of the discharge channel (T_{chan}) of the AE Kulon, model GL-206D on the power consumed from the rectifier powder source.

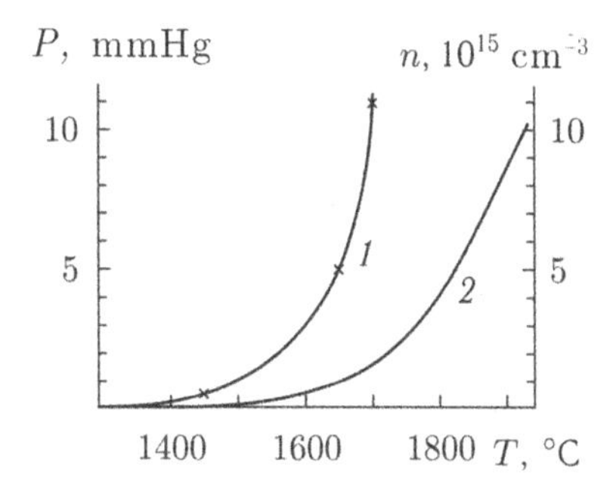

Fig. 3.30. The dependence of concentration (*1*) and pressure (*2*) of copper vapour on temperature.

presented in Table 3.3 correspond to the maximum values obtained when optimizing the CVL for the power consumption in the range of the operating PRF and in the steady (stationary) thermal regimes. For example, for the AE GL-206D at 15 kHz, the maximum power (P_{rad} = 15 W), which is also the working one, is set at the power consumption of the power source rectifier P_{rect} = 1.9–2.0 kW and the temperature of the discharge channel T_{chan} = 1600°C (Fig. 3.29).

Generation in the AE begins at P_{rect} = 1.4 kW, when T_{chan} = 1350°C. At a temperature T_{chan} = 1350°C, the concentration of copper vapour is $n = 10^{14}$ cm^{-3}, at T_{chan} = 1600°C – $n = 3 \cdot 10^{15}$cm^{-3} (see Fig. 3.30).

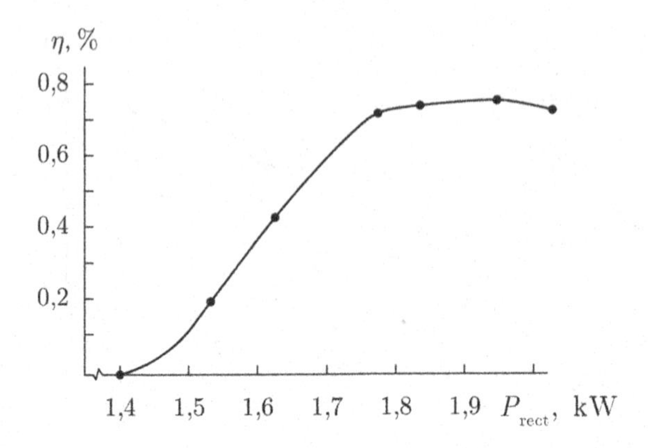

Fig. 3.31. Dependence of the efficiency of the laser AE Kulon, model GL-206E on the power consumed from the rectifier of the power source.

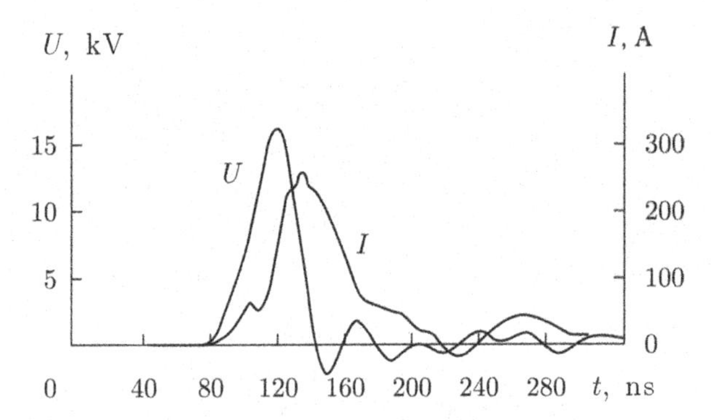

Fig. 3.32. Oscillograms of voltage pulses (*U*) and current (*I*) of the laser AE Kulon, model GL-206E.

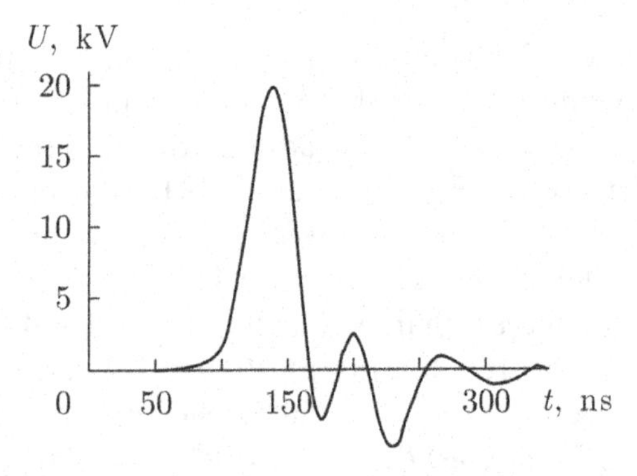

Fig. 3.33. Oscillogram of voltage pulses of the laser AE Kulon, model GL-206I.

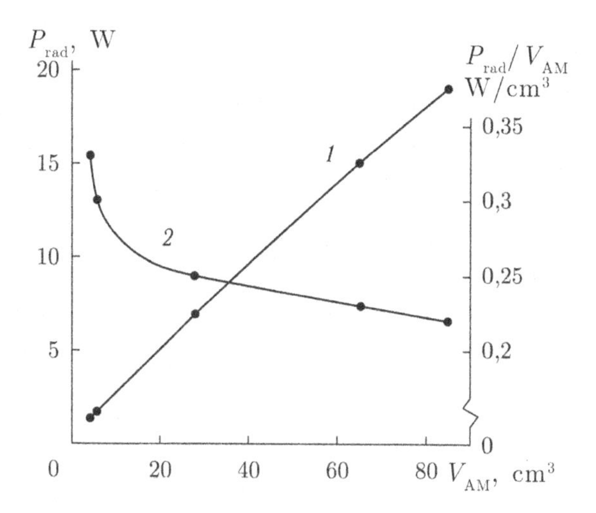

Fig. 3.34. Dependence of the average power (P_{rad}) (*1*) and specific (P_{rad}/V_{AC}) (*2*) radiation power of the industrial sealed-off laser AE of the Kulon series on the volume of the active medium (AM).

At a power consumption P_{rect} > 2.0 kW, when the temperature T_{chan} > 1600°C there is a sharp decrease in the radiation power, which is due to the thermal population of metastable levels of copper atoms. The maximum radiation power corresponds to a maximum efficiency equal to η = 0.8% (Fig. 3.31).

The pulses of the voltage and discharge current of the AE Kulon of the GL-206E model are shown in Fig. 3.32. With the optimum power consumption (P_{rect} = 1.9 kV) and the working PRF of 15 kHz the amplitude of the voltage pulses (*U*) is 16 kV for a base duration of ~70 ns, current (*J*) is 270 A for a duration of ~140 ns. The duration of the steep portion of the front of the current pulses is ~30 ns and corresponds to the measured values of the duration of the radiation pulses.

With an increase in the length of the discharge channel, in order to increase the efficiency of the AE, it is also necessary to increase the amplitude of the voltage pulses. In the longest AE GL 206I with a power up to 20 W, the voltage amplitude in the optimal operating mode is 21 kV (Fig. 3.33).

An important physical indicator of the efficiency of AE operation is the removal of power from a unit volume of the active medium. Figure 3.34 shows the dependences of the mean (P_{rad}) (*1*) and specific (P_{rad}/V_{AM}) (*2*) radiation power for industrial sealed-off AE of the

Kulon series (GL-206 according to TU) on the volume of the active medium are presented.

The radiation power increases with an increase in the volume of the active medium (curve 1), and the specific power density falls (curve 2), that is, it does not remain a constant value. If the AE GL-206A with the volume of AM of 4.2 cm^3 the specific power (power removal per unit volume) is 0.36 W/cm^3, then in GL-206I with an AM volume of 85 cm^3 is 0.24 W/cm^3 (1.5 times less). The latter, first of all, is associated with a decrease in the optimum operating temperature of the discharge channel and, correspondingly, the concentration of copper vapour in the AM. In the AE GL-206A and GL-206B, the working temperature of the discharge channel with an internal diameter of 7 mm reaches 1700°C ($n = 10^{16}$ cm^{-3}), in GL-206I and GL-206I with a channel diameter of 14 mm, the temperature is about 1600°C ($n = 3 \cdot 10^{15}$ cm^{-3}). The smaller the diameter of the discharge channel, the higher the rate of settling (decay) of the metastable levels of copper atoms in the interpulse period (due to the relative increase in the number of collisions of atoms with the channel wall), the higher the permissible operating temperature and the specific radiation power. Conversely, the larger the volume of speakers due to the increase in the diameter of the channel, the lower the rate of decomposition of metastable levels and, correspondingly, the operating temperature of the channel and the specific power.

The minimum (guaranteed) operating time of industrial sealed-off AEs of the GL-206A and GL-206B models of the pulsed CVL is at least 1000 h (T_{chan} = 1700°C), of the AE GL-206C, GL-206D, GL-206E and GL-206I – 1500 h T_{chan} = 1600°C), which is a quite acceptable level for this class of lasers and their wide application in modern technological and medical equipment. Limitation of the operating time of AEs on gold vapour GL-206F and GL-206G to 500 h is due to the increased optimum operating temperature of the discharge channel. (It is higher by 100–150°C than the AE on copper vapour.) During the guaranteed operating time, in accordance with the specification for AE, the reduction in the radiation power should not be more than 20%. Longevity tests and continuous operation of CVL show that the service life of sealed-off AEs on copper vapour is at least 3000 h. During this period of operation of the AE, the glow of the buffer gas of neon does not practically change in them, but the power of the radiation decreases. (The gas glow is checked by a high-frequency device of the Teslo type.) The first testifies to the preservation of the purity of the gaseous medium, and, accordingly,

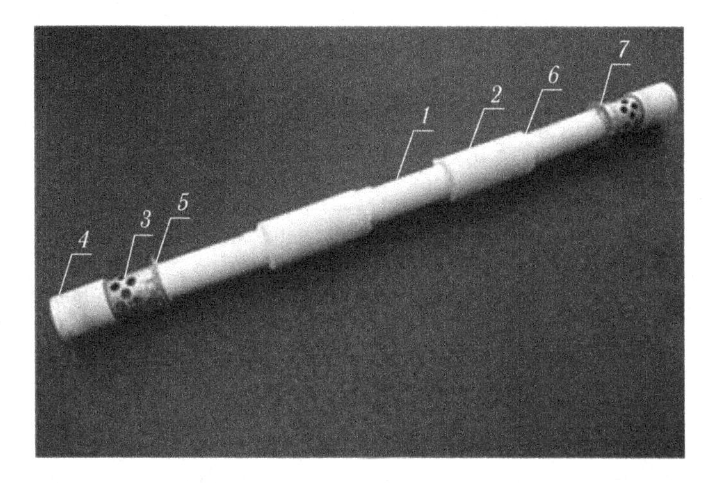

Fig. 3.35. The appearance of the sectional discharge channel of the laser AE GL-206C (Kulon) after 2000 hours of operation: *1* – the discharge channel tube; *2* – connecting sleeve; *3* – condenser; *4* – near-electrode tube; *5* – alumophosphate cement (APC); *6* – sealing cement; *7* – molybdenum ring and wire.

the high efficiency of the three-step technology for degassing and cleaning devices, and the second is about reducing the specific power take-off and the concentration of copper vapour in the AM. Devices that have been in operation for a long time (2000–3000 h) are usually analyzed for analysis of the state of its nodes and materials and their evaluation for reuse. A high-temperature partitioned discharge channel is usually retained by a solid node (Fig. 3.35), but can not be restored. Due to the strong sintering of the sealing cement (item 6), the ceramic channel is not disassembled into individual tubes (item 1) and bushings (item 2), which in turn does not allow replacement of the spent copper vapour generators for new ones.

Only the near-electrode ceramic tubes (item 4) without capacitors (item 3) can be reused. Molybdenum capacitors are in a state of embrittlement. The tungsten–barium cathode and the molybdenum anode are almost always suitable for reuse in new AEs since its service life is at least 6000 hours. The outer vacuum-tight metal-glass casing and the two end sections with optical windows, having a technological margin for the diameter of the welding of the metal cups, are suitable for two-three times repeated application. High-temperature heat insulators: fibrous grade VKV-1 and powder grade T can be re-applied only after removal of the sintered areas, which may constitute 15–20% of the total mass.

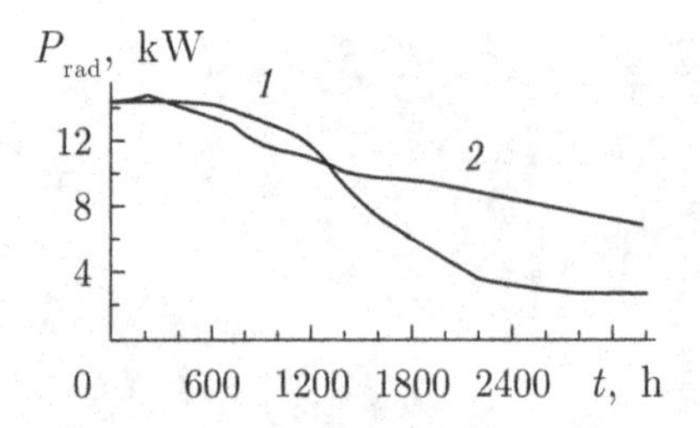

Fig. 3.36. Dependences of the average radiation power of the CVL with the Kulon AE, model GL-206E on the operating time: *1* – with standard copper vapour generators, *2* – with an increased supply of copper in the extreme copper vapour generators.

Figure 3.36 shows the dependence of the average radiation power for two AEs Kulon of the GL-206E model, differing in the reserve of the copper working substance. In the extreme copper vapour generators there was no substance, and in the only central one half of the initial mass remained. For the extreme generators, the calculated value of the service life (up to complete depletion of copper) is about 2000 h, the central one – 6000 h. At the same time, the path length of the diffusion escape of copper vapour for the extreme generators, determined by the distance to the capacitors is 90 mm, from a single central one 310 mm at both ends. Therefore, a more intensive loss of copper vapour and earlier depletion of metallic copper from the extreme generators leads in the process of operation to a decrease in the concentration of copper vapour in the AS and, accordingly, the radiation power. The latter is indicated, as already mentioned above, by the fall of curve 1 (Fig. 3.36) in the section of 1000–2200 h. With a runtime of more than 2200 h, the radiation power level is minimal, since the AE already operates in the regime with only one central generator, when the concentration of copper vapour in AM does not reach optimal values. The losses of radiation power from the dusting of the output windows does not have a noticeable effect on the course of curve 1, since they constitute about 3% of the total power.

In the AE Kulon GL-206E with a 2-fold increase in the copper reserve in the extreme copper vapour generators, the radiation power, as seen from curve 2 (Fig. 3.36), decreases monotonically and in 3200 h decreases from 14 to 7 W, to 50% level from the initial value. A portion of the steep power drop on curve 2, as in the case of the AE with standard generators (curve 1), is not present. But it became also clear that even with a large reserve of the active substance, while maintaining the purity of the buffer gas of neon and the conditions of excitation, there is still a decrease in power, albeit more delayed. This circumstance clearly indicates a decrease in the concentration of copper vapour in the AM and, correspondingly, the rate of evaporation from the surface of the molten copper in the generators during the operation of the AE. And at a time when the rate of evaporation of copper vapour from the generators becomes lower than the rate of vapour escape into the 'cold' ones, then a decrease in the radiation power becomes noticeable. This event usually occurs 300–500 hours from the beginning of the operation of the AE (see curves *1* and *2*).

To determine the reasons for the decrease in the rate of evaporation of copper vapour in the sealed-off AEs during the long-term operation of CVL, an analysis of the composition of the surface of metallic copper in standard samples of generators with a molybdenum substrate was carried out with the help of a secondary ion mass spectrometer MS-7201M (see Fig. 3.7) after working for 1600 and 2000 hours. In the mass spectrometer, methods were used to bombard the surface with a beam of Ar^+ ions with an energy of 5 keV (for knocking out ions) and quadrupole mass filtration. In this case, the surface of the samples was sprayed layer by layer to a depth of 56 nm. Based on the results obtained, the composition of the copper surface was evaluated. From the mass spectral data it follows that the surface of the remaining copper in the spent AE generators has practically an order of magnitude less pollution by carbon, nitrogen and oxygen than the original copper (Mob grade) before assembling the devices, which excludes their influence on the copper vapour pressure. The mass of carbon was 0.007%, nitrogen – 0.013%, oxygen – 0.004%. In the spectrum, insignificant traces of cuprous oxide Cu_2O, with a noticeable decomposition temperature only from 1800°C, were found. At the same time, despite the low solubility of molybdenum in copper, molybdenum is present in working copper with an operating time of 1600 hours, with a percentage content of about 5%, with an operating time of 2000 h to 12.5%. Molybdenum forms clusters in copper of various sizes, to microscopic (and probably nanoscale). The formation of molybdenum structures with a developed surface is confirmed by microscopic examination of samples the copper of which is etched with nitric acid. Segregation of copper on a large total surface of molybdenum clusters increases the enthalpy (necessary energy) of evaporation of copper from the system as compared to pure copper. This phenomenon naturally leads to a decrease in the rate of evaporation of copper vapour from the

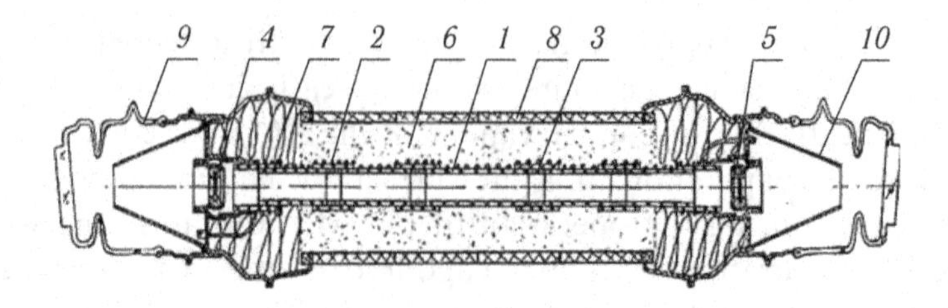

Fig. 3.37. The design of the sealed-off self-heating AE Kulon-20Sp of the pulsed CVL: *1* – a sectioned discharge channel; *2* – spiral from W–Re wire with a diameter of 1 mm; *3* – copper vapour generators; *4* – W–Ba cathode; *5* – molybdenum anode; *6, 8* – heat insulators; *8* – vacuum-tight shell; *9* – end section with an optical window; *10* – screen trap.

generators and their concentration in the AM and, as a consequence, to a decrease in the radiation power. In the future as a substrate material for copper vapour generators it is necessary to consider refractory rhenium (T_m = 2170°C) and its alloys with molybdenum and tungsten, which have a more favourable combination of the physical and chemical properties in comparison with molybdenum. These materials are plastic, their heat resistance is higher than that of molybdenum and is kept to a temperature 2000°C. At present, the AE Kulon" GL-205E with copper vapour generators on a rhenium substrate undergoes resource tests. The operating time is 500 h, no noticeable changes in the radiation power are observed.

Design and parameters of the AE Kulon-20Sp with a spiral heater of the discharge channel. With the purpose of increasing the reliability of the power source and decreasing the readiness time of the pulsed CVL, the sealed-off AE on copper vapour Kulon-20Sp with a spiral discharge channel heater [322] was developed and experimentally investigated on a laboratory test stand. The design of the AE of the Kulon-20Sp of the pulsed CVL is shown in Fig. 3.37.

The discharge channel (position 1) of the AE of the Kulon-20Sp, as in industrial AEs of the Kulon series, has a sectional construction [16, 42]. The channel consists of five ceramic tubes with an inner diameter of 20 mm and an outer diameter of 26 mm, interconnected by ceramic bushings with an internal diameter of 26 mm and an outer diameter of 32 mm. The ceramic material is aluminum oxide grade A-995 (99.8% Al_2O_3). On the inner surface of the connecting bushings there are four copper vapour generators (item 3) consisting of a cylindrical molybdenum substrate and an active copper substance (14 g each). The vacuum-tight shell (item 8) is cermet with glass end sections (item 9), in which optical windows for laser radiation exit are welded at an angle to the optical axis of the AE. The main part of the casing (item 8) is made of ceramics 22KhS which has an outer diameter of 72 mm, an inner diameter of 62 mm. Soldering of the middle seam of a ceramic cylinder was carried out with a PSr-72 solder with a melting point of 780°C. The space between the high-temperature discharge channel (item 1) and the vacuum-tight shell (item 8) is filled with a finely dispersed powder heat insulator (item 6) of the T grade from Al_2O_3 with the size of hollow microspheres of 20–200 μm, having a low thermal conductivity of 0.3 W/(m · K) The widened metal ends of the shell are filled with heat insulation from kaolin fiber (item 7) of grade VKV-1 (50% Al_2O_3 + 50% SiO_2) with fiber sizes 4 μm and thermal conductivity 0.27 W/(m · K). The maximum operating temperature of the thermal insulator of the T brand is 1600°C, grade VKV-1 – 1100°C.

The spiral heater (item 2) is wound directly onto the discharge channel. The spiral has 120 turns of tungsten–rhenium (W–Re) wire 1 mm in diameter, grade VR-20. The ends of the spiral, due to argon-arc welding, have reliable electrical contact with the electrode nodes of the AE. The resistance of the helix in the cold state is ~4.5 Ohm, in the state preheated to 1500°C it is 9 Ohm. The annular cathode 4, like in industrial AEs, is tungsten–barium, operates in the regime of auto-thermoemission at the localization of a pulsed arc discharge into a spot about 1 mm in size, the anode is molybdenum. The optical windows of the AE intended for output of laser generation

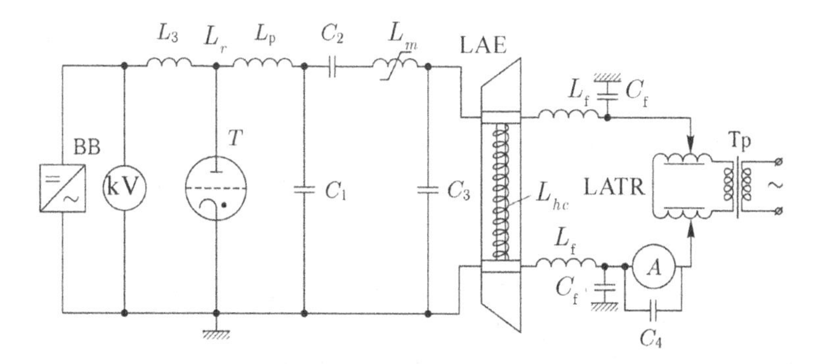

Fig. 3.38. The basic electrical circuit of a high-voltage pulsed power source: T – high-voltage pulsed hydrogen thyratron TGI1-2000/35, HR – high-voltage rectifier (0–5 kV), kV – kilovoltmeter, L_c – charge choke, L_r – resonant choke, $C_1 = C_2 = 1$ nF – storage capacitors, L_m – non-linear magnetic choke, $C_3 = 0.11$ nF – sharpening capacitor, LAE – laser active element, L_{hc} – heating coil of the discharge channel, $C_f = 1$ nF – filter capacitor, $L_f = 140$ μH - filter choke, A – ammeter (0–20 A), $C_4 =$ 4.7 nF – shunt capacitor, T_p – isolating transformer (380/165 V), LATR – laboratory autotransformer.

are installed at an angle of 82° to the axis of the device and are protected from dusting by screens (position10).

In the process of training AE, in its degassing in the mode of pumping neon, the maximum shell temperature was 520°C, the discharge channel 1580°C. Therefore, in the operating mode, the temperatures did not exceed the above values.

Figure 3.38 shows the principal circuit diagram of a high-voltage impulse puylsed power source [16, 42] with a circuit for feeding a spiral heater of the discharge channel of the AE.

The AE tests were carriedout at a neon buffer gas pressure of 300 mm Hg. At this pressure, the PRF of the pump current pulses of 12.5 kHz and the combined power supply (0.7 kW from the low-voltage spiral heater and 2 kW from the high-voltage pulse modulator), the average radiation power with a planar-spherical resonator (R_1 = 3 m, $R_2 = \infty$) was 17 W. In this mode, the amplitude of the pulses of the discharge current was 320 A with the duration of its front and the total duration of 35 and 100 ns, respectively (value of the voltage amplitude 15 kV). The shell temperature did not exceed 440°C. These results were obtained on a test bench without the use of a reverse coaxial conductor, under conditions of free convection and without an antireflection coating of the exit windows of the AE.

It should be emphasized that the use of an external spiral heater of the discharge channel allows to eliminate the danger of its destruction during the warm-up period by reducing the temperature difference between the external and internal walls, to shorten the heating time of the AE, to substantially increase the service life of the pulsed hydrogen thyratron (switch) and the reliability of the high voltage Power source as a whole for account for the decrease in the fraction of power invested in the gas discharge of the AE.

Basic parameters and characteristics of industrial sealed-off AE of the Kristall series. The main parameters of a new generation of industrial sealed-off self-heating AEs of the Kristall series of the pulsed CVLs are given in Table 3.4 [16, 41, 42, 323, 324]. They are obtained and optimized with a power source, whose high-voltage pulse modulator is made according to a capacitive voltage doubling scheme with a link of magnetic compression of nanosecond current pulses and an anode reactor (Fig. 3.27 *d*). Such a version of the thyratron IP today is the most reliable and effective source of excitation of AU on copper vapour. As a high-voltage pulse commutator in a modulator, a powerful hydrogen thyratron of the type TGI1-2500/50c with water cooling of the development of the

Table 3.4 The main parameters of the new generation of industrial sealed-off self-heating AE of the Kristall series

Parameter	Model						
	GL-205A	GL-205B	GL-205C	LT-50Cu-D	LT-75Cu	LT-100Cu	GL-205D
Laser environment	Copper vapour						Gold vapour
Radiation wavelength, nm	510.6; 578.2						627.8
Pressure of the buffer gas Ne, mm Hg	250	180	150	120	100	90	200
Diameter of the discharge channel, mm	20	20	32	32	45	45	20
Length of the discharge channel, mm	930	1230	1230	1520	1230	1520	930
Volume of active medium, cm^3	250	350	900	1200	1800	2200	250
Pulse repetition frequency (PRF): optimum operating range, kHz	10–12 8–20	10–12 8–20	9–12 8–20	9–12	9–11	9–11	14–17 10–20
Average radiation power (with optimal RPF), W: oscillator mode amplifier mode	30–35 40–45	40–45 55–60	50–55 70–75	60–80	74–97	90–117	4–6 –
Ratio of radiation powers *	1 : 1	1 : 1	1 : 1	1 : 1	1 : 1	1 : 1	–
Duration of radiation pulses, ns	25–30	25–30	30–35	30–35	35	35	15–20

Table 3.4

Divergence of radiation with a plane resonator, mrad	4	3	5	4	6	5	4
Divergence of radiation with a telescopic unstable resonator, mrad	0.07–0.1			–			0.1
Power consumed from the rectifier IP, kW	2.9–3.1	3.6–3.8	4.5–4.7	5.7	6.6	9.0	3.3–3.5
Readiness time (at optimum power consumption), min	60	60	80	80	100	100	60
Minimum (guaranteed) operating time, h	> 1500				> 1000		> 500
Service life not less than, h	3000				2000		1000

*P_{rad} (λ = 510.6 nm)/P_{rad} (λ = 578.2 nm).

Istok NPP [146] is used. The capacity of the storage capacitor is $2C$ = (1000 + 1000) pF, the sharpening capacitor is C_{ob} = 160 pF. A link of magnetic compression cooled by water includes 100–150 ferrite rings of the M1000NM brand with dimensions of 20 × 16 × 6 mm. The AEs of the Kristall series the volume of the active medium of which is 1–2 orders of magnitude greater than that of the AE of the Kulon series, when using an effective scheme of voltage doubling and magnetic compression of pulses have almost twice the radiation power than in the case of a direct scheme of the power source (Fig. 3.27 *a*). The optimization of the new generation of sealed-off AEs of the Kristall series of the pulsed CVL to achieve the maximum operating efficiency and power of radiation was carried out using, first of all, the results of unique experimental studies [16, 26].

The maximum values of the average radiation power, shown in Table 3.4, correspond to the AE with the antireflecting exit windows (transmittance τ = 98%). Currently, the AEs of the Kristall series are produced only with the antireflecting windows.

The construction and manufacturing technology of the AE of GL-205D of the pulsed gold vapour laser (GVL) are identical to the GL-205A model for the CVL and differ only in the composition of the active substance and, correspondingly, in the wavelengths of the radiation (see Table 3.4). The average radiation power of the

Table 3.5. Comparison of power and efficiency of old and new industrial models of AEs of the Kristall series

Parameter	Model AE Kristall					
	GL-201	GL-205A	GL-201E	GL-205B	GL-201 D32	GL-205C
	V_{AM} = 250 cm^3		V_{AM} = 350 cm^3		V_{AM} = 900 cm^3	
Average radiation power in generator mode, W	35	35	44	45	55	55
Power consumed from the rectifier of the power source, kW	3.6	3.0	4.4	3.6	5.5	4.6
Practical efficiency,%	1	1.17	1	1.2	1	1.2

AE GL-205D is 4–6 W, the guaranteed running time is 500 hours, which is 3 times less than that of the GL-205A, GL-205B and GL-205C models. The latter is explained by the fact that the optimal working temperature of the discharge channel of AE on gold vapour (T_{chan} ~ 1700°C) is 100–150°C higher than that of the AE on copper vapour. Application in the AE Kristall GL-205B and GL-205C of gold as an active substance allows to raise the level of radiation power at a wavelength λ = 578.2 nm to 10 W.

Comparative analysis of the efficiency of old and new models of AE of the Kristall series. The efficiency of old and new models of the AE of the Kristall series was estimated from the values of the average radiation power, the power consumed by the power source of the rectifier, and the practical efficiency in the optimum operating modes. The practical efficiency is defined as the ratio of the output radiation power to the power consumed by the rectifier of the power source. From the comparative Table 3.5 it follows that for the same values of the radiation power, the power consumption of the AE of the Kristall series of the new models GL-205A, GL-205B and GL-205C is approximately 1.2 times smaller, and the practical efficiency is correspondingly higher than for the old GL-201, GL-201E and GL-201E32. The decrease in the power consumption of the power source led to an increase in the service life of high-voltage pulse thyratrons which for a powerful TGI1-2500/50 is not less than 2000 hours. The service life of new AE models is determined practically only by the reserve of the active substance in copper vapour generators, the operating temperature of the discharge channel and buffer gas pressure neon, since the main causes of the destruction of the ceramic discharge channel are eliminated, overlapping of the channel aperture by drops of condensing copper, dusting of the output windows with copper vapour, electrode erosion products and other particles. At the same time, the minimum operating time of the instruments increased from 500 to 1500 h, ensuring high quality of the output radiation and reproducibility of the parameters. This level of basic parameters for industrial sealed-off AEs of the pulsed CVL is quite acceptable for wide application in modern technological equipment.

Influence of hydrogen on efficiency and power of radiation of industrial sealed-off self-heating AEs of the Kristall series. One of the ways to increase the radiation power and the efficiency of pulsed CVL is the addition of molecular hydrogen to the gas medium of AE [16–20]. In [126] it was shown for the first time that the addition of molecular hydrogen to the neon buffer gas significantly increases the efficiency of the CVL (up to 3%). In [20], the main experimental and theoretical studies are considered in which the influence of hydrogen on the operating mode of CVL and on its output characteristics is determined. Depending on the geometric dimensions of the discharge channel AE (D_{chan} = 1.8–10 cm, L_{chan} = 70–350 cm), purity of the active medium, buffer gas pressure (p_{Ne} = 24.5–79.6 mm Hg) and PRF (3.9–9.2 kHz), the optimal amount of hydrogen added is 0.5–3%. Hydrogen additives increase more significantly increase the efficiency of the CVL than the power of its radiation, as the electrical matching between the AE and the pumping modulator improves. The temperature of the discharge channel is increased by 50–100°C. The increase in the amplitude of the voltage pulses at the discharge gap of the AE reaches 30%, and the amplitude of the current pulses, on the contrary, decreases, and the duration of the radiation pulse increases. The latter, when using a telescopic unstable resonator, leads to an increase in the radiation power in a beam of diffraction quality. The optimal PRF corresponding to the maximum average radiation power is moved to the region of higher frequencies.

The authors [20] explain the increase in the efficiency and power of the laser radiation with addition of molecular hydrogen, in particular, by an intensive decrease in the electron temperature in the afterglow period due to elastic and inelastic collisions. 'Chilled' electrons speed up the recombination of electrons and ions, which reduces the pre-pulse concentration of the electrons. In this case, in the period between the current pulses, a faster and full transition of copper atoms to the ground state occurs and the pulse voltage applied to the AE increases.

Earlier, experimental studies of the effect of hydrogen on the efficiency of the sealed-off self-heating AEs of the Kristall series GL-201, GL-201E and GL-201E32 were considered. It was established that the addition of hydrogen to a partial pressure of 10 mm Hg into the active medium (neon pressure p_{Ne} = 50–250 mm Hg) leads to an increase in the radiation power by 1.2–1.5 times. Here, the influence of hydrogen additives on the operation of new models of AE on the copper vapour models GL-205A (Kristall LT-30Cu), GL-205B (Kristall LT-40Cu) and GL-205C (Kristall LT-50Cu) at the pressures of the neon buffer gas was studied (see Table 3.4) [99].

The working pressure of neon in the AE of GL-205A is 250 mm Hg, GL-205B – 180 mm Hg, GL-205C – 150 mm Hg. High purity neon was used with a volume fraction of 99.994% (impurities: 5 ·

10^{-3} % He + 67 · 10^{-5} % O_2 + 10^{-4} % N_2 + 10^{-4} % H_2 + 10^{-5} % CH_4 + 10^{-5} % CO_2 + 2 · 10^{-4} % H_2O), hydrogen of spectral purity with a volume fraction of 99.9999% (impurities: 2 · 10^{-5} % (O_2 + Ar) + 2 · 10^{-5} % N_2 + 3 · 10^{-5} % CH_4 + 2 · 10^{-4} % H_2O). The hydrogen was injected into the Kristall AE after finishing training for its degassing and purification with hydrogen and then filling it with pure neon to operating pressure and warming up to an operating temperature (~ 1600°C).

Figure 3.39 shows the dependence of the average radiation power of the AE GL-205B (LT-40Cu Kristall) of the CVL with an effective thyratron power source (Fig. 3.27 *e*) at 10.7 kHz from the hydrogen pressure added to the neon buffer gas. The exit windows of the AE did not have an antireflection coating. At the same time, the amplitude of the voltage pulses was 28.4 kV, and its duration on the base was 70 ns; for current pulses, the corresponding values are 0.4 kA and 140 ns. The hydrogen gas was usually introduced from the cathode side of the AE through a glass exhaust tube in the end section, its amount being controlled by a U-shaped oil manometer. In order to accelerate the process of mixing hydrogen with neon in the AE, the output valve of the glass exhaust tube was opened from the anode side and the gas was evacuated from the device until the initial pressure was established. In another case, the system was left for 12–16 hours; this time was sufficient for complete mixing of the gas mixture. In the absence of hydrogen additives, the average power of

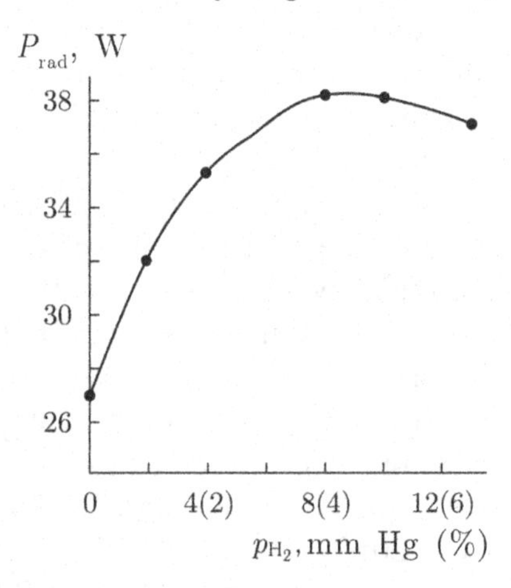

Fig. 3.39. Dependence of the average radiation power of the CVL with the AE Kristall of the GL-205B model with an effective thyratron power source with a PRF of 10.5 kHz on the hydrogen partial pressure (the value in the parentheses is the percentage of hydrogen).

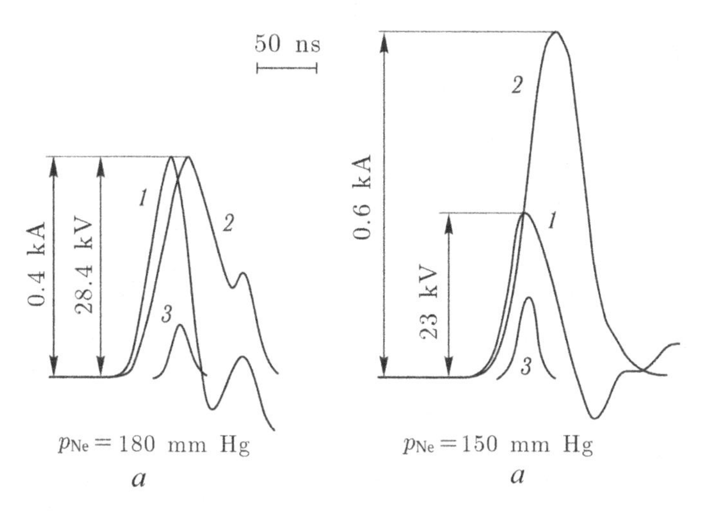

Fig. 3.40. The oscillograms of the pulses of voltage (1), current (2), and radiation (3) of laser AE GL-205A (LT-30Cu Kristall) (*a*) and GL-205C (LT-50Cu Kristall) (b) at PRF of 5 kHz.

the AE radiation was about 27 W at the optimum power consumption $P_{ext} \simeq 3.6$ kW (efficiency 0.75%). The maximum radiation power was achieved at the partial pressure of $p_{H_2} = 6–12$ mm Hg (3-6% of the gas mixture pressure), and was 37.5–38.5 W (see Fig. 3.39) at a power consumption of $P_{rect} \simeq 3.4$ kW (efficiency 1.12%). The temperature of the discharge channel increased from about 1550° to 1600°C. Thus, the addition of pure hydrogen to the GL-205B led to an increase in the radiation power by 1.4 times (from 27 to 38 W), practical efficiency by 1.5 times, and the discharge channel temperature by 50°C. The efficiency is increasing more noticeably than the radiation power, which is explained by the decrease in the power consumption from 3.6 to 3.4 kW. That is, the addition of hydrogen leads to an improvement in the matching of AE with elements of the pump circuit (first of all, losses in the thyratron are reduced). Approximately the same changes were detected in the characteristics of CVL with the addition of hydrogen to the gas environment of the AE GL-205A (Kristall LT-30Cu) and GL-205C (Kristall LT-50Cu). The oscillograms of the voltage pulses and discharge current for these AE are shown in Fig. 3.40 [42, 269–271]. The radiation power of the AE GL-205A increases from 25 to 31 W, i.e., by 1.24 times, the practical efficiency – from 0.75% to 1%, i.e., 1.3 times, for the AL GL-205C – from 34 to 52 W (1.53 times) and from 0.7% to 1.13% (1.6 times), respectively, from which it can be concluded that with an increase in the diameter of the discharge

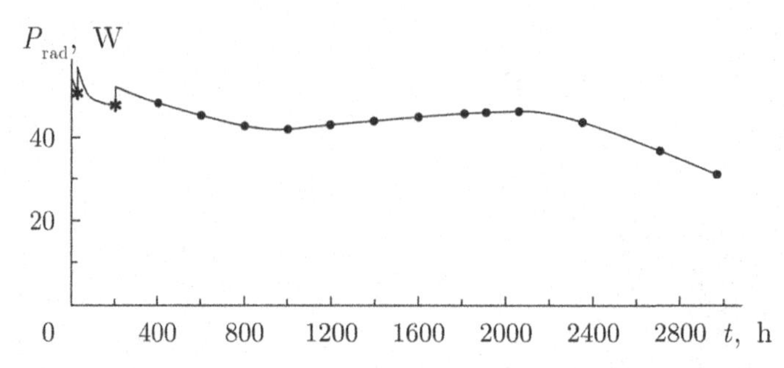

Fig. 3.41. Dependence of the average radiation power of CVL with AE Kristall LT-50Cu GL-205C with an effective thyratron power source on the time of operating. The asterisks mark the times at which the neon was pumped through the AE.

channel of the AE, the efficiency of the hydrogen additive increases. It should also be noted that the addition of hydrogen leads to an increase in the radiation power on the green line to a greater extent and in the total duration of the radiation pulses. For example, when using AL GL-205B without the addition of hydrogen, the power of emission on the green and yellow lines is approximately the same, and at the optimal hydrogen content the power of the emission on the green line was 62% of the total power, on yellow – 38%.

Figure 3.41 shows the results of tests of AE GL-205C (LT-50Cu Kristall) with hydrogen additives with a partial pressure of 12 mm Hg for 3000 hours in the sealed-off mode (neon pressure p_{Ne} = 150 mm Hg, exit windows of the AE – clarified).

At the initial moment, the radiation power was 55–56 W. After 25 hours of operation, the power of the AE decreased by 10%, and therefore through it a slow pumping (in the operating mode) of pure neon was performed to remove excess hydrogen. Most likely, after the regime of AE reduction by hydrogen, some hydrogen remained on its elements and at high operating temperatures it began to be released. The same technological procedure had to be done after 200 hours of operating time. In the period from 200 to 800 hours of operating time, a monotonic decrease in the radiation power from 52 to 42 W took place, which can be explained by the slow increase in the concentration of hydrogen in the active medium. Then the power slowly increased and at 2000 hours the operating time reached 46 W. This may be due to the subsequent slow decrease in the concentration of hydrogen in the active medium due to the prevailing processes of its absorption. Estimates show that the reserve of the active substance in copper vapour generators is sufficient to ensure the effective

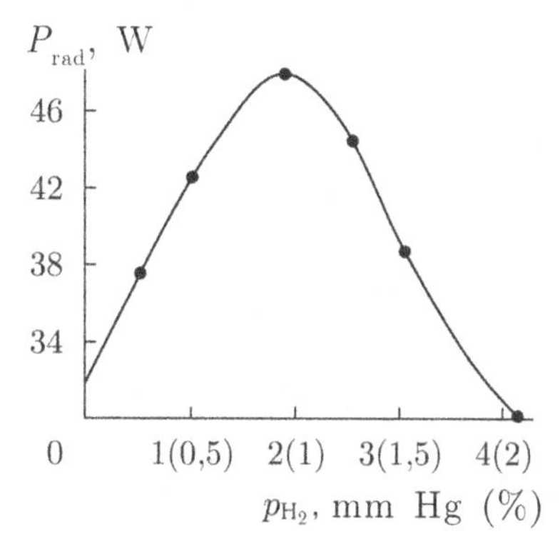

Fig. 3.42. Dependence of the average radiation power of the CVL with the LT-40Cu Kristall AE the LT-205B model and the lamp power source on the hydrogen partial pressure at 14.7 kHz PRF (values in the parentheses are the percentage of hydrogen).

operation of AE for more than 3000 hours. With the continuation of the AE tests to 3000 h, the radiation power was already P_{rad} = 34 W.

The dependence of the radiation power on the addition of hydrogen has a more pronounced character in the case of a lamp power source (Fig. 3.27 *b*) than for a thyratron (Fig. 3.27 *e*). When using the GL-205B AE, the maximum radiation power was achieved when p_{H_2} = 2 mm Hg. and was 48 W at the PRF = 14.7 kHz and P_{rect} = 4.3 kW (Fig. 3.42).

At partial pressures of hydrogen above 2 mm Hg the power falls off rather sharply. In the absence of hydrogen, the radiation power was equal to P = 32 W at a power consumption of 3.9 kW. The addition of hydrogen led to an increase in radiation power 1.5 times (from 32 to 48 W), practical efficiency 1.36 times (from 0.82% to 1.12%), power consumption 1.1 times (from 3.9 to 4.3 kW), the temperature of the discharge channel by 50°C (from 1550 to 1600°C).

If with the thyratron power supply, the addition of hydrogen causes an increase in efficiency greater than the increase in radiation power, due to a decrease in the power consumption (by 5%), then in the case of the lamp power source, on the contrary, a greater increase in the radiation power is detected (the power consumption increased by 10 %). These features when using the lamp power source can be explained only by shorter generated current pulses.

Figure 3.43 shows the dependence of the average radiation power of the AE GL-205B (LT-40Cu Kristall) on the PRF in the absence of an additive and with the addition of hydrogen (p_{H_2} = 2 mm Hg). Before recording of the curves preliminary purification (flushing) of the AE with hydrogen was carried out, which led to an increase in

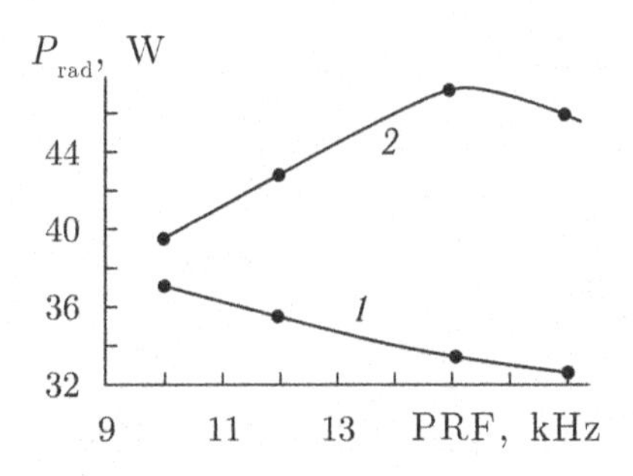

Fig. 3.43. Dependence of the average radiation power of CVL with AE Kristall LT-40Cu GL-205B and the lamp power source on the the PRF without additive (*1*) and with the addition of hydrogen at p_{H_2} = 2 mm Hg (*2*).

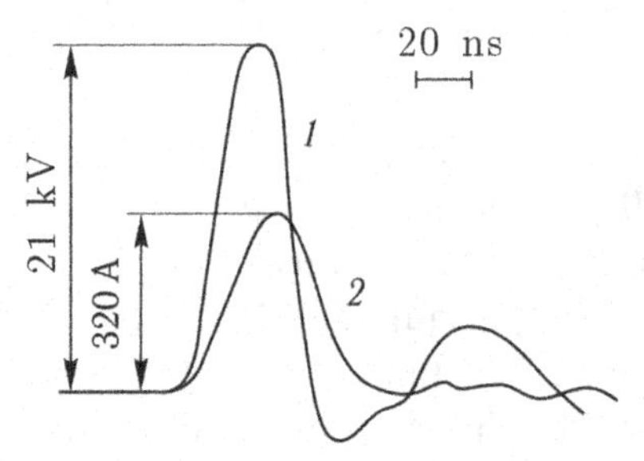

Fig. 3.44. The oscillograms of the voltage pulses (*1*) and the current (*2*) of the CVL with the AE Kristall LT-40Cu GL-205B and the lamp power source at 14.7 kHz PRF and p_{Ne} = 210 mm Hg. p_{H_2} = 2.5 mm Hg.

the radiation power by approximately 5%. This is most likely due to the reduction of copper oxide (CuO) in the near-surface layers of the copper vapour generators. The oxide reduces the rate of evaporation of copper in the generators, since it begins to decompose significantly only at a temperature of about 1800°C [277]. In the absence of hydrogen, the radiation power decreased from 37 to 32.3 W with an increase in the PRF from 10 to 17 kHz, and with the hydrogen addition it first increased from 39.5 to 47 W (at 14.7 kHz), then it slowly decreased (i.e., the addition of hydrogen to the CVL gas mixture leads to an increase in the optimal PRF). With an optimum 14.7 kHz PRF, the amplitude of the voltage pulse was 21 kV, the current is 320 A, and the duration of the leading edge of the current pulses is 25 ns (Figure 3.44).

To determine the rate of hydrogen escape from the active gaseous medium of the self-heating AE GL-205B (LT-40Cu Kristall) with a lamp power source, the dependence of the radiation power on the time of operating at the optimum hydrogen pressures was obtained

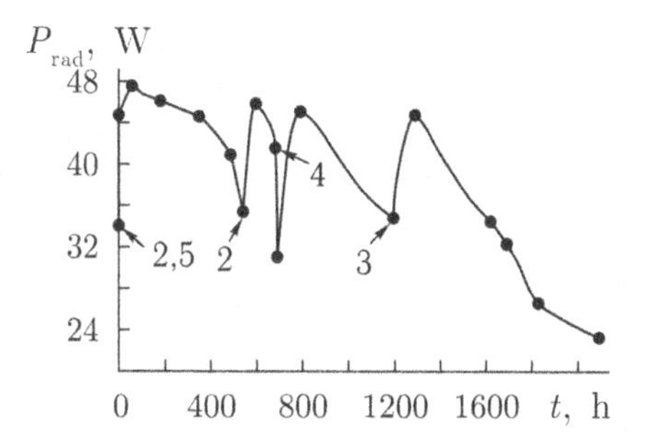

Fig. 3.45. Dependence of the average radiation power of CVL with AE Kristall LT-40Cu GL-205B and lamp power at 14.7 kHz PRF on the operating time. The value of the partial pressure of hydrogen (mm Hg) is indicated next to the corresponding points of the graph.

(Fig. 3.45). The tests were carried out at 14.7 kHz PRF. Before starting the tests, hydrogen was introduced into the AE to a partial pressure of 2.5–3 mm Hg. at which the radiation power reached 44 W.

Since the optimum value of the hydrogen pressure with a lamp-mounted power source was 2 mm Hg, the radiation power (within 70 hours) increased to 47.5 W, and then, during the next 500 hours, it decreased to 35 W. At this point (t = 570 h), hydrogen was injected into the AE (p_{He} = 2 mm Hg), and after 50 hours the radiation power increased to 46 W, and then decreased to 42 W for the next 80 hours. At the point t = 700 h, hydrogen was again allowed to reach p_{He} = 4 mm Hg), and the power dropped to 31.5 W. After 100 h after this (t = 800 h), the radiation power was restored to 45 W, and after another 400 h (t = 1200 h) it dropped to 34.5 W. After the addition of a new portion of hydrogen (p_{He} = 3 mm Hg), the complete escape of hydrogen was most likely to occur at the time t = 1700 h, when the power became 32 W.

An approximate hydrogen flow was estimated from the analysis of the experiment, taking into account possible errors in determining the pressure in the processes of hydrogen inlet and evacuation; the pressure p_{He} = 8–10 mm Hg is sufficient to provide 1000 hours of operation. A further decrease in the radiation power after 1700 h is due to the depletion of the copper reserve in the extreme (end) copper vapour generators. Therefore, in order to ensure that the time of guaranteed operating time exceeds 2000 h, it is necessary to increase the mass of copper in the extreme generators by at least 15%. During the total operating time of the AE t = 2100 h,

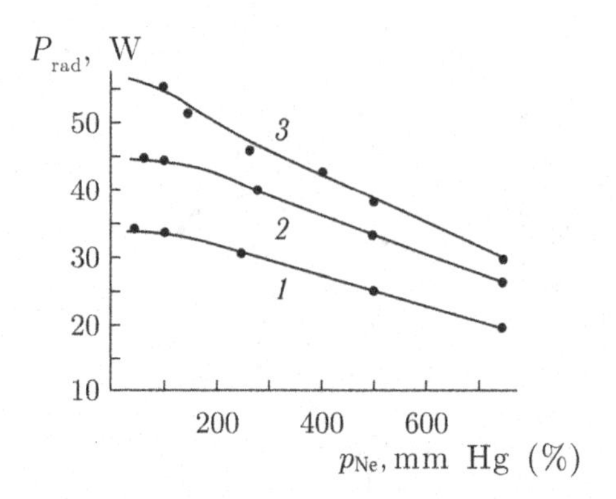

Fig. 3.46. The dependence of the average radiation power of the AE GL-205A (LT-30Cu Kristall) (*1*), GL-205B (LT-40CU Kristall) (*2*) and GL-205C (LT-50Cu) (*3*) with the thyratron power source on neon pressure at 10.7 kHz PRF.

approximately 270 switching cycles were made. At the same time, the output windows of the AE remained clean, and, as the control check showed, their transmittance was practically unchanged. The aperture of the discharge channel also remains unopened.

Dependence of the radiation power of the Kristall AE on the pressure of the neon buffer gas. Figure 3.46 shows the dependence of the average radiation power of the Kristall AE on the pressure of the neon buffer gas with the execution of the pump modulator according to the capacitive voltage doubling scheme with a magnetic compression link (Fig. 3.27 *e*).

When the pressure varies from 50 to 760 mm Hg the radiation power of the AE GL-205A (LT-30Cu Kristall) (curve *1*) decreased from 34 to 21.5 W (by 37%), GL-205B (LT-40Cu Kristall) (curve *2*) – from 44 up to 25 W (by 43%), GL-205C (LT-50Cu Kristall) (curve *3*) – from 56 to 30 W (by 47%). The decrease in the total radiation power is primarily due to the reduction in power on the green line (λ = 0.51 μm) due to the impairment of the characteristics of the pump pulses (the duration of the leading edge and the total duration of the current pulses increase and its amplitude decreases). It also follows from the comparative analysis that when the discharge channel elongates and its diameter increases, i.e., when the volume of the active medium increases (see curves 1–3 in Fig. 3.46), the relative power drop becomes more pronounced. Therefore, as a rule, the AEs with large volumes have lower operating pressures. Thus, in the AE GL-205A the working pressure is 250 mm Hg. and in the AE GL-205C – 150 mm Hg (see Table 3.4.)

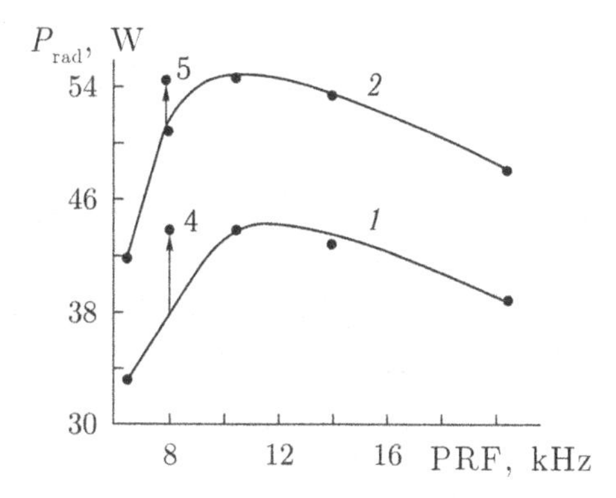

Fig. 3.47. Dependences of the average radiation power of the AE GL-205B (LT-40Cu Kristall) (*1*) and GL-205C (LT-50CU Kristall) (*2*) with the thyratron power source on the PRF.

Dependence of the radiation power of the Kristall AE on the pulse repetition frequency (PRF). The frequency characteristics of the Kristall AE (Fig. 3.47) were taken at the working neon pressures (see Table 3.4) with the same performance of the thyratron pump modulator of the power source (Fig. 3.27 *d*).

With an increase in the PRF from 10.5 to 20.5 kHz, the average radiation power of the AE GL-205B (LT-40Cu Kristall) decreased slightly – by 14%, AE GL-205C (LT-50Cu Kristall) – by 13%. In the case of a reduction in the PRF with respect to the optimal value (10.5 kHz), a relatively sharp decrease in the radiation power took place, which is explained by an increase in losses in the thyratron (due to the increase in the anode voltage), a decrease in the fraction of the power that is used to heat the AE and, a decrease in the temperature of the discharge channel. With a drop in the PRF from 10.5 to 6.5 kHz, the AE's power fell by 24%. When replacing the capacitance of the storage capacitor $2C_{cap}$ = (1000 + 1000) pF = 2000 pF and the capacitance C_{sh} = 160 pF by $2C_{cap}$ = (1500 + 1500) pF = 3000 pF and C_{sh} = 300 pF at 8 kHz, radiation power and efficiency (see points 4 and 5) werre almost the same as at 10.5 kHz.

Efficiency of industrial sealed-off AE Kristall. For industrial sealed-off self-heating AE GL-205A (LT-30Cu Kristall) with bleached exit windows at neon working pressure p_{Ne} = 250 mm Hg and the nominal power mode (see Table 3.4), the practical efficiency (determined from the power consumed from the power source rectifier) in the generator mode is about 1.17%, the efficiency of the AE (2.22% in the power input) is 2.23%, for AE GL-205B (Kristall LT-40Cu) with p_{Ne} = 180 mm Hg – the corresponding values

Table 3.6. The power consumed from the rectifier power source, and the radiation power of the CVL with the AE GL-205B (Kristall LT-40Cu) and the lamp IC

AE serial number	P_{rect}, kW	P_{rad}, W
127 001	4.2	48
127 002	4.1	48
127 003	4.1	46
127 009	4.2	46
127 012	4.12	46
127 014	4.2	47
127 015	4.15	48
127 128	4.2	46
127 129	4.2	47
127 130	4.25	48

are 1.2% and 2.4%, and for AE GL-205C (LT-50Cu Kristall) with p_{Ne} = 150 mm Hg – 1.2% and 2.4%. The efficiency of the AE is about twice the practical efficiency, since in the elements of the charging and discharge circuits approximately half of the power consumed by the rectifier is lost.

For the AE Kristall CVLs it is also important to know the efficiency in the power amplifier mode, since these AEs are mainly used in high-power laser systems such as the master oscillator – power amplifier (MO–PA) as the PA. In the CVL with AE GL-205A (LT-30Cu Kristall), the practical efficiency in the PA mode is about 1.5% (45 W/3 kW), the efficiency of the AE itself is 3% (2 times more), with AE GL-205B (Kristall LT-40Cu) the corresponding values are 1.7% (60 W/3.6 kW) and 3.4% and with AE GL-205C (LT-50Cu Kristall) – 1.63% (75 W/4.6 kW) and 3.3%. The radiation power and efficiency of the AE Kristall CVL in the PA mode is 1.3–1.4 times higher than in the MO regime.

Test results of the AE GL-205B (LT-40Cu Kristall) with a lamp power supply. The results of tests of the AE GL-205B (LT-40Cu Kristall) with a lamp power supply (Fig. 3.27 *b*), developed by Altek (Moscow), are analyzed. The power source with the GMI-29 A-1 modulator pulse lamp makes it possible to generate pump current pulses of 40–70 ns duration in a wide range of PRF and has high operational reliability (lamp life is about 2000 h). For testing,

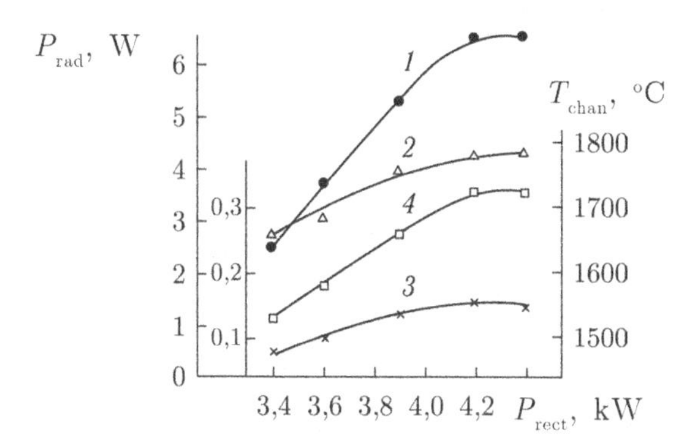

Fig. 3.48. The dependence of the average radiation power (*1*) of the AE GL-205D (LT-4AU Kristall) of the gold vapour laser, the temperature of its discharge channel (*2*), the practical efficiency (*3*), and the efficiency of the AE (*4*) on the power consumed by the power source rectifier under neon pressure 250 mm Hg and 16 kHz PRF.

10 serial devices from different lots were taken without special selection. The results of measurements for the same PRF – 12 kHz – are summarized in Table 3.6.

The spread of the values of the nominal power consumed by the power source rectifier and the radiation power was minimal (ΔP_{rect} ~ 150 W (i.e. 3.6%), ΔP_{rad} ~ 2 W (4.2%)). A plane-spherical resonator with a radius of curvature of the 'blind' mirror R = 3.5 m was used. The power of the AE emission with the lamp power source (48 W) is 1.2 times higher than with the thyratron (40 W). Using the antireflecting output windows, the radiation power was ~52 W with a practical efficiency of 1.25% and an efficiency of 2.5%. The efficiency of AE in the power amplifier mode reached 3.5%.

Industrial sealed-off self-heating AE of the pulsed gold vapour laser. The basic designs and technology of training for the AE of a pulsed GVL with a wavelength of λ = 0.628 μm are industrial pulsed CVLs, which use instead of the active substance of copper high purity gold Zl 999.9, with the optimal temperature increasing by 100–150°C (up to 1700–1800°C). For AE GL-206F (Kulon LT-1Au) with an average radiation power of not less than 1 W, the basic design is AE GL-206D, for AE GL-206G (Kulon LT-1.5Au) with a power of at least 1.5 W – GL-206E (see Table 3.3), for AE GL-205D (LT-4Au Kristall) with a power of at least 4 W – GL-205A (see Table 3.4). The radiation power of the GVL with the AE of these models is approximately six times smaller than the power of a copper vapour laser. Figure 3.48 shows the dependence of the average

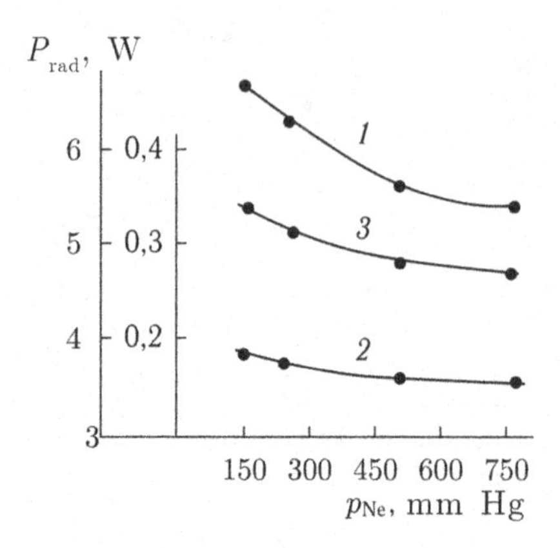

Fig. 3.49. Dependences of the average radiation power (*1*) of the AE GL-205D (LT-4AU Kristall) of the GVL, the practical efficiency (*2*), and the efficiency of the AE (*3*) on the neon pressure at 16 kHz PRF.

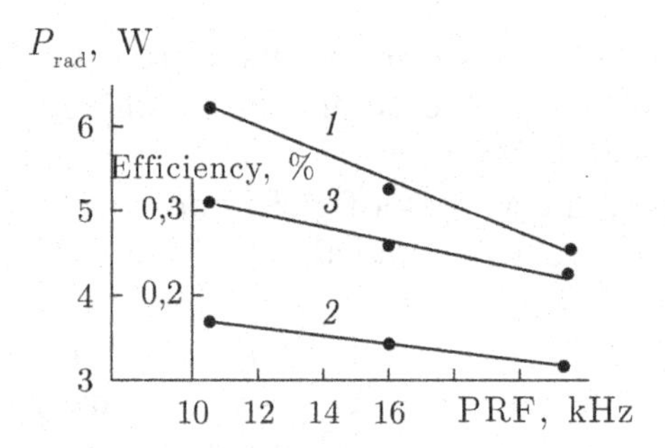

Fig. 3.50. Dependence of the average radiation power (*1*) of the AE GL-205D (LT-4AU Kristall) of the GVL, the practical efficiency (*2*), and the efficiency of the AE (*3*) on PRF at a neon pressure of 250 mm Hg.

radiation power of the AE GL-205D of the GVL, the temperature of its discharge channel of the practical efficiency and the efficiency of the AE on the power consumption. A power source with a high-voltage modulator was used which was made according to a scheme with a capacitive voltage doubling and a magnetic compression link (Fig. 3.27 *d*). The PRF was 16 kHz, the neon pressure in the AE was 250 mm Hg. The maximum radiation power (6–6.5 W) was achieved at the walls of the discharge channel at 1750–1800°C. A plane-spherical resonator with a radius of curvature of the 'blind' mirror of 3.5 m was used. At that, the practical efficiency was 0.15%, and the efficiency of the AE was 0.3%.

With the change in neon pressure from 150 to 760 mm Hg at 16.7 kHz, the radiation power decreased from 6.7 to 5.4 W (curve 1 in

Fig. 3.49), the practical efficiency – from 0.18% to 0.15% (curve 2), the efficiency of AE – from 0.34 % to 0.25% (curve 3). With an increase in the PRF, the decrease in the radiation power with increasing pressure becomes more abrupt. For example, at 21.7 kHz the power was reduced by 41%. At a constant neon pressure p_{Ne} = 250 mm Hg with an increase in frequency from 10.5 to 21.5 kHz, the radiation power decreased from 6.3 to 4.6 W (by 27%) (curve 1 in Fig. 3.50). The maximum radiation power was attained at a neon pressure p_{Ne} = 50 mm Hg and the hydrogen pressure p_{He} = 5 mm Hg and for optimized excitation conditions was 8.5 W. For AE GL-206F, the neon pressure change from 200 to 600 mm Hg led to a halving of power (from 1.5 to 0.75 W).

The obtained results also show that the pulsed GVL works rather effectively at neon pressures close to the atmospheric pressure; under these conditions, the service life of the commercial sealed-off AE on gold vapour is not less than 1000 h.

Dependences of the specific characteristics of industrial sealed-off AE of the CVLs on the volume of the active medium. The values presented in Tables 3.3 and 3.4 of the average radiation power for new models of industrial sealed-off self-heating AEs of the Kulon and Kristall series of the pulsed CVL were obtained under optimized pump modes with an effective thyratron power source (see Fig. 3.27 *d*). When choosing the operating pressure of the buffer gas, not only the output power of the radiation was taken into account, but also the service life of the AE. For all new industrial sealed-off models of AEs on copper vapors, the minimum operating time is at least 1500 h, which is 3 times more than that of the first (old) industrial models. During the minimum operating time, in accordance with the specification for AE, the reduction in the average radiation power should not exceed 20% of the nominal value.

Based on tabulated data, the dependences of the average and specific radiation power on the volume of the active medium (AM) have been constructed (Figs. 3.51 and 3.52), which are very important for evaluating the efficiency of sealed-off AEs.

These dependences indicate possible ways to increase the power and efficiency of AE with large volumes of AM. Figure 3.51 shows the curves corresponding to the radiation powers in the MO (*1*) and PA mode (*2*). When the volume of the active medium varies from V_{AM} ~ 4.2 cm^3 for AE GL-206A (Kulon LT-1Si) to V_{AM} ~ 900 cm^3 for AE GL-205C (LT-50Si Kristall), the average radiation power in the MO mode increased from 1.5 to 55 W, in the PA mode from 1.7 to 75

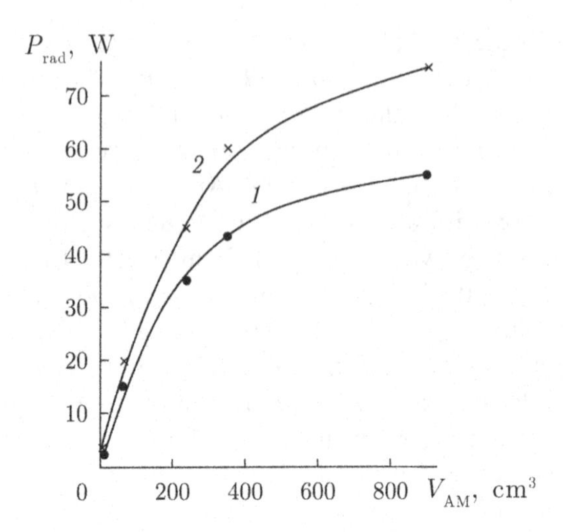

Fig. 3.51. Dependence of the average radiation power of industrial sealed-off self-heating AE pulsed CVLs in the generator mode (*1*) and the power amplifier mode (*2*) on the volume of the active medium.

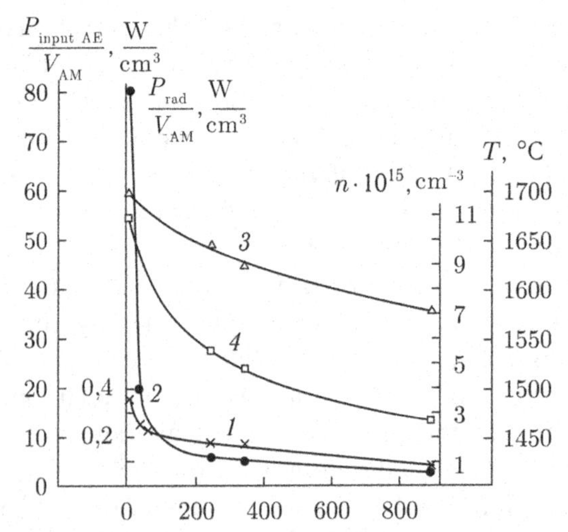

Fig. 3.52. Dependence of the specific radiation power (P_{rad}/V_{AM}) (*1*) of the industrial sealed-off self-heating AE of the pulsed CVL, the specific power input to the AE ($P_{input\ AE}/V_{AM}$) (*2*), the temperature of the discharge channel (*3*) and the concentration of copper atoms (*4*) on the volume of the AM.

W. If the volume has increased approximately 215 times (900/4.2), then the radiation power is only 44 times (75/1.7). Thus, the relative increase in power is about 4 times lower than the relative increase in the volume of the AM.

The values of the radiation power during operation of the AE in the PA mode and the power input in the AE are used to plot the curves reflecting the change in the specific radiation power (P_{input}/V_{AM}) and the specific input power (P_{input}/V_{AM}) in dependence on the volume of the AM (Fig. 3.52). If the specific power in the AE of the GL-206A with $V_{AM} \sim 4.2$ cm^3 (the power take-off from the volume unit of the AM) is 0.38 W/cm^3, then in the AE GL-205C with $V_{AM} =$ 900 cm^3 it is 0.085 W/cm^3, four times less, indicating that it is potentially possible to achieve a total power take-off from a GL-205C device (LT-50Si Kristall) to 300 W (75 × 4 = 300 W).To confirm this possibility, the curves of the dependence of the temperature of the wall of the discharge channel where the vapour generators of the active substance copper are located (curve *3* in Fig. 3.52), and the concentration of copper atoms (curve *4*) on the volume of the active medium of the AE was investigated. On the one hand, curve 2 indicates that in the active medium of AE with small volumes it is necessary to introduce a specific power of the order of 80–100 W/cm^3 in order to ensure the optimal regime. On the other hand, such a high level of specific power for AEs with large volumes is excessive, since even at a specific power higher than 3–4 W/cm^3, the radiation power decreases because of the superheating of the active medium. In AE GL-206A with V_{AM} = 4.2 cm^3, the operating temperature of the discharge channel is about 1700°C, which corresponds to a copper concentration of about 11 × 10^{15} cm^{-3}, and in GL-205C with V_{AM} =

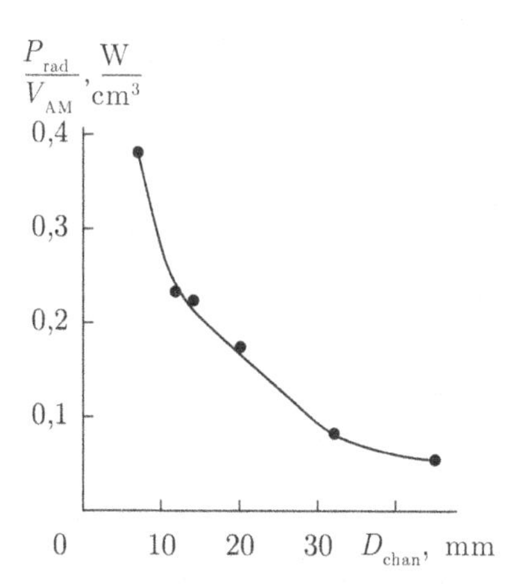

Fig. 3.53. Dependence of the specific radiation power of industrial sealed-off self-heating AE pulsed CVLs on the diameter of the discharge channel of the AE at optimum operating temperatures.

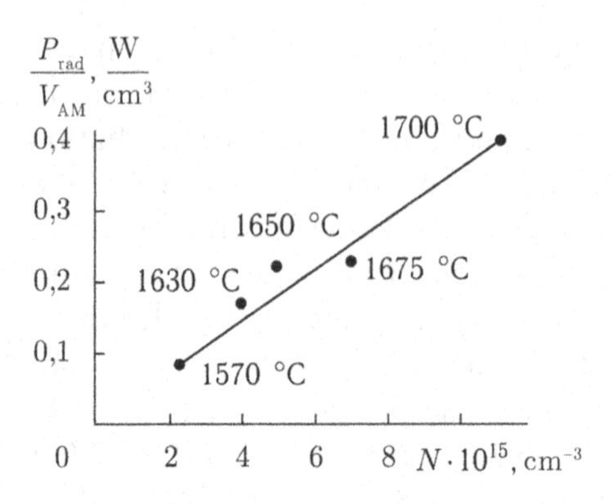

Fig. 3.54. Dependence of the specific radiation power of industrial sealed-off self-heating AE of the pulsed CVLs on the concentration of copper atoms in the AM at optimum operating temperatures.

900 cm^{-3}, the operating temperature is 1570°C, the concentration of copper atoms is 2.5×10^{15} cm^{-3}, that is, it is four times smaller (see curves *3* and *4*).

For additional analysis and evaluation of the performance of the industrial self-heating AEs of the CVL Figs. 3.53 and 3.54 show the dependences of the specific radiation power on the diameter of the discharge channel and on the concentration of copper vapour at optimal operating temperatures. The temperatures are indicated directly on the curve in Fig. 3.54. T_{chan} = 1570°C – optimum working temperature of AE GL-205C with D_{chan} = 32 mm, T_{chan} = 1630°C – GL-205B with D_{chan} = 20 mm, T_{chan} = 1650°C – GL-206E with D_{chan} = 14 mm, T_{chan} = 1670°C - GL-206C with D_{chan} = 12 mm and T_{chan} = 1700°C - GL-206A with D_{chan} = 7 mm.

As the diameter of the discharge increases, the probability of collision of copper atoms from the volume of the AM with the channel wall decreases and the rate of their reduction in the interimpulse period decreases from the state with a metastable level to the ground state. This process leads to an increase in the concentration of 'metastable' atoms in the AS and the need to reduce the optimal operating temperature of the channel (to avoid 'overheating' of the AM), which in turn leads to a decrease in the radiation power (Fig. 3.53). It turns out that the specific radiation power (ρ) and the concentration of copper vapour (N) are related to each other by a practically direct proportional dependence (Fig. 3.52 *b*), determined by the expression

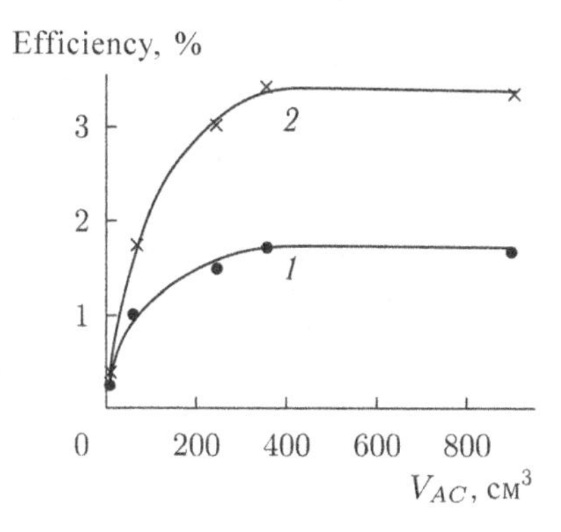

Fig. 3.55. Dependences of practical efficiency (*1*) and efficiency of AE (*2*) for industrial sealed-off AE of the pulsed CVL of the series Kulon and Kristall in the mode of the power amplifier on the volume of AM.

$$\rho = \rho_0 \cdot N / N_0,$$

where ρ_0 is the power density at the vapour concentration N_0. How many times the concentration of copper vapour decreases, the power take-off per unit volume of AM decreases as much as again, and vice versa.

The foregoing analysis of experimental studies clearly indicates that in order to achieve high values of radiation power in CVL with large volumes of the AM, the AE design and the excitation conditions should be such as to ensure the optimum operating temperature of the discharge channel at a level of 1700°C without overheating of the AM, i.e. would create optimal conditions for inversion of populations in the AM at a vapour concentration copper $n_{Cu} \sim 10^{16}$ cm^{-3}.

To estimate the maximum efficiency of the industrial AE of the CVL, curves were constructed for the dependence of the practical efficiency and efficiency of the AE on the volume of the AM (Fig. 3.55). When calculating the efficiency, the values of the radiation powers in the PA mode were used.

This is especially important for the AE of the Kristall series, since they are used mainly in multimodal CVLS 'MO–PA' for the formation of powerful radiation beams. The radiation power of the Kristall AE in the PA mode is approximately 1.3–1.4 times higher than in the MO regime. And, as can be seen from the course of the curves in Fig. 3.55, with a change in the volume of the active medium from 4.2 cm^3 (GL-206A) to 900 cm^3 (GL-205C), the practical efficiency increases from ~0.25% to ~1.7% (*1*), and the efficiency of the AE

Table 3.7. The main parameters of industrial sealed CVL produced by the Istok NPP and foreign analogs

AE model	Firm	AE type	Diameter of the discharge channel, mm	Frequency of repetition of pulses, kHz	Average radiation power, W	Power ratio*	Guaranteed minimum (minimum) operating time of the AE, h	Power consumed from the rectifier power source and from the grid, kW
GL-205A (Kristall LT-30Cu)	Istok»	Sealed-off	20	10–12	30–32	1 : 1	1500 > 2500	2,9–3,1 –
CVL-30	Israel, Nuclear Research Center	Partial pumping	30	5,2	30 ± 5	> 1,2 : 1	400** –	– 5
GL-205B (Kristall LT-40Cu)	Istok	Sealed-off	20	10–12	39–41 38–50 ***	1 : 1	1500 > 2500	3,6–3,8 –
AGL-45	England, Oxford Lasers	Pumping system	42	6	45	< 1,7 : 1	> 300** –	– 6
GL-206D (Kulon LT-5Cu)	Istok	Sealed-off	14	12–18	5–9	< 1,5 : 1	1500 > 2500	1,4–1,5 –

Table 3.7

SCuL05H	Bulgaria, Mashinoexport	Sealed-off	15	5–15	7	–	– 1000	– –
CVL-5W	USA, Lasers Now	Sealed-off	14	20	5	1,4 : 1	– > 800	– 1.2
GL-206D («Kulon LT-10Cu»)	SPE «Istok»	Sealed-off	14	14–17	10–15	< 1,5 : 1	1500 > 2500	1.8–1.9 –
Cu10-A	England, Oxford Lasers	Pumping system	25	8–14	10	2 : 1	> 300** –	– –
SCuL10H	Bulgaria, “Mashinoexport”	Sealed-off	25	5–15	10–12	–	– 500	– –
CVL-10W	USA, Lasers Now	-	20	20	10	–	– > 500	– –
CVL-10	Israel, Nuclear Research Center	Soldered	20	–	10	–	– 1000	– –

* P_{rad} (λ = 510.6 nm)/P_{rad} (λ = 587.2 nm). ** Operating time for one copper charge. *** With a lamp power source.

from 0.4% to 3.4% (*2*) with a decrease in the specific power take-off from 0.36 to 0.083 W/cm^3 (see curve 1 in Fig. 3.52).

Comparative analysis of the effectiveness of industrial sealed-off self-heating CVLs with foreign analogues. To assess the effectiveness of domestic pulsed CVLs on the basis of industrial sealed-off self-heating AEs of the Kulon and Kristall series produced by the Istok Company, a comparative analysis of the main parameters with foreign analogues close in terms of radiation power was carried out. From the data given it follows that the domestic AEs GL-205A (LT-30Si Kristall) has the same radiation power as the Israeli model CVL-30. If we compare the models by the diameter of the discharge channel, we can assume that the volume of the active medium of the CVL-30 is approximately twice as large as that of the GL-205A model (see Table 3.7) and the capacity of power take-off from a unit volume (efficiency) is lower by the same amount. The efficiency of the English AGL-45 model is about four times lower by the same criterion than the GL-205B model (LT-40Ci Kristall). Foreign analogues with a power of more than 10 W work mainly in the continuous pumping of the buffer gas, i.e., the laser is equipped with additional support elements. In addition, the CVL-30 and AGL-45 models require a new portion of working copper at certain intervals (300 and 400 hours respectively). Thus, the domestic devices of the Kristall series favourably differ from foreign ones with the same power level not only in efficiency, but also in that they have a sealed-off design of the AE. The latter increases the reliability of the laser as a whole and simplifies its operation.

The devices of the Kulon series also have an advantage over the analogues given. In particular, the efficiency of the GL-206E model (LT-10Si Kulon) is about twice the efficiency of the CVL-10W (USA) and CVL-10 (Israel) models and three times the efficiency of the CU10-A and SCuL10H models (Bulgaria). The CU10-A laser from Oxford Lasers works with neon pumping. The operating time of its AE on a single charge of copper is about 300 h. Lasers SCuL05H and SCuL10H of the firm Mashinoexport are manufactured both in the pumping system and in the sealed-off version. For sealed-off AEs of foreign lasers with a power of 5–10 W, the main reliability parameter is the service life (500–1000 h), which is 2–3 times less than the guaranteed operating time of the industrial sealed-off Kulon and Kristall. The most powerful Russian industrial sealed-off AE is the GL-205C (LT-50Si Kristall) produced by the Istok Company. Today, experimental samples in a sealed-off design with an output of

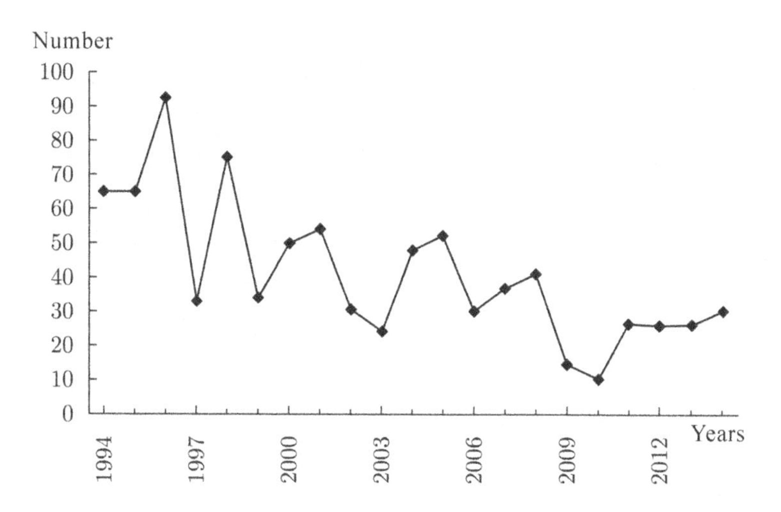

Fig. 3.56. The annual sales volume of industrial sealed-off AEs of the pulse CVLs of the Kulon and Kristall series.

80–120 W have been created at the Istok Scientific and Production Enterprise – LT-50-DSi Kristall, LT-70Si Kristall and LT-100Si Kristall. In the advanced foreign countries (USA, France, Japan, etc.), AEs with an average radiation power of 100–750 W have been developed (see Chapter 8 in [15, 20]) in connection with the implementation of the AVLIS program (laser isotope separation). In the American Lawrence Livermore National Laboratory a separate AE in the amplifier mode produced the radiation power of more than 750 W, the French firm CILAS – 400 W (1996), the Japanese company Toshiba – 615 W (1995).

The annual release of industrial sealed-off AEs and the main areas of their application. In recent years, the Istok Enterprise has modernized the production base, which makes it possible to produce up to 100 sealed-off AEs of all types per year. Figure 3.56 is a graph of the sales volume of the Kulon" and Kristall AEs for the last 20 years (from 1994 to 2014). During these 20 years, 560 AEs were sold, that is, an average of 31 units in year. The largest annual sales of AEs were in 1996 (92 units), the lowest – in 2009 and 2010 (14 and 10 units). A positive growth trend is forecast for the future. The volume of AE sales is determined mainly by the internal state of the Russian market and the financial situation of its main consumers. A number of industrial sealed-off AEs during this period were also sold abroad: China, South Korea, Iran, Switzerland, Finland, Italy, Armenia.

In Russia, the AEs of the CVLs are delivered to the Kurchatov Institute, P.N. Institute of Physics, the A.N. Prokhorov Institute of General Physics, Russian Academy of Sciences, (Moscow), The Institute of Spectroscopy of the Russian Academy of Sciences and the P.N. Lebedev KBBI of the Russian Academy of Sciences (Troitsk, Moscow region), The Medical sterilization systems" (Odintsovo, Moscow region); Federal State Unitary Enterprise RFYaTs-VNIIEF (Sorovo, Nizhny Novgorod region), the branch of the All-Union Electrotechnical Institute and NPP VELIT (Istra, Moscow region), The Research Institute of Precision Engineering (Zelenograd); St. Petersburg Technical University, The Institute of Electrophysics and Electricity (St. Petersburg), The Institute of Semiconductor Physics (Novosibirsk), the Institute of Atmospheric Optics (Tomsk), Tomsk State University, etc.

One of the important applications of pulsed CVLs is the laser isotope separation of different atoms (AVLIS technology), in which the difference in the absorption spectra of atoms of different isotopic composition is used. In the AVLIS technology, frequency-tunable lasers on dye solutions (DSL) pumped by a powerful CVLS are used. In such technological installations, CVLS operate according to the MO–PA scheme. In Russia, the development of isotope separation plants using sealed-off AE Kristall series is carried out at the Kurchatov Institute, Medical Sterilization Systems Ltd. (Odintsovo, Moscow Region) and the Institute of Semiconductor Physics (Novosibirsk) [15 –17, 20 22.23, 31, 147–153]. Another developing promising direction with the use of CVL is precision micromachining of materials, first of all thin-sheet metal and non-metallic materials (0.05–1.0 mm) for electronic products [15, 16, 20, 25, 33, 34, 108–113, 165, 169, 187–192, 364,366–375, 382–385, 389, 391, 394]. The spectrum of processed materials is practically unlimited: they are refractory materials (W, Mo, Ta), metals with high thermal conductivity (Cu, Ag, Al, Au), their alloys, semiconductors (Si, Ge, GaAs, SiC), dielectrics, graphites, diamonds and others. The so-called optical systems with intensified image brightness occupy a notable place in the field of application of CVL, in which active media of both CVL and GVL are used.

The CVLs and GVLs on the basis of sealed-off AE of the Kulon series separately and in combination with tunable wavelengths DSLs are widely used in medicine. Especially important is their use for the treatment of cancer by photodynamic therapy [16, 52–56, 166, 167, 172, 173, 182]. Another quite successful application of them can be considered treatment of vascular, pigmented and unpainted skin defects, for smoothing wrinkles, i.e. in dermatology and cosmetology [16,48–51]. Pulsed radiation of CVL with wavelengths $\lambda = 510$ and 582 nm and a pulse duration of 15–25 ns selectively coagulates skin defects without damaging the surrounding tissue and without causing pain (without anesthesia).

The CVLs in combination with DSLs, which allow to obtain effective and powerful tunable generation in the visible and near infrared regions of the spectrum, are also widely used for spectroscopic studies. The radiation of CVLs on the basis of sealed-off AEs of the Kristall series is efficiently transformed by means of non-linear crystals into the second harmonic, that is, into the ultraviolet region of the spectrum with an average output power of 1–9 W. The laser radiation is also used to pump a titanium–sapphire ($Al_2O_3 \cdot Ti^{+3}$) laser in order to obtain tunable lasing in the near-IR region, and when doubling the frequency, in the blue region of the spectrum [15, 16, 22, 23, 31, 104, 153–156]. Such multifrequency tunable laser systems with a large average generation power are unique.

In addition to technological applications, the CVLs and devices based on them have been used for many years for high-speed photography and holography, gas flow visualization, laser acceleration of microparticles, spectroscopic and mass spectroscopic studies, astronomical studies, and projection microscopy, cinema and television, navigation and location (see Chapter 1). The course of the development of research in these areas is widely discussed at numerous all-Russian and international conferences and seminars, which contributes to a stable interest in the CVL from scientific and production organizations.

3.6. Conclusions and results for chapter 3

1. Highly efficient and durable industrial sealed-off self-heating AEs of the pulsed CVL of a new generation operating in a mixture of copper, neon and hydrogen vapour, the Kulon series of small power (1–20 W) and the Kristall of medium power (30–100 W) have been developed.

2. High (maximum) efficiency, power, durability, quality and stability of output radiation parameters in the developed industrial sealed-off self-heating AEs of the pulsed CVL with a power of 1–100

W are achieved due to the implementation of a complex of scientific, technical and technological solutions:

– producing a sectioned ceramic discharge channel with blind grooves, in each of which a copper vapour generator is installed in the form of a molybdenum substrate with holes wetted by the active substance – molten copper and perforated end tubes (ceramics A-995–99.8% Al_2O_3 + 0.2% MgO);

– developed technology for degassing AE and restoring surface purity of copper vapour generators, electrode assemblies and other elements in a hydrogen atmosphere with neon at T_{work} = 1600°C after complete degassing of AE at T = 1700°C for 30–60 h (depending on the AE model);

– creation of an unheated auto thermoemission metal-porous tungsten–barium (W–Ba) cathode of a ring structure with an annular groove on the inner surface providing stable local combustion of a pulsed arc discharge (active substance – barium aluminosilicate of composition $3BaO \cdot Al_2O_3 \cdot 0.5CaO \cdot 0.5SiO_2$);

– a three-layer high-temperature heat insulator with low thermal conductivity (λ = 0.27–0.31 W/(m · K)) and a low specific gravity (ρ = 0.32–0.5 g/cm^2) based on Al_2O_3 and SiO_2 oxides, located in the space between the discharge channel with $T_{work} \cong 1600$°C, electrode nodes and the outer vacuum-tight shell with $T_{work} \cong 300$°C;

– application of output antireflecting windows with an angle of inclination to the optical axis of the AE, not exceeding the values

$$\alpha = \operatorname{arctg}\frac{(2-ab)}{(2a+b)},$$

where $a = D_{chan}/l_{chan}$ (D_{chan} is the diameter and l_{chan} is the length of the discharge channel), $b = D_{chan}/l_w$ (l_w is the distance from the end of the discharge channel to the window along the optical axis.

3. The service life of industrial sealed-off self-heating AEs of the new generation of the Kulon and Kristall series of the pulsed CVL, due to the high reliability of all functional units and effective protection against dusting of the output windows, is determined by only three factors: the mass of stored copper in copper vapour generators, the operating temperature of the discharge channel and the pressure of the buffer gas neon and is not less than 3000 hours. The reason for the reduction in power during the period after the guaranteed operating time (1500 hours) is the formation in molten copper of molybdenum structures (clusters) with a developed surface,

increasing the enthalpy (necessary energy) of evaporation of copper from the system.

4. Experimental studies on increasing the radiation power and the efficiency of pulsed CVLs in dependence on the conditions of excitation of AM with the industrial sealed-off AEs of the Kulon and Kristall series were carried out at a pulse repetition frequency in the range of 6–25 kHz, the pressures of the neon buffer gas of 50–760 mm Hg and the partial pressure of hydrogen up to 10 mm Hg.

5. The power source with a high-voltage pulse modulator of pumping, produced according to the scheme of capacitive voltage doubling with links of magnetic compression of nanosecond current pulses and an anode reactor, remains the most reliable and efficient pulsed CVL pumping generator.

6. The addition of hydrogen to the active medium of the CVL with a partial pressure of 2–10 mm Hg leads to an increase in radiation power up to 2 times and an efficiency of 1.5 times, depending on the parameters of the pump current pulse, pulse repetition frequency, neon pressure and the diameter of the discharge channel. The hydrogen is introduced into the AE after a complete cycle of degassing and purification of the AE and at optimal operating temperatures (~1600°C).

7. As the diameter and length of the discharge channel of the AE and, correspondingly, the volume of the active medium increase, the relative decrease in the radiation power with increasing pressure of the neon buffer gas becomes sharper because of the deterioration in the characteristics of the pump current pulses. When the neon pressure varies from 50 to 760 mm Hg and the 10 kHz pulse repetition frequency, the average radiation power of the CVL with the Kulon AE, model GL-206E with a diameter and length of the discharge channel of 14 and 490 mm decreases from 15 to 10 W (~33%), the Kristall AE, models GL-205A with a diameter and the length of the channel 20 and 930 mm – from 34 to 21 W (~38%), GL-205B with a diameter and length of the channel 20 and 1230 mm – from 44 to 25 W (~43%), GL-205D with a diameter and the length of the channel 20 and 1230 mm – from 56 to 30 W (~47%), GL-205D with a diameter and channel length of 20 and 930 mm (in gold vapours) – from 8.5 to 5.4 W (by ~36%).

In the pulse repetition frequency range of 6–20.5 kHz, the average radiation power of the Kristall AEs of the GL-205A, GL-205B, and GL-205C models at operating neon pressures and optimized pumping parameters varies within 25%.

8. The efficiency of pulsed CVLs with the industrial sealed-off AE of the Kulon series with a radiation power of 1–20 W with diameters of the discharge channel D_{chan} = 7, 12 and 14 mm and the volume of AM (V_{AM}) from 4 to 85 cm^3 is 0.2–1%, the Kristall series with a power of 30–55 W with D_{chan} = 20 and 32 mm and V_{AM} = 250–900 cm^3 – 1.1–1.2% and with a power level of 60–100 W with D_{chan} = 32 and 45 mm and V_{AM} = 1200 –2200 cm^3 – 1.0–1.1%.

9. More important parameters of the AEs of the Kristall series are those when operating them in the power amplifier mode, since they are mainly used in high-power CVLs of the MO–PA type as the PA. In the CVLS, when using in the MO of the AE in the GL-205 model, the average power of the radiation and the efficiency of the CVLS and the efficiency of the AE GL-205A as the PA was 45 W, 1.5% and 3.0%, respectively, with GL-205B as the PA – 60 W, 1.7% and 3.4%, with the GL-205C as the PA – 75 W, 1.63% and 3.3%, with LT-50Cu-E as the PA – 80 W, 1.4 % and 2.7%, with LT-75Cu as the PA – 97 W, 1.5% and 3.0%, with LT-100C as the PA – 117 W, 1.3% and 2.6%, which in 1.2–1.3 times more in comparison with the generator mode.

10. It is established that in the industrial sealed-off self-heating AEs of the Kulon and Kristall series with an increase in the diameter of the discharge channel from D_{chan} = 7 mm with the volume of the active medium V_{AM} ~ 4.2 cm^3 (GL-206A) to D_{chan} = 32 mm with V_{AM} ~ 900 cm^3 (GL-205C) the radiation power of the CVL when operating in the generator mode increases from 1.4 to 55 W, in the PA mode from 1.7 to 75 W. At the same time, the volume of AM increased approximately 215 times (900/4.2), and the radiation power – only 44 times (75/1.7). This means that the relative increase in the radiation power is approximately 4 times lower than the relative increase in the volume of the AM, which in turn indicates a potential possibility of achieving power take-off from one AE of the GL-205Ctype to 300 W (75 × 4 = 300 W).

11. It was found that at optimal operating temperatures of the discharge channel of the AE in the range 1500–1700°C, the specific power release and the concentration of copper vapour in the AM are related to each other by a direct proportional dependence determined by the expression $\rho = \rho_0 \cdot N/N_0$, where ρ_0 is the specific power at a vapour concentration of N_0. How many times the concentration of copper vapour decreases, the power take-off per unit volume of AM decreases as much as again, and vice versa. If the optimum working temperature of the channel of the AE of GL-206A (D_{chan} = 7 mm) is

about 1700°C, then for the AE of the GL-205C (D_{chan} = 32 mm) is lower by about 130°C – 1570°C, which corresponds to a decrease in the concentration of copper atoms in the AM from $11 \cdot 10^{15}$ cm^{-3} to $2.5 \cdot 10^{15}$ cm^{-3}, that is 4 times. The specific power take-off has decreased by the same factor – from 0.38 W/cm^3 to 0.085 W/cm^3.

12. An important conclusion follows from the analysis of the experimental and theoretical studies carried out in this paper: in order to achieve high efficiency of CVL with large volumes of AM, the AE design and the excitation conditions should be such as to ensure the operating temperature of the discharge channel with copper vapour generators at the level 1700°C without superheating of the AM, i.e., optimal conditions were created for inversion of the populations in the AM at a copper vapour concentration $N_{Cu} \cong 10^{16}$ cm^{-3}.

13. The industrial sealed-off self-heating AEs of the Kulon series with a radiation power of 1–20 W and of the Kristall series with a power of 30–100 W are preferred as regards the efficiency, guaranteed running time and operating conditions in comparison with the similar foreign analogues. The removal of the radiation power from a unit volume of the active medium of the AE of the Kulon series is approximately 2 times higher, in the Kristall AEs it is 4 times, the minimum operating time is 4–5 times larger, and the sealed-off design of the AE does not require additional operational support elements.

14. The total volume of sales of the new generation of industrial sealed-off AEs of the Kulon and Kristall pulsed CVLs for the period from 1994 to 2014 (for 20 years) amounted to about 560 units, that is, an average of about 30 units/year. In the following years it is expected that the sales will increase.

4

Highly selective optical systems for the formation of single-beam radiation of diffraction quality with stable parameters in copper vapour lasers and copper vapour laser systems

Modern fields of application of pulsed copper vapour lasers (CVL) and the more powerful copper vapour laser systems (CVLS) based on them require not only an accurate knowledge of the radiation characteristics, but also the possibility of forming radiation beams with high (necessary) quality and controlled parameters [227–233]. These areas include advanced technologies for precision microprocessing of materials for electronic components, separation of isotopes and the production of pure substances for the needs of nuclear power engineering, medical technologies, pumping wavelength-tunable lasers based on dye solutions (DSL) and non-linear crystals (NC), analysis of the composition of substances, location, nanotechnology, etc. [12–57].

The present chapter is devoted to the study and development of optical systems for the formation of high-quality single-beam radiation in CVL and CVLS: diffraction divergence and with stable parameters. To achieve this goal, experimental and theoretical studies of the dynamics of the formation of the structure, spatial, temporal, and energy characteristics of the pulsed radiation of CVL and CVLS with optical systems possessing high spatial selectivity were carried out: with one convex mirror, with an unstable resonator (UR) with two convex mirrors and a telescopic UR and by the definition of the conditions for the formation in a resonator of single-beam radiation with diffraction divergence and stable parameters. The laser radiation of this quality is focused with the help of short-focus lenses (50–200 mm) into a spot with clear boundaries 5–20 μm in diameter and a peak power density of 10^9–10^{12} W/cm^2, sufficient for efficient microprocessing of metallic materials and a large range of semiconductors and dielectrics [15, 16, 20, 33, 34]. Since the processing is carried out in the evaporation mode in a small section (5–20 μm) and with small values of pulsed energy (0.1–1 mJ) and high repetition rates (5–30 kHz), a high quality of the cut is achieved: the heat-affected zone of $\leq$ 5–10 microns, surface roughness $\leq$ 1–2 microns and an order of magnitude higher productivity in comparison with traditional methods of processing [16, 20, 34].

4.1. Distinctive properties and features of the formation of radiation in a pulsed CVL

One of the main differences between pulsed CVL and other types of lasers is the combination of a short time of existence of population inversion (τ = 20–40 ns), commensurate with the radiation path in the resonator (L = 0.5–2.0 m), with a large gain of the active medium (AM) (k = 10–10^2 dB/m). Due to high amplification, the CVL can work in the superradiance mode: without mirrors or with one mirror, but the radiation is incoherent. In the operating mode of CVL with an optical resonator, during the existence of a population inversion (τ), laser radiation has time to make only a few double passes in the resonator ($n = \tau/(2L/c)$, where L is the resonator length, c is the speed of light) and the modes in their usual understanding, formed as a result of hundreds of passes, can not be formed. With conventional plane-parallel, planar and other resonators, the divergence of the laser beam remains two orders of magnitude greater than the diffraction (minimum) limit, and therefore the beam can not be focused into a spot of micron sizes and with a high peak power density >10^9 W/cm^2) necessary for efficient microprocessing of materials.

Therefore, in this work, with reference to CVL with nanosecond duration, studies were carried out with optical systems possessing potentially high spatial selectivity: with one convex mirror, UR with two convex mirrors and a telescopic UR. For the first time, a telescopic resonator in CVL was used and investigated in [74] and then in [91–94], where it was shown that at large magnifications of the resonator (M = 100–300) a beam with a diffraction divergence is formed toward the end of the pulse. But the authors of these works did not fully disclose the dynamics of the formation and structure of the output radiation beam, which turned out to be the most important for practical applications. Somewhat later, in my works and together with other authors we presented the results of large-scale experimental and theoretical studies of the structure and characteristics of the radiation of CVL with different types of resonators and the application of the first, most powerful for that time, industrial sealed-off active elements (AE) of the Kristall series the GL-201 model (20–25 W) and the GL-201E model (40–45 W) [16, 26, 230–233]. And it was established that the output radiation of CVL with the telescopic UR has a multibeam and discrete structure, in which each beam, in the process of formation, competing with each other in power, acquires its spatial, temporal and energy characteristics. The diffraction beam is always preceded by several beams with greater divergence, and the instabilities of the axis of the beam pattern of the diffraction beam are commensurate with its divergence. In this form, the output radiation was simply unsuitable for high-quality microprocessing of materials and that prevented the creation of technological equipment on the basis of CVL.

In the same papers, in order to get rid of the multibeam radiation structure, the optical variant alternative to the resonator was first proposed and investigated – the operating mode of CVL with a single convex mirror (single-mirror regime) at small radii of curvature [8,9]. In the experiments we used the same AE Kristall: models GL-201 and GL-201E. In this mode, the structure of the output radiation is two-beam: an incoherent beam of superradiance formed by the geometric aperture of the active element (AE), and a beam with high spatial coherence formed by the mirror and the output aperture of the AE. The divergence of the second useful beam can be varied within wide limits by changing the radius of curvature of the convex mirror. Since this beam is formed with the participation of a single mirror, it also possesses a high stability of the axis of the radiation pattern. At radii of the convex mirror one or two orders of magnitude smaller than the distance from the mirror to the output aperture of the AE, the divergence of the beam becomes close to diffraction. At R = 1−3 cm, the divergence of the CVL beam was θ = 0.2−0.3 mrad (3–4 times the diffraction limit), which was sufficient to achieve a density of peak power of 10^9–10^{10} W/cm^2 in a focused spot with a diameter d = 20–50 μm and, accordingly, for the processing of thin-sheet materials.

The radiation characteristics of CVL with UR with two convex mirrors have not been investigated before and have been investigated only in the framework of this paper, where its capabilities for the technology of microprocessing of electronic component materials have been determined.

To date, a series of small-sized modern, high-reliability and effective CVL facilities has been created on the basis of a new generation of industrial sealed-off Kulon AEs with an average radiation power of 1 to 20 W (see Chapter 1) [16, 26, 157–165]. They are already used for pumping lasers tunable over wavelengths on dye solutions, analysis of the composition of substances, nanotechnology, medicine, etc. But the question of the possibility of their use for microprocessing of materials remained open. Therefore, it made sense to conduct research of optical systems with high spatial selectivity with a new generation of industrial CVLs.

4.2. Experimental settings and research methods

The experimental facilities for studying the structure and characteristics of the output beam of pulsed CVL radiation are presented in Figs. 4.1 *a* and *b*. The CVL studies were carried out with optical systems possessing high spatial selectivity: in the regime with one convex mirror, the single-mirror mode (Fig. 4.1 *a*), with UR with two convex mirrors and telescopic UR (Fig. 4.1 *b*). The most powerful industrial sealed-off AE from the Kulon series was used as a source of generation in the CVL: the GL-206E model with the power of 15 W and the GL-206I model with a power of 20 W (see Chapter 1).

Figure 4.2 shows their appearance. Pumping (warming up and excitation) of the AE of the CVL was produced by a high-voltage pulsed power source with a thyratron modulator, produced according to the capacitive voltage doubling scheme with two links of magnetic compression of nanosecond current pulses and an anode reactor (Fig. 4.3) [16, 158–161, 164]. The power source with this electrical circuit of the modulator remains the most reliable and efficient pump generator.

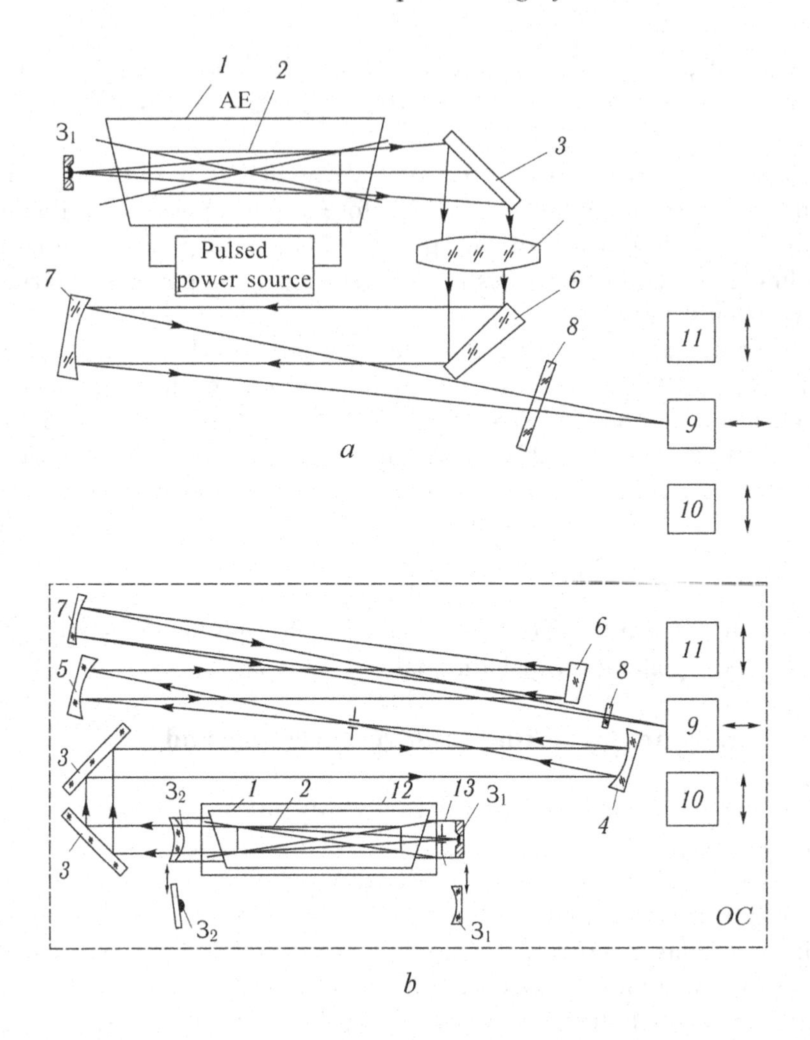

Fig. 4.1. Experimental facilities for studying the structure and characteristics of the output radiation of pulsed CVL with optical systems with high spatial selectivity (*a* and *b*): OC – optical honeycomb table type 1NT10-20-20 with pneumatic insulators AP-500 and AP-1000 located between the table and four its supports. model 1TS065-12-06 (firm Standa); *1* – AE; *2* – AE discharge channel; *A* – collimating lens with focal length F = 1.6 m; $S_1(a)$ is a convex mirror; $S_1(b)$–$S_2(b)$ – UR with two convex mirrors; $S_1'(b)$–$S_2'(b)$ – the telescopic UR; *3* – flat rotary mirrors; D_1 – diaphragm with aperture d = 0.1 mm; D_2 – the diaphragm of the spatial filter-collimator (SFC); 4 and *5* – input and output concave mirrors of a collimator SFC with a radius of curvature R = 1.25 m; *6* – beam splitter plate with reflection coefficient ρ = 4%; *7* – focusing mirror with a radius of curvature R = 15 m; *8* – neutral filter; *9* – a radiation power meter (millivoltmeter M136 with a converter of laser radiation power TI-3); *10* – digital oscilloscope GDS-840S with a photocell FEK-14K; *11* – BeamStar-FX beam analyzer; *12* – water-cooled heat sink; *13* – fluoroplastic (sealing) tubes

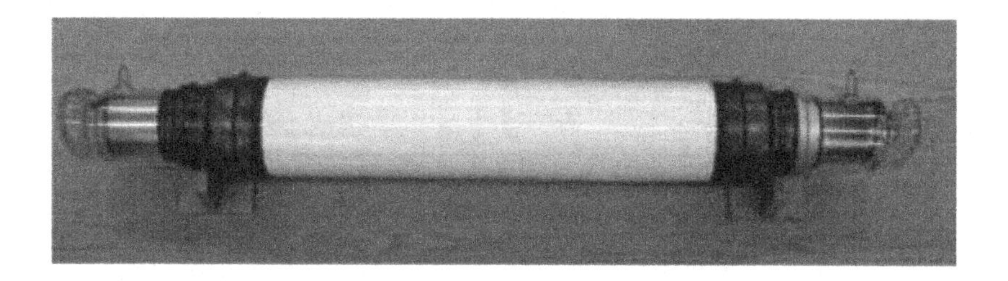

Fig. 4.2. The appearance of industrial sealed-off AE Kulon, models GL-206E (15W) and GL-206I (20W).

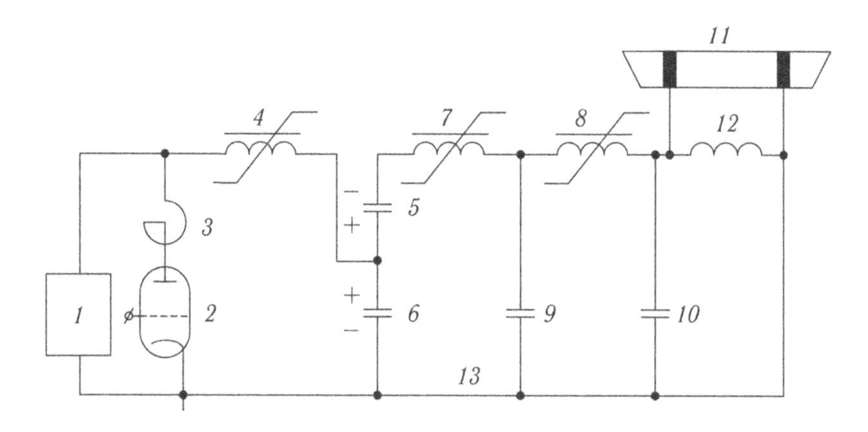

Fig. 4.3. The basic electrical scheme of the high-voltage pulse modulator of the power source: *1* – high-voltage rectifier-charger; *2* – thyratron switch TGI1-1000/25 or TGI2-1000/25; *3* – anode reactor, *4*, *7* and *8* – first, second and third non-linear inductances; *5* and *6* – series-connected storage capacitors of capacitance (1000+1000) pF; *9* – storage capacitor with a capacitance of 500 pF; *10* – a sharpening capacitor with a capacity of 110 pF; *11* – AE with parallel inductance 12; *13* – common ground bus.

Table 4.1. The geometric dimensions of the AE of the models of the GL-206E and GL-206I

Geometric dimensions / AE model	l_{AE}, mm	l_{chan}, mm	l_{AM}, mm	l_1, mm	l'_1, mm	l, mm
GL-206E	770	515	440	50	220	692
				250	420	892
GL-206I	900	640	565	50	220	816
				250	420	1016

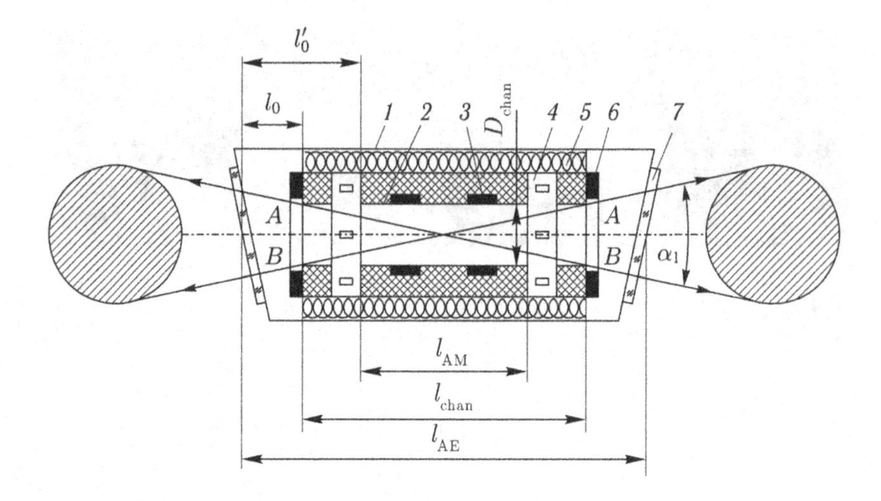

Fig. 4.4. Schematic representation of AE Kulon CVL without mirrors: *1* – vacuum-tight shell; *2* – discharge channel; *3* – copper vapour generators; *4* – condensers of copper vapour; *5* – heat insulator; *6* – electrodes; *7* – output windows; l_{AE} is the length of the AE; D_{chan} and l_{chan} are the diameter and length of the discharge channel; l_{AM} is the length of the active medium; l_0 and l_0' are the distances from the output window to the discharge channel and the active medium, respectively; α_1 is the angle of propagation of the beam of superradiance.

Figure 4.4 is a schematic representation of the AE Kulon without mirrors. The main functional nodes, overall dimensions of the AE and its discharge channel, necessary for analysis and calculation of the space-time radiation characteristics are shown: l_0 – distance from the output window (position 7) to the end of the discharge channel (item 2); l_0' – the distance from the output window (item 7) to the active medium of the discharge channel (item 2); l_{chan} is the length of the discharge channel (the distance between the electrodes (item 6)); l_{AM} – length of the active medium (distance between the condensers of copper vapour (item 4)). AB is the output aperture of the AE, equal to the internal diameter D_{chan} of the discharge channel, α_1 is the propagation angle of the beam of superradiance formed by the total geometric aperture of the discharge channel (D_{chan}/l_{chan}) from the amplifying emission. Within this angle, approximately 95% of the energy of the output radiation is concentrated.

The geometric dimensions of the AE of the models GL-206E and GL-206I, which are necessary for calculating the divergence of radiation, are given in Table 4.1.

Table 4.1 also gives the distances that determine the location of the convex mirror S_1 with respect to the AE and its elements; l_1, l_1'

and l are the distances from the mirror S_1 to the AE, its AM, and the output aperture.

A dull convex mirror S_1 in the optical circuits a) and b) is installed in a blackened metal frame with a conical surface absorbing the radiation. The output mirror S_2 (b) of the UR is an antireflecting positive lens in the form of a convex-concave meniscus, on the convex surface of which a mirror spot with a diameter of 1–1.5 mm is applied (patent No. 2 432 652 RF) [371]. The focus of the mirror lens $S_2(b)$ is aligned with the focus of the blind mirror $S_1(b)$, so that the diverging beam formed by UR is converted (collimated) into a parallel (cylindrical) beam with minimum divergence and that is convenient for practical applications. In the first experiments, an output convex mirror with a diameter of 1.5 mm was glued to a plane-parallel, antireflecting glass plate ($S'_2(b)$). The angle between the optical axis of this mirror and the plate was 86°, which eliminated the inverse 'parasitic' connection from the plate with the active medium of the AE. In this case, the beam from the resonator comes out divergent and an additional optical element is required to collimate it into a parallel beam. In the telescopic UR, the focus of a deaf concave mirror $S'_1(b)$ is aligned with the focus of the output convex mirror $S'_2(b)$ which is the basic condition for the formation of a parallel beam with a plane wave directly in the resonator.

For the convenience of conducting research on the characteristics of a divergent qualitative beam of radiation produced in a CVL in a single-mirror mode (a), it was first collimated into a cylindrical beam using a lens with a focal length F = 1.6 m (L), and then focusing with a spherical concave mirror with R = 15 m (item 7). When recording the oscillograms of pulses and removing the intensity distribution in the constriction of the focused radiation beam, the power was previously attenuated by introducing into the light flux a beam splitter plate (position 6) and neutral light filters (position 8).

In experimental installations, the average power of the radiation was measured with a millivoltmeter M136 with a TI-3 laser radiation power converter connected to it (item 9), recording radiation pulses with an oscilloscope of the GDS-840S type with a photocell FEK-14K (item 10), and studying the distribution of radiation intensity and measurement of the diameter in the focal plane (practically in the waist of $1/e^2$) of the focused beam – BeamStar-FX beam analyzer (item 11). Since it is assumed that the intensity distribution in the focal plane corresponds to the far-field distribution ($>D^2/\lambda$, where D is the diameter of the beam, λ is the wavelength of the radiation), then for the practical determination of the divergence (θ) a simple method is the focal spot method:

$$\theta = \frac{d_0}{F}, \tag{4.1}$$

where d_0 is the diameter of the focused radiation beam in the focal plane (in the neck), F is the focal length of the focusing optical element.

In all the cases considered below, the divergence of the laser radiation was reduced to the diameter of the aperture of its discharge channel (D_{chan} = 14 mm).

An important technological parameter in microprocessing by pulsed laser radiation is the density of peak power in a focused spot that determines the quality and productivity of processing. The peak power density is determined by the formula

$$\rho = \frac{P_{rad}}{f \cdot \tau\pi \cdot r^2}, \tag{4.2}$$

where P_{rad} is the average radiation power, f is pulse repetition frequency (PRF), τ is the half-height pulse duration, r is the radius of the spot of the focused beam of radiation. In our experiments, the working PRF of the CVL was f = 14–15 kHz, the pulse duration at half-height was τ = 10–13 ns.

4.3. Structure and characteristics of radiation of CVL in single-mirror mode. Conditions for the formation of single-beam radiation with high quality

Experimental studies of the characteristics of pulsed CVL radiation in a single-mirror regime, as mentioned above, were carried out in the regime with a single convex mirror in the setup shown in Fig. 4.1 *a*. Figure 4.5 shows separately the optical scheme for this mode.

Investigations and calculations were carried out for two cases of the arrangement of the convex mirror S_1 relative to the AE: at l_1 = 50 mm and l_1 = 250 mm (see Fig. 4.5 and Table 4.1).

In the single-mirror mode, in accordance with the results of our investigations in [16, 232, 233] and in the present work, the output radiation of the CVL has a strictly two-beam structure: an incoherent beam of superradiance I (Fig. 4.5), formed from amplified

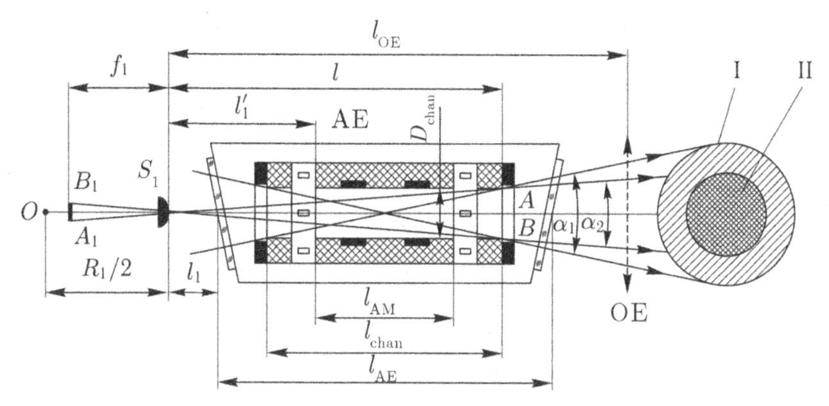

Fig. 4.5. The optical scheme of CVL with the Kulon AE in single-mirror mode (with one convex mirror): l_{AE} – length of AE; l_{chan} and D_{chan} – length and diameter of the discharge channel; l_{AM} is the length of the active medium; S_1 – convex mirror with radius of curvature R_1; l_{OE} is the distance from the mirror S_1 to the collimating or focusing optical element of the OE; A_1B_1 – image of the output aperture of the discharge channel AB in the mirror Z_1; l_1 and l_1'– distance from mirror S_1 to AE and to AM; l is the distance from mirror S_1 to the output aperture AB; f_1 – distance from the mirror S_1 to the image A_1B_1; α_1 – angle of propagation of the beam of superradiance formed by the total aperture of the discharge channel; α_2 is the angle of propagation of the beam of superradiance formed by the mirror S_1 and the output aperture of the discharge channel AB; I and II – the intensity distribution of the beams of superradiance in the near zone.

spontaneous emission by the total geometric aperture of the AE discharge channel, and a beam II with high spatial coherence formed with the participation of mirror S_1 and the output aperture of the discharge channel AB. Hereinafter, the first incoherent beam will be referred to simply as a background beam, as unsuitable for practical applications, the second with a high coherence – a qualitative beam.

Based on the consideration of the geometric path of the rays in the CVL with a single convex mirror (see Fig. 4.5), the divergence of the collimated qualitative radiation beam, i.e., transformed into a cylindrical beam with a minimum divergence [232, 233], is calculated. The condition for collimating this beam is the alignment of the focus of the optical element (OE) with the plane of the image A_1B_1 of the output aperture AB of the discharge channel in the convex mirror S_1 (Fig. 4.5). Image A_1B_1 physically represents an imaginary radiation source equivalent to AE with a single mirror. The divergence of the collimated beam is determined by the ratio of the image size A_1B_1 in the mirror to the distance from the image to the OE:

$$\theta = \frac{A_1B_1}{f+L}.$$

To determine the location (f) and the size (A_1B_1) of the image, we use formulas (4.3) and (4.4) for a convex spherical mirror:

$$\frac{1}{l} - \frac{1}{f} = -\frac{R}{2}, \quad (4.3)$$

$$\frac{H}{h} = \frac{f}{l}, \quad (4.4)$$

where l is the distance from the object (AB) to the mirror (Fig. 4.5); f is the distance from the image (A_1B_1) of the object in the mirror to the mirror (S_1); h and H are the sizes of the object (AB) and the image (A_1B_1) of the object in the mirror (S_1); $AB = A'B' = D_{chan}$ is the diameter of the aperture of the discharge channel of the AE. Wherein

$$f = \frac{A_1B_1 \cdot L}{D},$$

$$A_1B_1(H) = \frac{D_{chan} R/2}{R/2 + l}.$$

Then the minimal (limiting) divergence of the radiation beam (without allowance for diffraction) is

$$\theta = \frac{RD_{chan}}{l}\left[R + \frac{L}{l}(R+2l)\right]. \quad (4.5)$$

As the value of R tends to zero, the divergence θ also tends to zero. Actually, because of the diffraction on the output aperture, the AE θ tends to the diffraction limit $\theta_{diff} = 2.44\lambda/D$ ($D = D_{chan}L/l$ is the diameter of the radiation beam at the OE). Taking this into account, the expression for the divergence is written in the form

$$\theta_{lim} = \frac{RD_{chan}}{l}\left[R + \frac{L}{l}(2R+2l)\right] + \theta_{dif}. \quad (4.6)$$

For $L = l$ we obtain

$$\theta = \frac{R \cdot D_{chan}}{2l(R+l)} + 2.44\frac{\lambda}{D_{chan}}. \quad (4.7)$$

The divergence determined by formula (4.7) corresponds to the divergence of the radiation beam, reduced to the diameter D_{chan} of the aperture of the discharge channel AE.

If the focal length F of the optical element is less than L, then the radiation beam is focused into a spot with a diameter

$$d = \theta\left\{\frac{1}{F} - \frac{1}{l}\left[\frac{L}{l} + \frac{R}{R+2l}\right]\right\}. \tag{4.8}$$

Then, to determine the radiation power density in a focused spot, we have the following formula:

$$\rho = 4P_{rad} \cdot \frac{\left\{\frac{1}{F} - \frac{1}{l}\left[\frac{L}{l} + \frac{R}{R+2l}\right]\right\}^2}{\pi\theta_{lim}^2}, \tag{4.9}$$

where P_{rad} is the average radiation power of a qualitative radiation beam.

Analysis of formulas (4.7) and (4.9) implies that the divergence and radiation power density can be varied within wide limits, changing the curvature radius R of the convex mirror and the distance l from the mirror to the output aperture AB. When R is one or two orders of magnitude smaller than the distance l, the beam divergence θ becomes close to the diffraction limit: $\theta = 2–3\theta_{dif}$. An increase in the distance l is not always expedient, since in this case, due to the limited time of existence of population inversion, the radiation power is substantially reduced.

Figure 4.6 shows the dependence of the calculated divergence, Fig. 4.7 – the average radiation power with a change in the radius of the convex mirror S_1 within the range of 0.6–3 cm for the Kulon AE of the models GL-206E (*a*) and GL-206I (*b*). For AE GL-206E and GL-206I with a channel aperture diameter D_{chan} = 14 mm, the diffraction divergence is θ_{difr} = 0.1 mrad.

Curves *1* and *3* in Fig. 4.6 are calculated for the distance from the mirror to the AE l_1 = 50 mm, *2* and *4* for l_1 = 250 mm. The larger the distance l_1 and the longer the discharge channel l_{chan} and, correspondingly, the distance from the mirror to the output aperture AB (l) AE, the smaller the divergence of the qualitative beam (curve *4*). The AE of the Kulon series, model GL-206I, is longer than the AE of the GL-206E) by 130 mm.

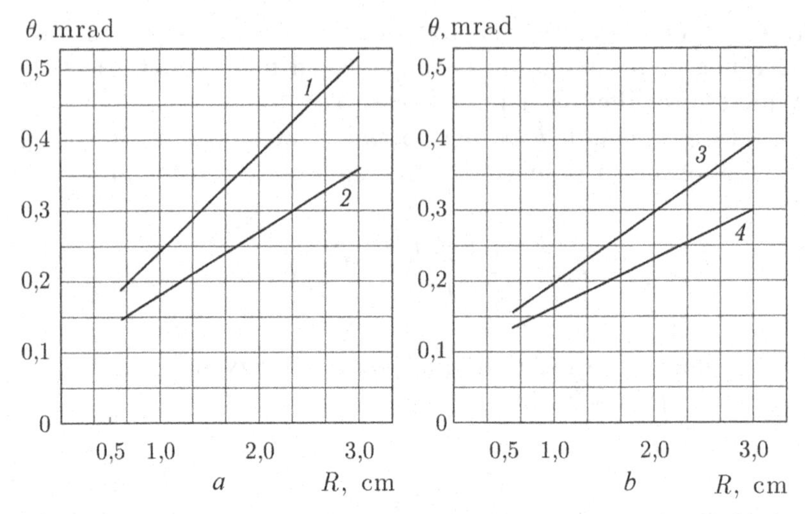

Fig. 4.6. Dependence of the calculated divergence of the qualitative radiation beam of the CVL with the Kulon AE of the models GL-206E (*a*) and GL-206I (*b*) in the single-mirror regime (*a*) on the radius of curvature of the convex mirror S_1. Curves 1 and *3* are calculated for l_1 = 50 mm, curves *2* and *4* for l_1 = 250 mm.

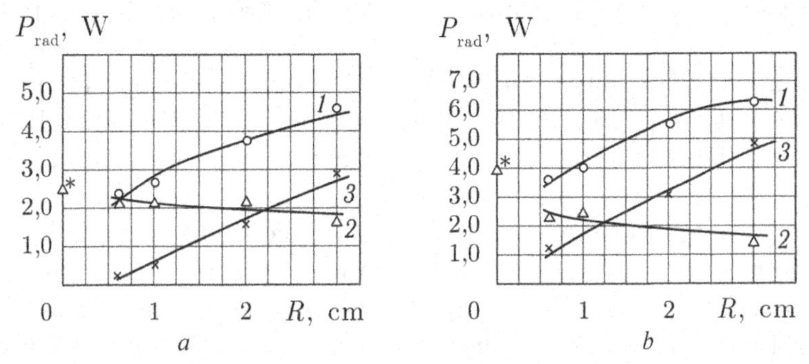

Fig. 4.7. The dependence of the average power in the total (*1*), background (*2*) and qualitative (*3*) radiation beams of the CVL with the AE Kulon GL-206E (*a*) and GL-206I (*b*) in the single-mirror mode on the radius of curvature of the convex mirror. l_1 = 50 mm, *x* – power of the beam of superradiance without mirrors.

The experimental values of the divergence of the qualitative radiation beam were determined from the results of measurements of the minimum diameter in the constriction of the focused beam (Figure 4.1, item 11) and use of formula (2.3). For example, in Fig. 4.8 (color insertion) shows the distribution of the intensity of radiation in the constriction in the operating mode of CVL with AE GL-206I and a convex mirror with R = 3 cm at l_1 = 250 mm.

The following notations are indicated on the intensity distribution field horizontally: W, 2W, FWHM, Correlation, PeakHeight. W is

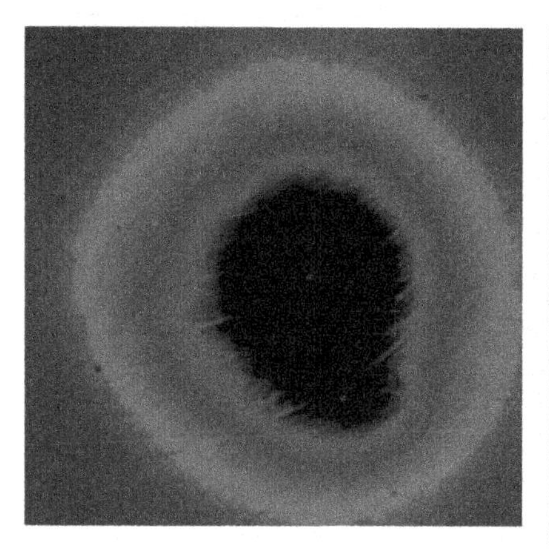

2D distribution

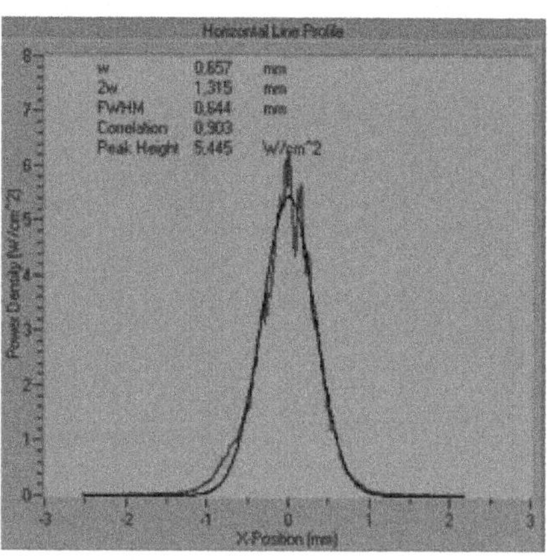

Horizontal distribution

Fig. 4.8. Distribution of the intensity of the output radiation beam of the CVL with the GL-206I AE in the single-mirror mode in the focal plane (waist) of the mirror with the curvature radius R = 15 m, R_1 = 30 mm, l_1 = 50 mm.

the size of the waist, defined as the half-width of the profile with respect to the $1/e^2$ level from the maximum intensity of the beam. 2W is the total width of the profile along the $1/e^2$ level from the maximum intensity of the beam. FWHM is the total width of the profile at the half maximum beam intensity. Correlation is the best correlation correspondence between the beam profile and the ideal Gaussian beam. PeakHeight is the height of the beam peak W/cm^2.

The diameter of the waist at the level $1/e^2$ (2W in Fig. 4.8) from the maximum intensity was 1.315 mm, the divergence of the beam in accordance with the formula (4.1) – $\theta \cong 0.25$ mrad, which agrees well with the calculated value – $\theta = 0.3$ mrad (see curve 4 in Fig. 4.6). It follows from the horizontal intensity distribution that the degree of beam correlation in the single-mirror mode with the Gaussian beam is 0.903.

With an increase in l_1 from 50 to 250 mm, the radiation power decreased insignificantly – by 5...6%, which is insignificant in the contribution to the power density in comparison with the decrease in the divergence.

Figure 4.9 shows examples of oscillograms of the background incoherent (*1*) and qualitative (*2*) radiation beams of CVL with the AE Kulon GL-206I at radii of curvature of the convex mirror R = 0.6; 1; 2 and 3 cm and the distance from the mirror to the AE l_1 = 250 mm. At l_1 = 250 mm, the time of double transmission by the radiation of the distance from the mirror to the active medium is

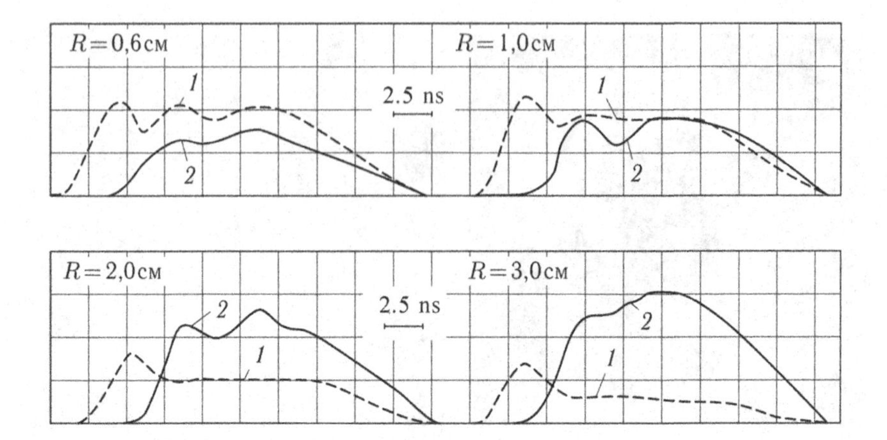

Fig. 4.9. The oscillograms of the pulses of the background (*1*) and qualitative (*2*) radiation beams of the CVL with the AE Kulon GL-206I in the single-mirror mode for different radii of curvature of the convex mirror and l_1 = 250 mm.

$$\Delta t = \frac{2l_1'}{c} = 2.8 \text{ ns}, \tag{4.10}$$

where l_1' = 0.42 m, $c = 3 \cdot 10^8$ m/s. The total radiation pulse duration for the base is 23–25 ns. Time Δt = 2.8 ns, as can be seen from the oscillograms, corresponds to the time lag of the beginning of the pulses of the qualitative beam (*2*) from the pulses of the background beam (*1*).

As the distance l_1' increases, the delay time interval Δt increases and the duration of the pulses of the qualitative beam decreases, which in principle leads to an improvement in the processing quality due to a decrease in the interaction time with the material. With increasing Δt, the divergence of the qualitative beam decreases, but the radiation power also decreases. Depending on the problem posed, a change in the distance from the mirror to the AE can optimize the density of the peak radiation power in the focused spot, both in the operation of the CVL in the mode of an individual generator, and in the CVLS operating as a master oscillator – a spatial filter-collimator – a power amplifier (MO–SFC–PA).

In technological installations using a two-wave CVL, the radiation is focused using achromatic objectives with F = 30–150 mm. Figure 4.10 shows the dependences of the minimum spot diameter (d) and the peak power density (ρ) in the waist of the focused beam of laser radiation in single-mirror mode from the radius of curvature of the convex mirror S_1 (R_1) with an achromatic objective with F = 70 mm.

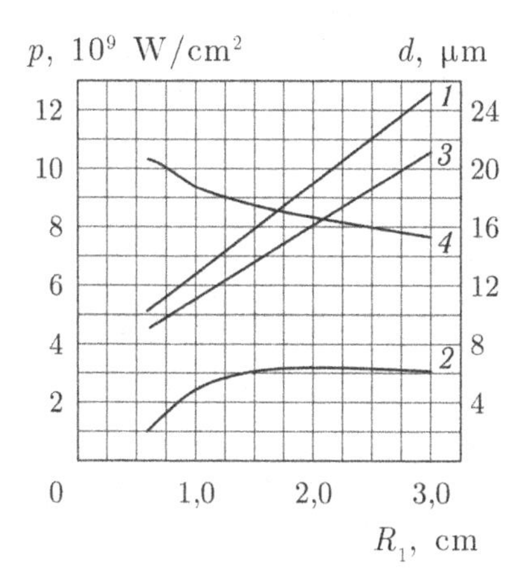

Fig. 4.10. Dependence of the spot diameter and the peak power density in the waist of the focused beam of the laser radiation of the CVL with the Kulon AE GL-206E (curves *1* and *2*) and GL-206I (curves *3* and *4*) on the radius of curvature of the convex mirror with the achromatic objective F = 70 mm.

The course of the curves in Fig. 4.10 shows that the change in the radius of curvature of the mirror within R_1 = 0.6–3 cm corresponds to an increase in the diameter of the focused spot with AE GL-206D within d = 10.5–25 μm (curve *1*) and with a longer GL-206I in (curve *3*), peak power density ρ = (0.96–3.2) · 10^9 W/cm^2 (curve *2*) and (10–7.7) · 10^9 W/cm^2 (the curve *4*), respectively. But at the same time high processing capacity is not ensured, since the energy in the pulse is insignificant – 0.05–0.2 mJ/cm^2.

An increase in the radiation power of the CVL in the regime with one convex mirror. In order to increase the power of a qualitative beam of radiation (superradiance) of the CVL in the mode of operation with a single convex mirror, the optical scheme shown in Fig. 4.11 is used. It is protected by the Author's certificate No. 1 438 549 of the USSR 'Pulsed resonator-free laser' [314]. The laser contains an AE (item 1) with a discharge channel (item 2), at one end of which there is a convex mirror (item 3), and on the other – a focusing optical element (item 4), a diaphragm (position 5) and a concave mirror (item 6). A collimating or focusing optical element (item 7) is installed behind the mirror (item 3). The diaphragm (item 5) is set at a distance l from the focusing element (item 4), namely in the focusing plane of the qualitative beam. The optical centre of the concave mirror (6) is in the plane of the diaphragm, and the normal to the surface of this mirror forms an angle α with the optical axis.

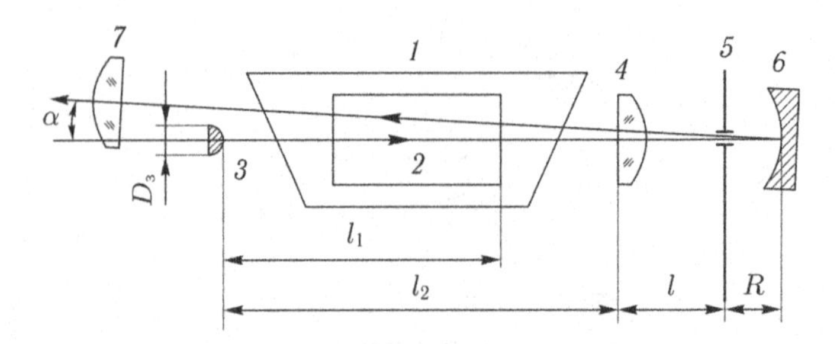

Fig. 4.11. Optical scheme for increasing the output power of CVL in the regime with one convex mirror: *1* – AE; *2* – discharge channel; *3* – convex spherical mirror; *4* – focusing optical element; *5* – the diaphragm; *6* – concave spherical mirror; *7* – collimating or focusing optical element.

The optical axis of the collimating or focusing element (item 7) behind the mirror (item 3) is aligned with the mirror axis (item 6).

The laser works as follows. In the discharge channel (position 2) of the AE, amplifying spontaneous 'seeds' form, leading to the formation of two beams of superradiance (see the beginning of Section 4.3 and Fig. 4.5). The beam of superradiance, formed with the participation of a convex mirror (item 3), is of high quality. This beam is focused by the optical element (item 4) on the diaphragm (item 5) and is extracted from the background of the beam with low coherence. Using the formula of a thin lens to determine the distance l, we obtain

$$l = \left[\frac{1}{F_2} - \frac{1}{\dfrac{F_1 l_1}{l_1 + F_1} + l_2} \right]^{-1},$$

where l_1 and l_2 are the distances from the mirror (item 3) to the far end of the discharge channel (item 2) and to the focusing element (item 4), respectively; F_1 and F_2 are the focal length of the mirror (item 3) and the focusing element (item 4).

The quality beam separated by the diaphragm (item 5) is reflected from the concave mirror (6) at an angle

$$\alpha = \frac{D_m}{l_2 + l + R}$$

to the optical axis (D_m is the diameter of the mirror (item 3), $l_2 + l + R$ is the distance from the mirror (item 3) to the mirror (item 6) with the radius of curvature R and is focused again into the plane of the diaphragm. close to the size of the spot from the quality beam.Through the focusing element (item 4), the beam reflected from the mirror (position 6) returns as a converging beam back to the discharge channel (position 2), where it is further amplified in power and leaves the mirror (item 3). The diameter of the working surface of the mirror (item 3) is close to the diameter of the whole mirror. When the radius of curvature of the convex mirror (item 3) is 1–2 orders of magnitude smaller than the distance l_1, the diameter of its working surface is a fraction of a millimeter behind the mirror (position 3) (in the image plane of the discharge channel) the beam is again focused, and then the element (item 7) is converted to a beam with the desired quality.When the collimating element (position 7) is installed at the output of the AE, the divergent laser beam is transformed into a beam with a minimum divergence. In the case of the presence of a focusing element at the output (position 7), the beam is focused into a spot with a high intensity, so that, without introducing an additional optical element, it is possible, for example, to microprocess materials. This optical scheme was implemented using of the industrial sealed-off AE Kristall GL-201. The distance l_1 from the mirror (item 3) to the far end of the discharge channel (item 2) was 110 cm, the distance l_2 from the mirror (item 3) to the optical element (item 4) was 140 cm. The mirror (item 3) had a diameter D_m = 0.15 cm and a focal length F_1 = 0.75 cm (or 0.3 cm), the mirror (item 6) – a diameter of 3.5 cm and a radius of curvature R = 10 cm. The focusing element (position 4) was an enlarged lens with a focal length F_2 = 23 cm. The diaphragm (position 5) was set at a distance l = 28 cm from the element (item 4) and had a hole diameter of ~0.01 cm, due to which the separation of a qualitative beam from a background with low coherence was conducted. The angle of inclination α of the mirror (6) to the optical axis was 1.7 mrad. At the output of the AE (item 1), a lens with a focal length of 110 cm was installed behind the mirror (item 3) as a collimating element (position 7).

In the layout of the CVL, in which the mirror (item 3) had a focal length of 0.75 cm, the average radiation power was 6 W, with F_1 = 0.3 cm – 3 W, which is 6 and 15 times more than in the mode with one convex mirror. The beam divergence at F_1 = 0.3 cm (θ = 0.15 mrad) was only twice as large as the diffraction limit θ_{diff} = 0.07 mrad, and at F_1 = 0.75 cm (θ_{real} = 0.2 mrad) – three times.

Thus, the proposed optical scheme (Fig. 4.11) allows several times and even by an order of magnitude to increase the radiation power

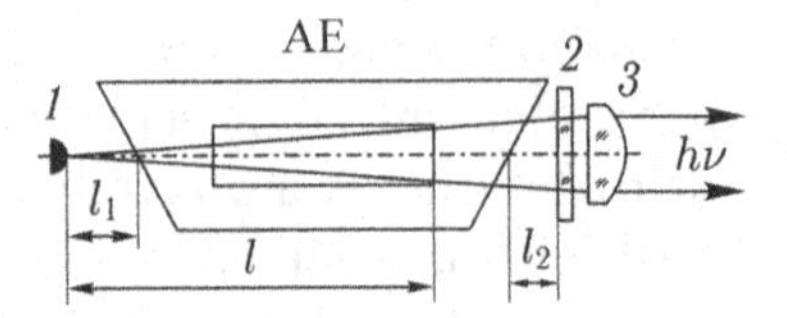

Fig. 4.12. Optical scheme for reducing the divergence of the laser radiation in the regime with one convex mirror: *1* – convex spherical mirror; *2* – plane-parallel plate; *3* – collimating lens

of the CVL. But since this circuit in relation to the single-mirror regime additionally includes four optical elements, the instabilities of the position of the axis of the beam pattern of the radiation beam and the pulsed energy increase noticeably. In addition, the adjustment of the entire laser system is also complicated.

Reduction of the divergence of radiation of the CVL in the regime with one convex mirror. The divergence of the qualitative radiation beam (superradiance) in the operation of a linear displacement laser in the regime with a single convex mirror can be reduced using the optical scheme shown in Fig. 4.12. It is protected by a patent for No. 44004 RF 'Pulsed laser' [395].

This optical scheme (Fig. 4.12) differs from the basic circuit (Fig. 4.5) in that a plane-parallel reflective plate (position 2) is installed between the collimating element (position 3) and the AE at a distance $l_2 \leq l_1$ from the AE perpendicular to the optical axis of the laser transparent for laser radiation. In this case, the cathode in the discharge channel of the AE is located on the laser radiation exit side, that is, on the side of the plane-parallel plate.

The proposed device works as follows. When the discharge is turned on in the discharge channel of the AE near the cathode, there are spontaneous seeds growing in both directions – two incoherent waves of superradiance. One wave is directed towards the convex mirror (item 1), the second – on the plane-parallel plate (item 2) with partial transparency for laser radiation. Since the plate (position 2) is located much closer to the cathode region of the AM and has a larger working surface than the mirror (item 1), the beginning of the process of forming a qualitative radiation beam is completely determined by the characteristics of this plate. In this case, part of the superradiance passes through the plate and forms a low coherent background beam with a large divergence determined by the total geometric aperture of the discharge channel. The other part of it, reflected, again falls into the amplifying AM of the discharge channel and spreads along the channel, having greater spatial coherence than before reflection from the plate. This wave, amplified in the active medium, arrives at the area opposite to the cathode of the discharge channel (anode region) earlier than the reflected part of the first wave reflected by the convex mirror, substantially reducing the population inversion of the AM of the discharge channel. Therefore, the relatively weak first wave, which is not able to amplify, is damped from the convex mirror (position 1), and the counter reinforced wave reflects from the convex mirror, increasing its coherence. In an active medium, this divergent wave, with high spatial coherence, is once again amplified, partially reflected from the plate, partially passes through it to the collimating lens (position 3), forming a cylindrical (plane) radiation beam with a divergence smaller than with the base optical circuit (Fig. 4.5). As a result, two spatially separated radiation beams are observed behind the lens: a superluminal beam formed by the aperture of the discharge channel and a second qualitative beam with practically a diffraction divergence.

The efficiency of the proposed optical scheme was checked with a CVL with a pulse repetition frequency of 10 kHz, in which the convex mirror had a radius of curvature of $R = 0.6$ cm. The CVL used a Kristall AE, model GL-201, with a diameter and a discharge channel length of 2 and 93 cm,, the distance between the convex mirror and the plane-parallel plate was 150 cm and $l_1 = l_2$ (Fig. 4.12). At the output of the CVL (after collimation), a strongly diverging incoherent background beam of superradiance and a beam with a diffraction divergence (0.07 mrad) of 0.5 W power were recorded. For the prototype, the divergence value was 0.1 mrad at a radiation power of 0.35 W. When the diffraction beam was focused by an achromatic lens with a focal length of 100 mm, the diameter in the focused spot was 7–8 μm, and the peak power density was $9 \cdot 10^9$ W/cm^2. In the prototype, these values were 10 μm and $3 \cdot 10^9$ W/cm^2.

Thus, the use of an additional plane-parallel plate in the side of the radiation exit in the CVL with one convex mirror led to a decrease in the divergence of the qualitative radiation beam by a factor of 1.5 (practically to the diffraction limit) and to an increase in the peak power density in the focused spot by a factor of three.

It is necessary to emphasize that from the practical point of view, when operating the CVL in the single-mirror mode, it is preferable to use long AEs relatively with a small diameter of the discharge channel (D_{chan} = 10–20 mm) and an increased distance from the mirror to the AE. But it is undesirable that the distance from the mirror to the active medium of the AE is greater than 0.7–1 m, which corresponds to a double-pass time by radiation of this distance Δt = 5–7 ns, since with a general pulse duration of 20–25 ns leads to a noticeable decrease of power in a qualitative radiation beam. As the length of the AE increases, the beam divergence decreases and tends to the diffraction limit, and the radiation power increases, which together leads to a sharp increase in the peak power density. Therefore, from this point of view, the industrial AE Kristall of the GL-205A and GL-205B models with a discharge channel length of 0.93 and 1.23 m and a diameter of 20 mm (see Table 3.4) are more effective [16, 26]. To the merits of the single-mirror CVL regime, in addition to the formation of a single qualitative beam, it is necessary to attribute the high stability of the axis position of the beam pattern of this beam and its pulse energy, to a disadvantage – the divergence is 2–4 times greater than the diffraction limit, which reduces the power density in the 5–15 times.

The main application of CVL in the single-mirror mode has been the pumping of wavelength-tunable lasers on dye solutions (DSL), since a peak power density of 10^6–10^7 W/cm^2 is sufficient for effective DSL pumping. For example, CVL with the Kristall AE and one convex mirror with a radius of curvature R = 5 cm (θ = 0.5 mrad, which is 7 times greater than θ_{diff} = 0.07 mrad), is used as a master oscillator in a 300 W CVLS operating according to the effective scheme, the master oscillator – the spatial filter-collimator – the power amplifier (MO–SFC–PA). In turn, the 300 W CVLS in combination with the 150 W DSL pumped by it, which also operates according to the MO–PA scheme, is used in unique modern technological equipment for isotope separation.

As the practical experience shows, the level of the peak power density of 10^9 W/cm^2 achieved in the single-mirror mode is sufficient for productive microprocessing of foil materials and cutting of solders (0.02–0.1 mm). In order to process thicker materials, the CVL in the single-mirror mode is used as the MO in the CVLS of the MO–SFC–PA, when using the Kristall AEs of the GL-205A and GL-205B models as PA, the radiation power increases by more than an order of magnitude (30–60 W) [16, 26]. At the same power levels, to maximize the productivity and quality of the material microprocessing, it is necessary that single-beam radiation has a diffraction divergence and stable parameters.

4.4. Structure and characteristics of the laser radiation in the regime with an unstable resonator with two convex mirrors

Conditions for the formation of single-beam radiation with diffraction divergence and stable parameters

In order to obtain single-beam radiation with a diffraction divergence and high stability of the axis position of the radiation pattern and pulsed energy in CVL, studies were made of the spatial, temporal and energy characteristics of the laser radiation in the regime with an unstable resonator (UR) with two convex mirrors. The experimental setup is shown in Fig. 4.1 *b* and separately Fig. 4.13 shows the optical scheme with the indication of the geometric dimensions necessary for the calculation of the divergence and time parameters of the radiation pulses.

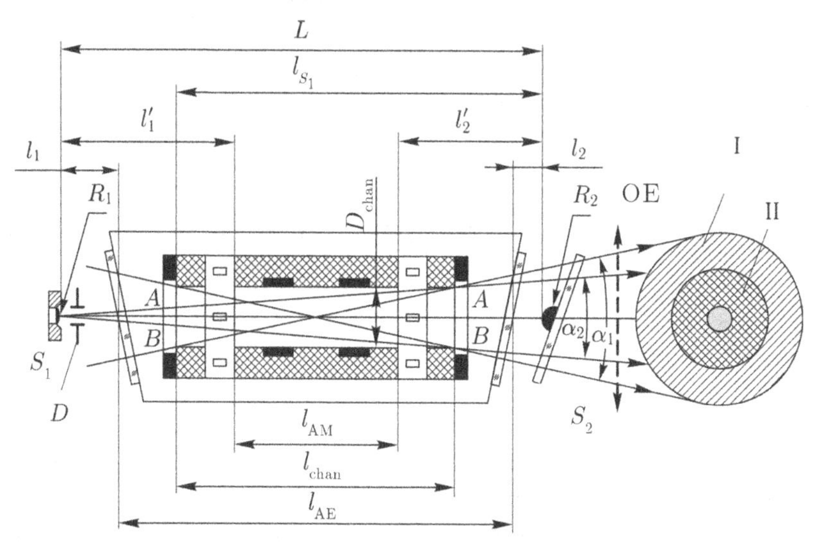

Fig. 4.13. The optical scheme of the CVL with the AE Kulon in the mode with UR with two convex mirrors: S_1 and S_2 are the 'dumb' and output mirrors of the UR with curvature radii R_1 and R_2; D – diaphragm with $d = 0.1$ mm; L – length of the resonator, l_{AE} – length AE; l_{chan} and D_{chan} – length and diameter of the discharge channel; l_{AM} – length of the AM; l_1 and l'_1 – distance from mirror 3 to AE and to AM; l_2 and l'_2 – distance from mirror 3 to AE and to AM; l_{S_2} is the distance from the mirror S_2 to the output aperture AB of the discharge channel; α_1 is the angle of propagation of the incoherent superradiance beam formed by the total geometric aperture of the discharge channel; α_2 is the propagation angle of the resonant (qualitative) radiation beam; I and II – the distribution of the intensity of the superradiance beam and the resonator radiation beam in the near zone. OE – collimating or focusing optical element or optical system.

The unstable resonator (UR) with two convex mirrors in comparison with the known telescopic UR and the single-mirror mode potentially has maximum spatial selectivity and, accordingly, the possibility of forming a single-beam radiation with a minimum (diffraction) divergence.

A careful analysis of earlier studies on the dynamics of the formation of a multibeam structure of the output radiation in pulsed CVLs with optical resonators [16, 91–94, 230–233] and our experimental studies and calculations with UR allowed us to establish that in the UR with two convex mirrors a strictly single-beam radiation of diffraction quality can form ($\alpha_2 = D_{chan}/l$) directly on the background of an incoherent aperture superradiance beam ($\alpha_1 = 2D_{chan}/l_{chan}$), but with the obligatory fulfillment of three interrelated conditions:

1. A blind mirror S_1 of the UR must be installed at a distance from the AS equal to at least half the length of the AM and the distance from the output mirror S_2 to the AM and not more than half the distance traveled by the radiation during the lifetime of the inversion and the sum of the length of the AS and the distance from the output mirror S_2 to the AM, that is,

$$\frac{l_{AM}}{2} + l_1' < l_1' < \frac{\tau_{inv} \cdot c}{2} - (l_{AM} + l_2'). \tag{4.11}$$

This is a very important position, since it allows us to begin the process of forming a resonator beam from the output mirror S_2. And then the beam formation in the resonator proceeds in the following sequence: AM → output mirror S_2 → AM → a blind mirror S_1 → AM → an output (from the side S_2). At the initial stage, a part of the radiation from an aperture superradiance beam with $\alpha_1 = 2D_{chan}/l_{chan}$ is reflected by the mirror S_2 is reflected back to the AM of the discharge channel. Then, amplified in the AM of the discharge channel, it manages to leave earlier from the end which is near to the blind mirror S_1 before the part of the radiation reflected from the same aperture beam of superradiance reaches the AM. In this case, the reflected part of the radiation from the mirror S_1 from the superradiance beam with $\alpha_1 = 2D_{chan}/l_{chan}$ does not receive the preferential amplification in the AM and decays. That is, conditions are practically created that prevent the appearance of a second beam, which is 'parasitic' for our case, with a divergence close to the diffraction one.

2. The output mirror S_2 of the UR should be as close to the AE as possible and have a radius of curvature greater than that of the blind mirror S_1. In order to ensure that the selectivity of the radiation in the resonator is maximal, the blind mirror must have a small radius of curvature (R_1 = 0.6–3 cm) (as in the single-mirror mode) – by 1–2 orders of magnitude less than the distance from the mirror S_1 to the output aperture of the discharge channel AB ($R_1 < l/10–100$). Under these conditions, the feedback from the mirror S_2 with the AM begins earlier and from a larger working surface, which is in addition to the first condition that prevents the second unwanted beam from coming from the mirror S_1 with a divergence close to the diffraction one.

3. Under the conditions of sections 1 and 2, the parameters of UR must satisfy the requirements ensuring the formation of a diffraction beam of radiation already for the first double pass of radiation in the resonator, otherwise the practical application of UR with two convex mirrors is lost.

To fulfill the third condition, it was necessary to establish the dependence of the divergence of the resonator beam (θ) formed in it for the first double pass of the radiation, on the parameters of the UR, geometric dimensions of the discharge channel (aperture) of the AE and the dimensions connecting them with each other. To this end, it was necessary to derive a formula for the calculation of divergence, to perform calculations and experimental studies for small radii of curvature of mirrors. To derive the formula, using the laws of geometric optics, an optical circuit was constructed (Fig. 4.14) for successive displacements of the image of the aperture of the AM in the convex mirrors of the resonator in the received beam formation direction (see section 1) and indicating the necessary carrying out calculations of geometric dimensions. When in order to determine the location of the image and its dimensions, we used formulas (4.12) and (4.13) for a convex spherical mirror

$$\frac{1}{l}-\frac{1}{f}=-\frac{R}{2} \tag{4.12}$$

and

$$\frac{H}{h}=\frac{f}{l}, \tag{4.13}$$

where l is the distance from the object to the mirror (in Fig. 4.14 – l_{S_1} and l_{S_2}); f – distance from the image of the object in the mirror to the mirror (in Fig. 4.14 –f_1 and f_2); h and H are the size of the object and the image of the object in the mirror (in Fig. 4.14 – AB, $A'B'$ and A_1B_1, A_2B_2). $AB = A'B' = D_{\text{chan}}$ is the diameter of the aperture of

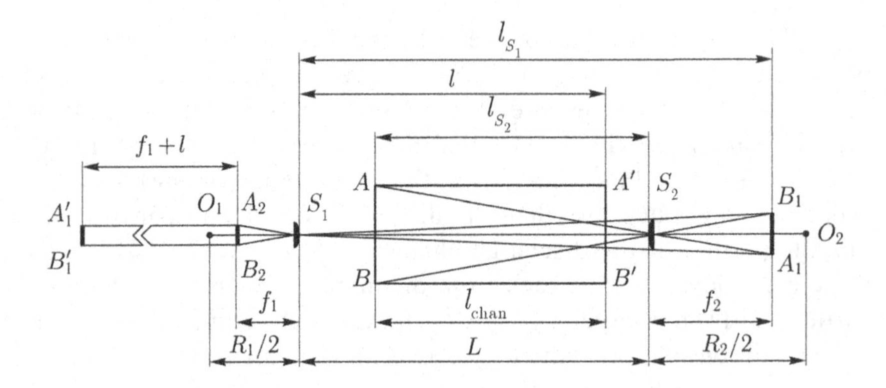

Fig. 4.14. Optical scheme of sequential moving of the image of the aperture of AE in convex mirrors of UR: $AB = A'B'$ – output apertures of the discharge channel of the AE; l_{chan} is the length of the AE discharge channel; L is the resonator length; S_1 and S_2 are the 'blind' and output convex mirrors of UR with the radius of curvature of R_1 and R_2; l is the distance from the mirror S_1 to the output aperture $A'B'$; l_{S_2} is the distance from the mirror S_2 to the output aperture AB; A_1B_1 is the image in the mirror S_2 of the output aperture AB; f_2 isd the distance from the mirror S_2 to the image A_1B_1; l_{S_1} is the distance from the mirror S_1 to the image A_1B_1; A_2B_2 is the image of the image A_1B_1 in the mirror S_1; f_1 is the distance from the mirror S_1 to the image A_2B_2; $A_1'B_1'$ is the image of output aperture $A'B'$ in image A_2B_2.

the discharge channel of the AE.

The formula obtained for calculating the divergence of the radiation beam (with allowance for the diffraction limit) has the following form:

$$\theta = \frac{D_{chan} \cdot R_1 \cdot R_2}{4 \cdot (R_1 + 2f_2 + 2L)(R_2 + 2l_{3_2})(l + f_1)} + \frac{2.44\lambda}{D_{chan}}, \tag{4.14}$$

where

$$f_2 = \frac{R_2 \cdot l_{S_2}}{R_2 + 2l_{S_2}}, \quad l_{S_2} = \frac{l_{AE} + l}{2} + l_2,$$

$$f_1 = \frac{R_1 \cdot l_{S_1}}{R_1 + 2l_{S_1}} = \frac{R_1(L + f_2)}{R_1 + 2(L + f_2)}.$$

Tables 4.2 and 4.3 show the results of calculating the divergence of the output radiation beam using the formula (4.14) for the CVL with the AE Kulon of the model GL-206E for different radiuses of curvature of the output mirror S_2 UR – R_2 = 1.5; 3; 5; 12 and 25 cm, the radius of curvature of the blind mirror is S_1R_1 = 3 cm and the

Table 4.2. GL-206E: l_1 = 5 cm, l_2 = 10 cm, l_{S_2} = 74 cm, l = 69.2 cm, L = 92 cm

R_2, cm	1.5	3	5	12	25
$\theta \times 10^{-3}$, rad	0.1016	0.1031	0.1051	0.1112	0.1206

Table 4.3. GL-206E: l_1 = 25 cm, l_2 = 10 cm, l_{S_2} = 74 cm, l = 89.2 cm, L = 112 cm

R_2, cm	1.5	3	5	12	25
$\theta \times 10^{-3}$, rad	0.1070	0.1020	0.1033	0.1073	0.1135

Table 4.4. GL-206I: l_1 = 5 cm, l_2 = 10 cm, l_{S_2} = 83 cm, l = 81,6 cm, L = 105 cm

R_2, cm	1.5	3	5	12	25
$\theta \times 10^{-3}$, rad	0.1011	0.1021	0.1034	0.1076	0.1141

Table 4.5. GL-206I: l_1 = 25 cm, l_2 = 10 cm, l_{S_2} = 83 cm, l = 101.6 cm, L = 125 cm

R_2, cm	1.5	3	5	12	25
$\theta \times 10^{-3}$, rad	0.1007	0.1014	0.1023	0.1052	0.1097

distances from the blind S_1 and the exit S_2 mirrors to AE, respectively, are l_1 = 5 and 25 cm and l_2 = 10 cm (Fig. 4.14). The diameter of the aperture of the AE D_{chan} = 14 mm. Experimental studies were also carried out at the same radii of mirrors and distances to the AE.

The results of calculating the divergence of the radiation from the CVL with the AE GL-206I are presented in Tables 4.4 and 4.5.

From the analysis of Tables 4.2–4.5 it follows that with UR with two convex mirrors, when the radius of curvature of the blind R_1 (3 cm) and the output R_2 (1.5, 3 and 5 cm) mirrors is 1–2 orders of magnitude smaller than the resonator length L (92, 112, 105 and 125 cm), the divergence of the beam formed for the first double pass of radiation in the resonator becomes equal to the diffraction limit ($\theta_{dif} = 0.1 \cdot 10^{-3}$ rad). Thus, the calculations show that the use of highly selective UR with two convex mirrors in CVL, in comparison with the known types of resonators, makes it possible to form a beam of diffraction quality already for the first double pass of radiation in the resonator and directly from the incoherent aperture beam of superradiance.

Tables 4.6 and 4.7 show the measured values of the total average power (P) and power in the diffraction beam of radiation (P_{difr}) of the CVL with the AE Kulon GL-206E for different values of the

Table 4.6. The radiation power of the total and the diffraction beam of the CVL with the AE Kulon GL-206E at l_1 = 5 cm

R_2, cm	0	3	5	12
ΣP, W	2.6	1.4	2.3	1.8
P_{difr}, W	0.72	0.6	1.0	0.93

Table 4.7. The radiation power of the total and the diffraction beam of the CVL with the AE Kulon GL-206E at l_1 = 25 cm

R_2, cm	0	3	5	12
ΣP, W	2.6	1.54	2.4	2.3
P_{difr}, W	0.66	0.7	1.05	1.08

radius of curvature of the output mirror S_2: R_2 = 0, 3, 5 and 12 cm and the constant radius of curvature of the blind mirror S_1: R_1 = 3 cm and the distance from the output mirror S_2 to the AE l_2 = 10 cm. For R_2 = 0, when the mode is single-mirror, i.e. there is no output mirror S_2, the qualitative beam is not strictly diffractive, since its divergence is 4–5 times that of the diffraction limit. Table 4.6 shows the power values at a distance from a blind mirror S_1 to AE l_1 = 5 cm, in Table. 4.7 – l_1 = 25 cm.

As can be seen from the tables, the value of the average radiation power in the diffraction beam is in the range P_{difr} = 0.6–1.1 W. The power at l_1 = 25 cm is slightly higher than at l_1 = 5 cm, since the conditions for the non-competitive formation in the resonator of the diffraction beam l_1 = 25 cm are closer to the fulfillment of the basic first condition.

Figure 4.15 shows examples of oscillograms of the background incoherent 1 and diffraction 2 radiation beams.

The total duration of the radiation pulse along the base was $\tau \cong 25$ ns. As can be seen from the oscillograms, the beginning of the pulse of the diffraction beam of radiation 2 lags behind the background beam of superradiance 1 by approximately 4–6 ns, which practically corresponds to one double pass of the radiation in the resonator ($\Delta t = 2L/c$) and the logic of the process of formation of the diffraction beam in the direction of the AM → S_2 → AM → S_1 → AM → output (condition 1).

In experimental studies (in the laboratory) of the intensity distribution in the spot of the focused beam of radiation (in the

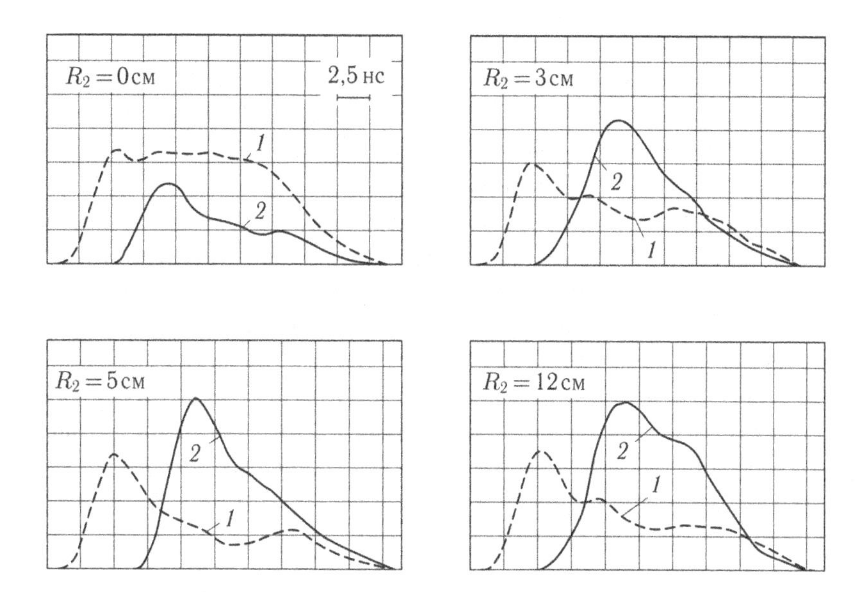

Fig. 4.15. Oscillograms of pulses of background (*1*) and diffraction (*2*) radiation beams of the CVL with AE Kulon GL-206E in the regime with UR with two convex mirrors. Radius of curvature of a blind mirror R_1 = 3 cm, output mirror R_2 = 0; 3; 5 and 12 cm, the diameter of the bounding diaphragm on the blind mirror is d = 0.1 mm. The distance from a blind mirror to an AE is l_1 = 25 cm.

waist), in order to determine the minimum diameter of the waist to calculate the beam divergence from formula (3.1), continuous chaotic oscillations of this spot were observed due to the instability of the position of the axis of the beam pattern. Therefore, it was not possible to accurately measure the diameter of the chaotically moving spot. In addition, with such a spot, it was practically impossible to conduct high-quality microprocessing of materials. It was found that the instability of the spatial position of the axis of the radiation pattern of the diffraction beam of laser radiation of CVL with UR is due to the high sensitivity to external influences of the elements and the field of the unstable resonator (mechanical, thermal and acoustic), since the diameter of the working region of its convex mirrors is about 100 μm. To eliminate the effect of mechanical floor vibrations and acoustic vibrations on the stability of the output radiation parameters, all the necessary elements of the optical system with UR were installed on a honeycomb optical table of type 1NT10-20-20 of Standa company (Fig. 4.1 *b*). Between the table and its four supports of the 1TS065-12-06 there were the pheumatic insulator AP-1000, and between the supports and the floor – rubber insulators.

To eliminate the influence of air–heat fluxes on the radiation parameters, the field of the resonator with the AE must be closed. For this purpose, the heat-loaded AE Kulon was installed in a cylindrical water-cooled heat sink, and the space between the mirrors and the heat sink was sealed with insulating tubes with an internal diameter larger than the diameter of the radiation beam. At the same time, the temperature of the water-cooled heat sink, due to feedback, was maintained at the same temperature as the optical table. In addition, the temperature of the air in the working room was also kept constant (20–21°C).

An additional bounding diaphragm (*D* in Fig. 4.13) was installed in addition to fixing the spatial position of the axis of the radiation pattern of the diffraction radiation beam just before the blind mirror S_1 of the resonator. The size of the aperture opening should be commensurable with the diffraction spot on the mirror

$$D \leqslant \frac{2.44 \cdot l}{D_{\text{chan}}}, \tag{4.15}$$

where l is the distance from the blind mirror S_1 to the output aperture of the AE. For example, for AE GL-206E with l = 69.2 cm (Table 4.3), the diameter of the diaphragm bounding the diffraction spot on the blind mirror S_1, in accordance with (*9*) is no more than 69 μm, with l = 89.2 cm (Table 4.3) – 89 μm, for AE GL-206I with l = 81.6 cm (Table 4.4) and l = 101.6 cm (Table 4.5), respectively, 82 and 102 μm.

These measures have resulted in almost complete elimination of instabilities in the position of the axis of the radiation beam pattern (by more than two orders of magnitude, $\Delta\theta < \theta_{\text{difr}}/100$), as well as pulsed energy, i.e., necessary conditions for qualitative experimental research and efficient microprocessing of materials.

Figure 4.16 presents an example of the distribution of the radiation intensity of the CVL in the regime with UR with two convex mirrors in the constriction of the focused beam.

In this case, the radiation intensity distribution has a very high degree of correlation with the Gaussian beam – 0.915. The diameter of the beam waist is $2W$ = 0.571 mm, which corresponds to the diffraction divergence of the beam – θ_{diff} = 0.1 mrad. The average radiation power in the diffraction beam was equal to P_{diff} = 1.3 W. The instabilities of the position of the axis of the beam pattern of the diffraction beam, as well as the pulsed energy, were practically

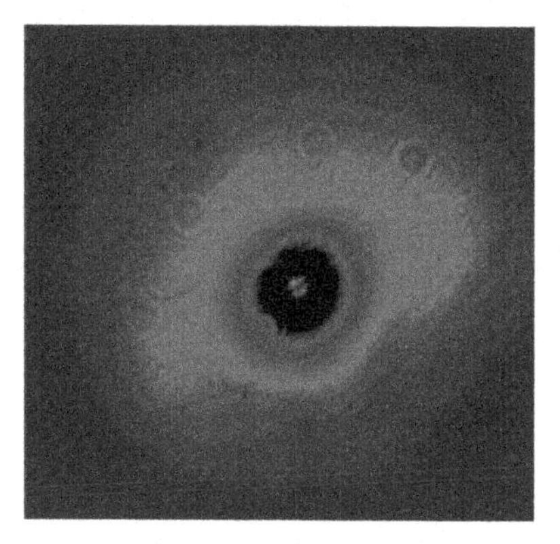

2D-distribution

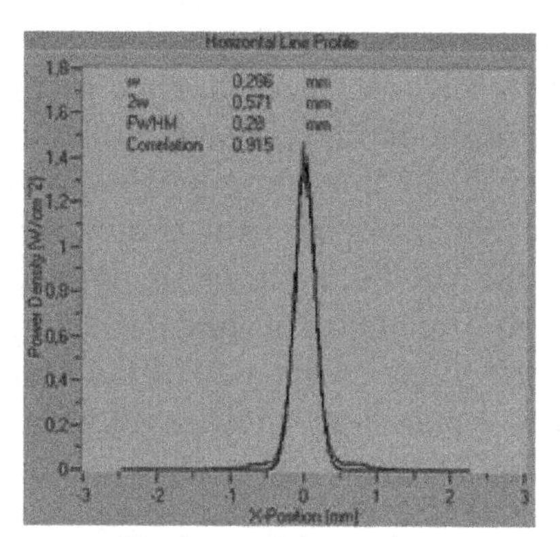

Horizontal intensity distribution

3D-distribution

Fig. 4.16. The distribution of the intensity of the output beam of radiation from the CVL with the AE Kulon GL-206D in the regime with UR with two convex mirrors in the focal plane (waist) of the focusing mirror with a radius of curvature $R = 15$ m, $R_1 = 30$ mm. The output mirror was a positive meniscus with a radius of curvature of the convex working surface $R_2 = 70$ mm ($F = 0.93$ m); $l_1 = 75$ mm, $l_2 = 80$ mm, $L = 920$ mm, diameter of diaphragm of the resonator $D_1 = 0.1$ mm; diameter of the diaphragm of spatial filter–collimator $D_2 = 0.15$ mm; $R_{rad} = 1.3$ W.

not observed, as evidenced by the stable position of the focused radiation spot on the screen.

The analysis of the experimental and theoretical studies discussed above leads to the following main conclusion. In the pulsed CVL in the mode with UR with two convex mirrors, the formation of single-beam radiation with diffraction divergence is ensured with the obligatory performance of three interrelated conditions and the radiuses of the curvature of the mirrors R = 1–10 cm. The use of an optical honeycomb table with vibration damping supports, the placement of the heat-loaded AE in a cooled heat sink, the sealing of the UR field and the maintenance of a constant temperature in the design of the experimental setup made it possible to get rid of the instabilities of the axis of the radiation pattern q and a pulsed energy. These measures significantly improved the quality of microprocessing materials.

Since in CVL when operating in a mode with UR with two convex mirrors, the power in the diffraction beam has small values – 1–1.5 W, which is about 10% of the total radiation power, then its most promising application is the use as master oscillator (MO) in the CVLS of the MO–SFC–PA type. Structurally, it is simpler and convenient for the spatial matching of the diffraction beam of the MO with the aperture of the active medium PA to use the MO in the UR as the exit mirror of the positive convex concave meniscus (Fig. 4.1 *b*) [372]. The average radiation power and the density of the peak power of the diffraction beam in the CVLS with the use of industrial sealed AEs as PAs (see Chapter 1) [16] can be increased by 1–2 orders of magnitude: from the AE Kulon of the model GL206E and GL206I to 15–25 W and up to 10^{11} W/cm^2, respectively, with AE Kristall of the models GL-205A, GL-205B and GL-205C up to 30–100 W and up to 10^{12} W/cm^2, which significantly increases the productivity of microprocessing and the thickness of the processed material to 1–2 mm.

CVLS of the type MO–SFC–PA, in which the pulsed CVL with AE Kulon GL206E and UR with two convex mirrors was used as the MO, was the basis for the creation of the most powerful modern automated laser technological installation (ALTI) Karavella-1M (20–25 W) for the productive and high-quality manufacture of precision electronic parts made of metallic materials up to 0.5 mm thick and non-metallic ones up to 0.7–1 mm (see Chapters 7 and 8).

4.5. Structure and characteristics of the radiation of CVL in the regime with telescopic UR. Conditions for the formation and separation of a radiation beam with diffraction divergence

The experiments used a telescopic UR, considered for the first time

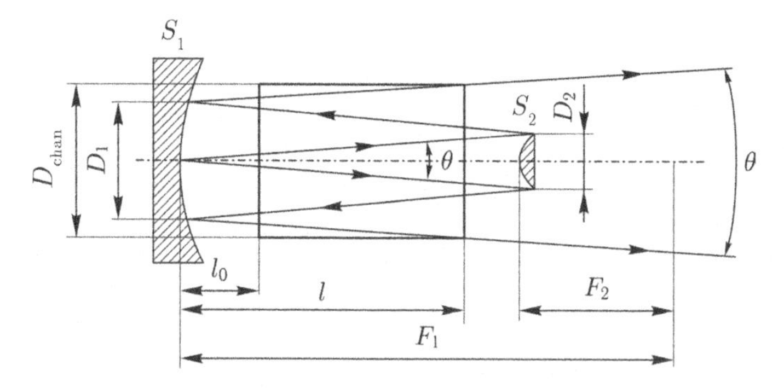

Fig. 4.17. Optical scheme of the AE with a telescopic UR.

in [303], and at the initial stage of the development of pulsed CVLs, were most widely studied in [91–94]. The telescopic resonator is schematically shown in Fig. 4.17. It consists of concave and convex mirrors with coincident foci. In such a resonator, an idealized parallel beam (without allowance for diffraction) after reflection from two mirrors is converted again into a parallel one, but its diameter increases or decreases (depending on the direction of radiation propagation) by a factor of M. The increase in the resonator is determined according to the expression $M = F_1/F_2$, where F_1 and F_2 are the focal lengths of mirrors S_1 and S_2. The diameters of the mirrors satisfy the relation $D_1/D_2 = F_1/F_2 = M$ ($D_1 = D_{chan}$ is the diameter of the discharge channel), the output of the radiation beam is from the side of the mirror S_2. In the intensity distribution at the output of the resonator there is an annular gap due to the geometric shadow from the mirror S_2. It was shown in [91] that the number of double passes of radiation in the resonator, necessary for the formation of a diffraction beam, is related to the parameters of the resonator by the following relation

$$m = 1 + \frac{\ln M_0}{\ln M}, \tag{4.16}$$

where $M_0 = D_1^2 / 2\lambda F_1$ is the Fresnel number.

The first large-scale experimental studies of the telescopic UR in CVLs were carried out with an industrial sealed-off AE Kristall of the GL-201 model with a direct design of a pumping modulator of the power source and a PRF of 8 kHz, when the radiation pulse duration at the base was approximately 40 ns. The increase in M of the telescopic UR varied within $5 \leq M \leq 300$. In the $60 \leq M \leq 300$ range, the output radiation had a four-beam structure: two incoherent

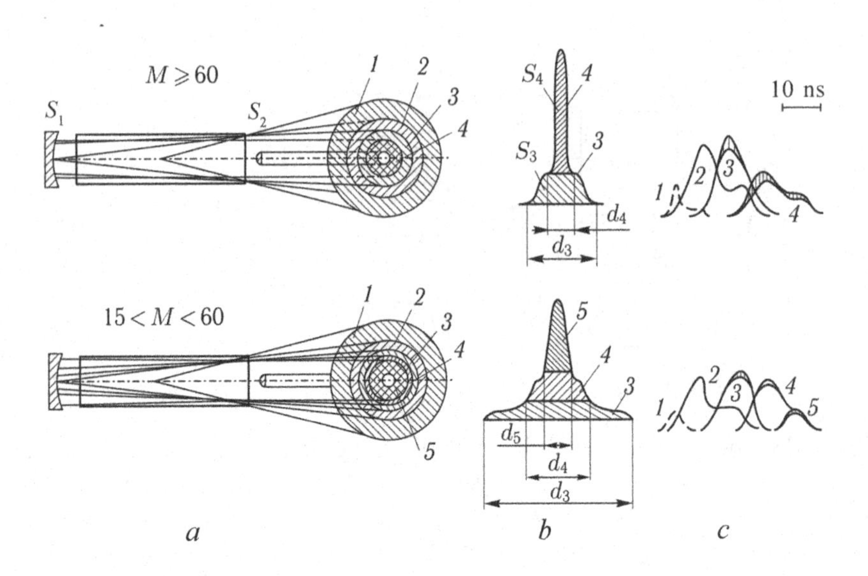

Fig. 4.18. The distribution of the intensity of the output radiation of the CVL in the regime with the telescopic UR in the near (*a*) and far (*b*) zones and the oscillogram of the radiation beams (in): *1*, *2* – superradiance beams; *3*, *4*, *5* – resonator beams.

superradiance beams and two resonator beams, at $15 \leq M < 60$ the structure was five-beam: a third resonator beam appeared (Fig. 4.18 *a*). Resonator beams were clearly observed when the radiation was focused by a mirror with a radius $R = 15$ m. The distribution of the intensity of the output radiation in the far zone (the focusing plane) has a clearly pronounced step-like character (Fig. 4.18 *b*). For such a distribution it is not difficult to estimate the diameters of focused resonator beams and calculate their divergence, respectively, and from the oscillograms (Fig. 4.18 *c*) determine the origin and duration of the pulses of radiation of these beams. In addition, according to the area under the curves of the intensity distribution of the beams (*b*) and the area under the oscillograms of their pulses, the percentage of power in each beam and its absolute value can be calculated. At $M \geq 60$, the second resonator beam (*4*) has a diffraction divergence, which is defined by the expression $\theta_{dif} = 2.44\lambda/D$, where D is the diameter of the beam, and λ is the wavelength of the radiation. For industrial AE GL-201 with a diameter of the discharge channel $D_{chan} = 20$ mm, we obtain θ_{dif} ($\lambda = 0.51$ µm) $= 0.062$ mrad and θ_{dif} ($\lambda = 0.58$ µm) $= 0.07$ mrad. At $5 \leq M \leq 15$, beams with a diffraction divergence do not have time to form, and at $15 < M < 60$ divergence of the third resonator beam (*5*) reaches a diffraction limit.

Peak power densities in the radiation focusing spot when working with the telescopic UR reach 10^{12}–10^{13} W/cm^2, which is 1–2 orders of magnitude greater than when working with a single convex mirror. However, oscillations of the spots are observed in the focusing plane, especially strongly from the beam with diffraction divergence, and on the oscillogram the spikes of the radiation pulses of these beams are blurred, that is, the positions of the axis of the radiation pattern and the impulse energy are unstable. With increases in the resonator of the order of 10^2, the displacements of the spot of the diffraction beam can be commensurate with its divergence, and the instability of the pulsed energy reaches values of 20–30%. These instabilities are due to the high sensitivity of UR to external mechanical and acoustic influences, to air and heat fluxes, to dust, and also, possibly, instability and heterogeneity of discharge burning, etc.

To determine the divergence of the radiation beams, a system of three simple equations is compiled on the basis of the geometric path of the rays in telescopic UR (Fig. 4.18) and taking into account its properties:

$$\begin{cases} D_2 / L = 0 \\ (D_{\text{chan}} - D_1) / l = \theta, \\ D_1 / D_2 = M. \end{cases}$$

The solution of the system of equations gives the following expression

$$\theta = \frac{D_{\text{chan}}}{(ML + l)M^{n-1}}.$$

The expression for the divergence of beams formed through n double passes of radiation in the resonator will have the form

$$\theta = \frac{D_{\text{chan}}}{ML + l}.$$

but with the allowance for the diffraction limit

$$\theta = \frac{D_{\text{chan}}}{(ML + l)M^{n-1}} + \frac{2.44\lambda}{D_{\text{chan}}}. \tag{4.17}$$

At large magnitudes of the resonator ($M \geq 50$), the formula for calculating the divergence is simplified:

$$\theta = \frac{D_{\text{chan}}}{M^n L} + \frac{2.44\lambda}{D_{\text{chan}}}. \tag{4.18}$$

The results of the calculations show that for the formation of the diffraction beam at $M = 5$, five double passes of radiation in the resonator ($n = 5$) are necessary, at $M = 10$ and 30 four and three ($n = 4$ and 3) respectively, and starting with $M \geq 60$ – two double passes ($n = 2$). These values of n coincide with the number of passes m calculated by formula (4.16).

To date, a large volume of CVL studies with telescopic URs has been carried out using the most powerful industrial sealed AE Kulon model GL-206I (see Chapter 2) in the experimental setup shown in Fig. 4.1 *b* ($S'_1(b)$–$S'_2(b)$). The telescopic resonator was used with an increase in $M = 220$. The blind concave spherical mirror S_1 had a radius of curvature of the surface $R_1 = 2200$ mm, the output convex spherical mirror $R_2 = 10$ mm, so that the resonator length (L) was practically 1000 mm. The dynamics of the formation and structure of the output radiation of the CVL with the given UR is described above. At $M = 220$, the structure of the output radiation of the CVL is four-beam (Fig. 4.18) – two beams of superradiance (*1* and *2*) and two resonator beams (3 and 4). The resonator beam *3* is formed as the first double pass of the radiation in the resonator ($2L/c = 7$ ns), the beam *4* – for the second double pass and diffraction divergence. Since the total duration of laser pulses is ~25 ns, radiation in the resonator has time to make a third double pass. The beam formed in the third pass is also diffracted and enhances the beam power *4*. The first resonator beam *3* lags the $2L/c$ beam of superradiance *2*, while the resonator beams are also lagging behind each other.

The divergence of the resonator beams with telescopic UR was calculated by formula (2.20), which is in good agreement with the experimental data. The calculated value of the divergence for the first resonator beam ($n = 1$) is $\theta = 0.16$ mrad, which is 1.6 times the diffraction limit – $\theta_{\text{difr}} = 0.1$ mrad. For the second and third beams ($n = 2$ and 3), the divergence is of the diffraction type. Therefore, for high stability of the position of the axes of the beam pattern of these beams, it is impossible to separate them from each other either spatially or in time, and as a single diffraction beam (beam *4* in Fig. 4.18) appears to form. Figure 4.18 shows that the intensity distribution in the far zone (*b*) has a step-like character due to the presence of a beam *3* with a divergence larger than the diffraction limit. The radiation with this distribution was practically unsuitable for high-quality microprocessing of materials.

In our experiments, due to the high stability of the position of the axes of the radiation pattern of the radiation beams of the CVL, achieved by eliminating mechanical and thermal effects on the operation of the optical system, a spatial filter–collimator (SFC) was used to extract the diffraction component from the total output radiation. The SFC consists of two concave mirrors with a radius of curvature $R = 1.25$ m (items 4 and 5 in Fig. 4.19 *b*) and a diaphragm D_2 located in the focal plane of these mirrors. To study the intensity distribution of the radiation beam emitted by the diaphragm D_2, in the far zone the beam was focused by a mirror with a radius $R = 15$ m (position 7 in Fig. 4.19 *b*). The investigations were carried out for different diameters of the diaphragm of the SFC: $D_2 = 1$; 0.25; 0.1 and 0.06 mm. Figure 4.19 shows the intensity distribution for diaphragms with $D_2 = 0.25$ and 0.06 mm. At $D_2 = 0.25$ mm (Fig. 4.19 *a*) at the base of the intensity distribution there are wings corresponding to the divergence span of the first resonator beam (beam *3* with $\theta = 0.16$ mrad), with $D_2 = 0.06$ mm (Fig. 4.19 *b*) the wings disappear and the radiation corresponds to a given distribution with a single-beam structure.

The divergence of the single-beam radiation (θ) was calculated by the focal spot method, that is, by the formula (2.3). In this case, the diameter of the beam cross section (d_0) in the focal plane (in the waist) in our case was determined as the arithmetic mean of two dimensions: the size of the waist from the horizontal intensity distribution is $2w_x$ and the vertical dimension is $2w_y$. $d_0 = (2w_x + 2w_y)/2$. The fact that d_0 is somewhat different horizontally and vertically is due to the displacement at a small angle of the optical axes of the mirrors of the collimator (*4* and *5*) and the focusing mirror (*7*) relative to the incident beam of radiation being incident on them. Figure 4.19 *b* shows the horizontal dimension $2w_x = 0.593$ mm, vertical $2w_y = 0.538$ mm, then $d_0 = (0.593$ mm $+ 0.538$ mm$)/2 = 0.556$ mm, which corresponds to beam divergence $\theta = 0.556$ mm/7500 mm $= 0.074$ mrad, i.e., to the diffraction limit ($\theta_{dif} = 0.1$ mrad).

With a decrease in the diameter of the diaphragm D_2 in the SFC from 1 to 0.06 mm, the average power of the radiation beam emitted by the diaphragm from the total radiation decreased from 9.3 to 5 W (Fig. 4.20).

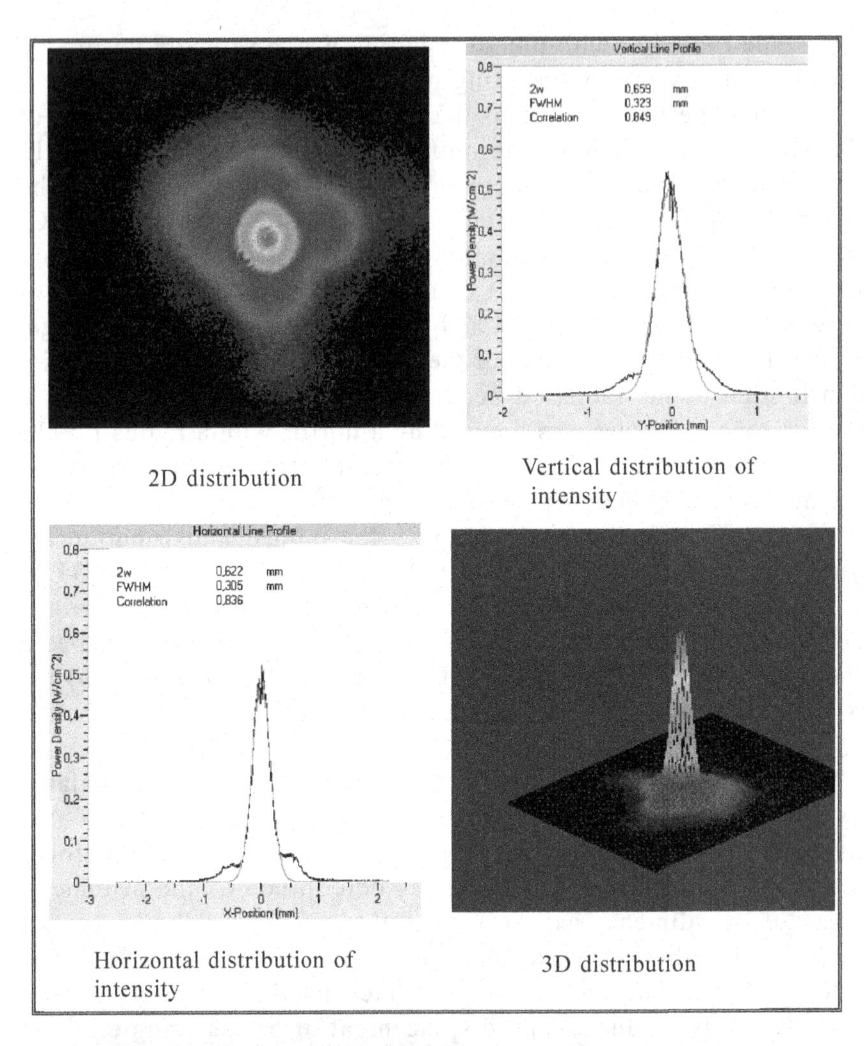

2D distribution

Vertical distribution of intensity

Horizontal distribution of intensity

3D distribution

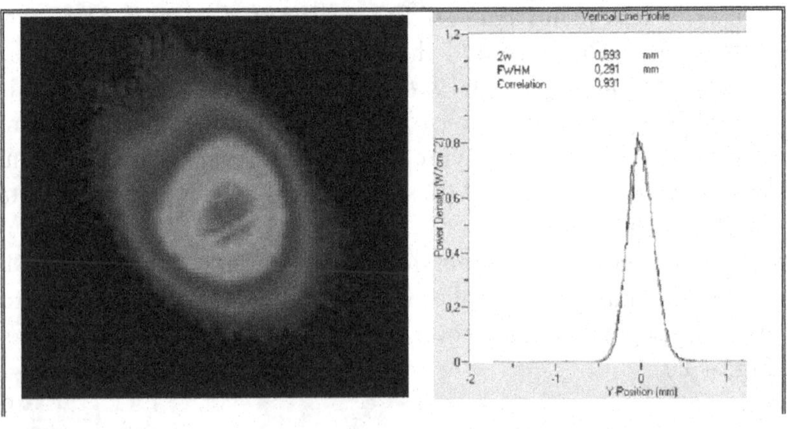

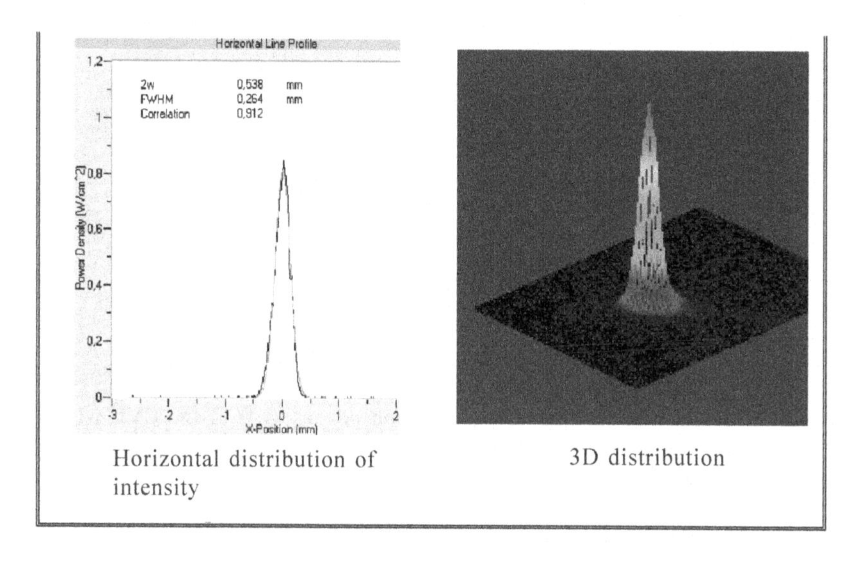

Horizontal distribution of intensity

3D distribution

Fig. 4.19. The intensity distribution of the output beam of laser radiation from the AE Kulon GL-206I in the regime with a telescopic UR with an increase of $M = 220$ ($R_1 = 2200$ mm, $R_2 = 10$ mm) in the focal plane (waist) of the mirror with a radius of curvature $R = 15$ m.

When the diameters of the diaphragm of the SFC are $D_2 = 1.0$ and 0.25 mm, the resonator beams 3 and 4 pass through these holes completely and have a total average radiation power of 9–10 W. At $D_2 = 0.06$ mm, a strictly diffractive component is distinguished, since this diameter corresponds to the minimum dimension of the constriction of the diffraction beam focused by the input mirror (*4*) of the SFC ($d_0 = R_4/2 \times \theta_{\text{dif}} = 1250 \text{ mm}/2 \cdot 0.1 \cdot 10^{-3} = 0.062$ mm). In this case, as can be seen from the intensity distribution in Fig. 4.19 *b*, the correlation with the Gaussian beam is high (0.93), which practically indicates a coincidence of the angular and energy divergences.

The average power of the diffraction beam is 5–6 W (which is about 50% of the total radiation power), as practical experience shows, it is sufficient for productive and high-quality microprocessing of metallic materials up to 0.3 mm thick, dielectrics and semiconductors up to 0.5–0,7 mm. CVL in the mode of a single generator with sealed-off AE GL-206I and the telescopic UR with the use of SFC became the basis for the development of modern compact industrial ALTI Karavella-2 and Karavella-2M with a laser radiation power of 5–8 W for manufacturing precision metal and non-metallic

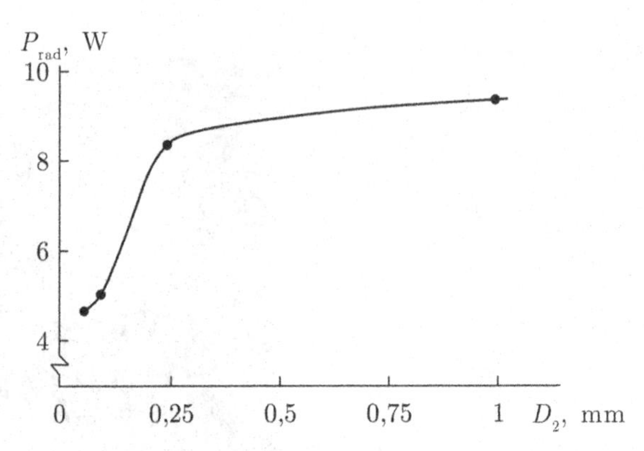

Fig. 4.20. Dependence of the average radiation power (P_{rad}) of the CVL with the AE Kulon GL-206I in the telescopic mode with an increase in M = 220 on the diameter of the diaphragm (D_2) of the SFC.

parts of electronic components, in particular, microwave devices (see Chapters 6 and 8).

The use of CVL in the mode of operation with the telescopic UR as a MO in the CVLS of the type MO–SFC–PA was the basis for the creation of a modern industrial ALTI Karavella-1 with a radiation power of 10–15 W for precision microprocessing of metallic materials up to 0.5 mm thick and nonmetallic up to 1–1.2 mm of the IET (see Chapters 7 and 8).

4.6. Investigation of the conditions for the formation of a powerful single-beam radiation with a diffraction divergence in a CVLS of the MO–PA type

Investigation of the conditions for the formation of a powerful single-beam radiation with a diffraction divergence in the CVLS of the MO–PA type. The greatest practical value in a pulsed CVL is represented by radiation beams with a diffraction and a divergence close to it, formed in the operating mode with an unstable resonator (UR) or with a single convex mirror. But the radiation power concentrated in qualitative beams and, first of all, with the diffraction divergence makes up a smaller part of the total output power – 10–40% (see section 4.2), which is a significant drawback of the CVL operation in the regime of a separate generator. The most effective way to increase the power and efficiency of a pulsed CVL, as has been shown in many works, including ours, is the use of the CVLS operating according to the effective MO–PA scheme [15–23, 32–34,

41, 42, 99, 101, 103, 227–233, 253–264, 304–306, 323, 324, 365, 366, 368, 370, 372, 375, 378, 386, 389, 400]. The main feature of such pulsed laser systems is that the saturation regime in the PA occurs at relatively low power levels of the input signals from the MO. The first CVLS of the type MO–PA was created at the Lawrence Livermore National Laboratory (LLNL) in the USA in 1976 in the framework of the AVLIS program for the separation of uranium isotopes [15]. By 1979, such a system of 21 CVL modules with a total output power of radiation 260 W was constructed. In 1991, a new generation of PA allowed to obtain power of 1.5 kW in a chain of three PA and one MO. Separate amplifiers in such a chain could generate radiation with a power of more than 750 W at an efficiency of ~1%. Research and development of this class of high-power laser systems is being carried out in several other countries: Japan, France, England, China, Israel, Russia, India, and others (see Chapter 1).

This section of the chapter presents the results of studies of the spatial, temporal, and energy characteristics of CVLS radiation, which also operates in accordance with the MO–PA scheme, using the new sealed-off AE of the Kulon series in new models GL-206E and 206I and old Kristall GL-201 and GL-201E and GL-206E models, and new models GL-205A, GL-205B and GL-205C (see Tables 4.3–4.5) [16, 32–34, 41, 42, 99, 103, 227–233, 253–264, 304–306, 323, 324, 365, 366, 386]. The purpose of the research is to determine the conditions necessary for the formation in CVLS of powerful single-beam radiation with a diffraction divergence with stable parameters sufficient for efficient microprocessing of materials up to 1–2 mm thick.

Investigation of CVLS type MO–SFC–PA in the working mode of MO with telescopic UR and industrial sealed AE Kristall GL-201 (GL-205 A). Experimental installation, methods and means of measurement. The scheme of the experimental CVLS is shown in Fig. 4.21 [16, 231]. The CVLS studied by us differs from the previously used CVLS of the MO–PA type in that a spatial filter collimator (SFC) is installed between the MO and the PA. As an AE in MO and PA, the first experiments used industrial sealed-off self-heating AE Kristall of the old model GL-201, and then a new GL-205A (see Tables 4.4 and 4.5) with the same power level and the discharge channel diameter D_{chan} = 20 mm and the volume of the AM V_{AM} = 250 cm^3.

In the telescopic UR of the MO with the AE Kristall GL-201 the blind mirror had a radius of curvature R = 3 m, diameter –

D_3 = 35 mm (position 3). With an increase in the telescopic UR M = 200, a convex mirror with R = 15 mm (D_3 = 1.5 mm) was used as the output mirror (item 4), and at M = 30 and 100 – glass menisci without coating with D_3 = 35 mm. A convex mirror with D_3 = 1.5 mm is glued to an antireflecting thin-walled glass substrate with a diameter of 35 mm. The reflection coefficient of mirrors having a multilayer dielectric coating is 99%. A blind concave mirror with a radius of curvature R = 1.8 m was used in the MO with AE Kulon GL-206D, the output convex mirror with R = 10 mm (M = 180). Heating and excitation of the AE of the emitter were provided by a two-channel synchronized power source containing two identical high-voltage rectifiers (item 5) and two high-voltage nanosecond pump pulse modulators (position 6) based on hydrogen thyratrons of the type TG1-2000/35 and TGI1-2500/50 for the radiator with the AE Kristall, on the basis of transistor switches for the radiator with the AE Kulon GL-206E and GL-206I. The heating voltage of a hydrogen generator and a cathode of thyratrons is stabilized. To improve the excitation efficiency of high-voltage exciters, high-voltage modulators were designed according to the scheme of transformer or capacitive voltage doubling with magnetic links of current pulse compression [16, 225]. The modulators were started from a common master pulse

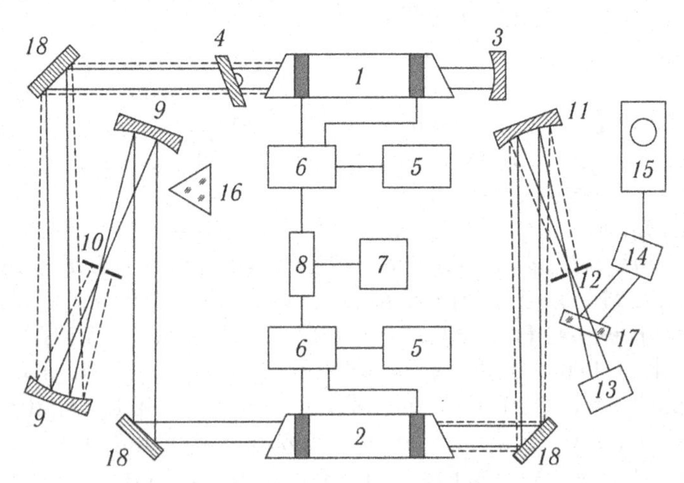

Fig. 4.21. Experimental CVLS of the type MO–SFC–PA with telescopic UR in MO and industrial sealed-off Kristall: *1* – AE MO; *2* – AE PA; *3* and *4* – blind concave and exit convex mirrors of the UR, *5* – high-voltage rectifiers of the power source; *6* – high-voltage modulators of nanosecond pump pulses; *7* – master pulse generator; *8* – nanosecond delay line of the MO and PA channels, *9* – SFC mirrors; *10* and *12* – selection diaphragms, *11* – focusing mirror; *13* – millivoltmeter with a converter of laser radiation power TI-3; *14* – photocell FEK-14 K; *15* – oscilloscope C1-75 or GDS-840S; *16* – glass prism; *17* – beam splitter plate; *18* – flat rotary mirrors.

generator (item 7), equipped with an adjustable nanosecond delay line at the output (key 8). The latter made it possible to shift the pulses of the discharge current of the AE of the MO (item 1) and the PA (positioin 2) with the thyratron power source within a range of ±50 ns in time relative to each other. With the transistor power source, the dissynchronization of the pulses of the MO current with respect to the pulses of the PA current was made within ±1000 ns.

In order to isolate the diffraction beam of MO radiation from its background component (from incoherent superradiance beams) and resonator beams with a divergence greater than the diffraction and spatial matching of the extracted beam with the aperture of the discharge channel of the PA, a SFC was placed between the MO and the PA. SFC consisted of a mirror collimator formed by two mirrors (item 9) with a radius of curvature R = 1.6 m (D_3 = 35 mm) for a radiator with an AE in the MO of the GL-201 model and with R = 1.2 m for a radiator with AE in MO GL-206 and a diaphragm (position 10) located in the focal plane of the entrance mirror of the collimator. A spatial filter formed by a focusing mirror (position 11) and a diaphragm (item 12), separated a diffraction beam at the output of the PA.

The average radiation power in the CVLS was measured with a M-136 millivoltmeter with a TI-3 laser radiation power converter connected to it (item 13). To register the radiation pulses, a photoelectric cell FEK-14K (item 14) was used, on which the radiation was removed by a beam splitter plate (position 17), and oscilloscopes of the type C1-75 or C7-10A and digital GDS-840S (item 15) were also used. The radiation intensity distribution along the beam cross-section at the input and output of the PA was investigated with the help of the FD-24K photodiode, the receiving surface of which was limited to a diaphragm 0.3 mm in diameter. The radiation was attenuated in order to ensure a linear operating mode of the photodetectors. To determine the dependence of the average radiation power on the output of the PA from the power at the input, the input power was varied with the help of a set of neutral calibrated light filters. The divergence of the output radiation beam was estimated from the diameter of the spot in the focal plane of the mirror (item 11) with R = 15 m and D_3 = 50 mm. Investigations of the CVLS emission characteristics were carried out at the steady-state optimum operating temperature of the AE. The power consumption from the power source rectifier (item 5) for the AE GL-201 with a working PRF of 8 kHz was 3.5 kW, for the AL GL-206E and GL-206I at 14 kHz, respectively, 2 and 2.2 kW.

The results of research of CVLS and their analysis. In the CVLS with a two-channel thyratron power source, the duration of the radiation pulses of MO and PA, in which the GL-201 or GL-205A were used, was the same and amounted to about 30 ns. The emission output of the MO with the telescopic UR had four beam structures: two incoherent superradiance beams with a geometric divergence of 50 and 18 mrad (background beams) and two resonator beams with small divergence [227, 228, 230, 231]. When the telescopic resonator $M \geq 60$ increases, the second resonator beam has a diffraction divergence ($\theta_{dif} = 2.44 \cdot \lambda/D_{chan} = 0.07$ mrad). As M increases, the power in the background and second resonator (diffraction) beams increases, while in the first resonator beam it decreases. In this case, the divergence of the first resonator beam tends to diffraction (not reaching the diffraction limit), which makes its spatial separation difficult. Figure 4.22 *a* shows the oscillograms of the pulses of the output radiation of the MO with an increase in $M = 200$: the total beam (*1*), the background beam with $\theta = 18$ mrad (2) and the total resonator beam (3) consisting of beams with $\theta = 0.14$ mrad and $\theta_{dif} = 0.07$ mrad. The oscillogram of the background beam pulse with $\theta_{geom} = 50$ mrad was not recorded because of its low intensity, since its power is approximately three to four orders of magnitude less than the total radiation power of the MO. The radiation pulses of the beams partially overlap in time, and the origin of the pulses of the resonator beam with $\theta = 0.14$ mrad lags behind the background with $\theta_{geom} = 18$ mrad by about 10 ns, diffraction by 20 ns. The interval of 10 ns corresponds to the time of one double passage, and 20 ns corresponds to two double passes of radiation in the resonator. At $M = 200$, the total output power of the generator was 18 W, of which 9 W per the background beam *2*, 9.4 W for the resonant beam, $\theta = 0.14$ mrad, 5.4 W, and the diffraction beam – 3.6 W [231].

An analysis of the temporal, spatial, and energy characteristics of the output radiation of the MO allows us to conclude that when all of its components are fed to the input of the PA, all its components will be amplified, competing in power with each other. Therefore, to obtain a powerful single-beam radiation with a diffraction divergence at the output of the PA, it is necessary to get rid of the influence on its operation from other radiation beams. For this, as follows from the analysis of curve 1 in Fig. 4.23, the power density of

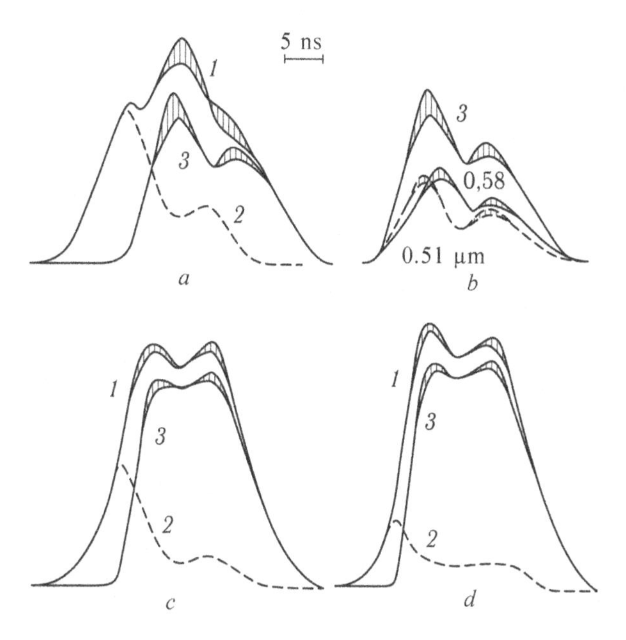

Fig. 4.22. The oscillograms of the radiation pulses of the total (*1*), background with θ_{geom} = 18 mrad (*2*) and the qualitative (*3*) beams with the increase in the telescopic UR M = 200 at the output of the MO (*a*) and at the output of the PA without a diaphragm in the SFC (*c*) and with the diaphragm with a hole diameter of 0.5 mm (*d*). The resonator beam (*3*) is composed of two beams: with θ_{geom} = 0.14 mrad and θ_{dif} = 0.07 mrad. Oscillograms of radiation pulses with λ = 0.51 µm and λ = 0.58 µm of the resonator beam (*3*) of the MO (*b*).

background beams at the input of the PA must be substantially reduced (to a value less than 0.15 mW/cm²). The results on which curve *1* is plotted are in good agreement with the results of [305]. The effective gain of the active medium of the PA reached values 10^3–10^4 (curve *2*), when the input power of the radiation was unity and the fraction of milliwatts.

The power density of the background could be reduced by increasing the length of the optical path from MO to PA; for our case, it should have been about 70 m. In the experimental laser system MO–SFC–PA (see Fig. 4.21), the optical path length is 7 m. The maximum output radiation power was achieved by means of an adjustable delay line (item 8). The total radiation power at the CVLS output with increasing the resonator M = 200 in the absence of the diaphragm (position 12) in the SFC was 38 W (about 60% of the power was at λ = 0.58 µm), while in the background beam – 9.5 W, in the first resonator with θ = 0.14 mrad – 15.5 W and in

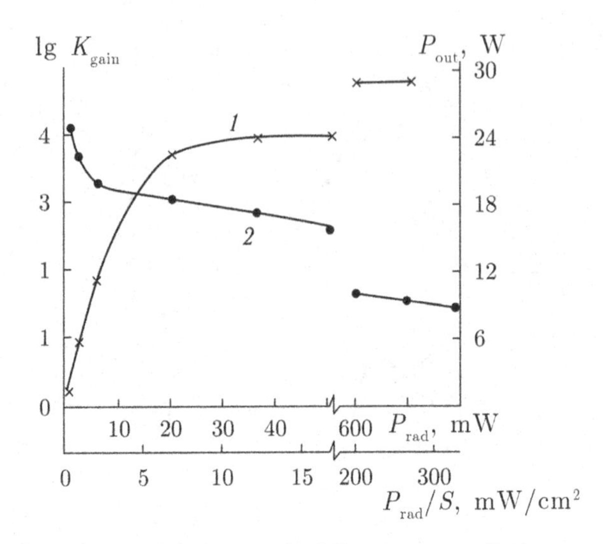

Fig. 4.23. The dependence of the removal of the average radiation power in the CVLS with the PA (*1*) and its gain factor (*2*) on the input power (power density).

the diffraction beam (θ_{dif} = 0.07 mrad) – 13 W. The power of the background radiation of the MO at the input of the PA was about 1 W. The oscillograms of the radiation pulses of the laser system are shown in Fig. 4.22 *c* and *d*. The blurring of the vertex of the pulses of radiation from beams with a small divergence at the output of the MO (see Fig. 4.22 *a*) indicates instability of the pulse energy in the range of 10–15%. This instability is due to the high sensitivity of UR to mechanical influences and air-heat fluxes. At the output of the PA (see Fig. 4.22 *c*, *d*), the instability of the pulse energy is noticeably smaller than at the output of the MO (Fig. 4.22 *a*, *b*). This is explained, firstly, by the relatively small contribution of the generator radiation to the total power of the CVLS and, secondly, by the smoothing action of the active medium of the PA in its saturation operation. The smaller the increase in the resonator of the MO, the greater the divergence and power of the first resonator beam. The corresponding changes also occur at the output of the PA. Thus, for example, with M = 30 using a diaphragm in the SFC with a hole diameter of 0.7 mm, the majority of the output power of the CVLS was concentrated in a beam with θ = 0.5 mrad and was about 35 W at a total power of 38.5 W.

Figure 4.24 shows the dependence of the average radiation power in the total (*1*), background (*2*), and diffraction beams (*3*) on the temporal detuning Δt of the CVLS channels with increasing telescopic resonator of the MO $M = 200$. The positive values of Δt correspond to the lead, and the negative values correspond to the lag of the light signal (radiation pulse) of the MO in relation to the light signal of the PA. When the MO signal was advancing by 3–4 ns, the power in the diffraction beam (curve *3*) reached a maximum value (31.5 W). With the lag, the absolute value and the fraction of power in the qualitative beam decreased relatively rapidly. When the signal of the MO is behind by 6–8 ns, the power in the background beam (curve *2*) reached the highest value (20 W). Such a redistribution of power between the beams in the case of detuning the pulses of MO and PA radiation is explained by the fact that in the input signal the pulses of the beam radiation overlap in time only partially, which makes it possible to introduce one or another component of the MF signal into the effective amplification zone and thereby optimize the laser system. The introduction of a diaphragm (item 12 in Fig. 4.21) with a hole diameter of 0.3 mm into the focal plane of the collimator led to a decrease in the fraction of power in the background beam at zero detuning from 25% to 15% (see Fig. 4.21 *c*, *d*, curves *2*) and a slight decrease in the power of the system (from 38 to 37 W). At the same time, the power in the qualitative beam increased from 28.5 to 31 watts. The detuning of the input signal from the MO (advance with respect to the pulse of the PA) made it possible to increase this power to 34–35 W, which is approximately four times the power of the low-diverging beams of a separate MO. The practical efficiency for the entire laser system of the MO–SFC–PA type was approximately 0.5%, the efficiency of the PA based on the

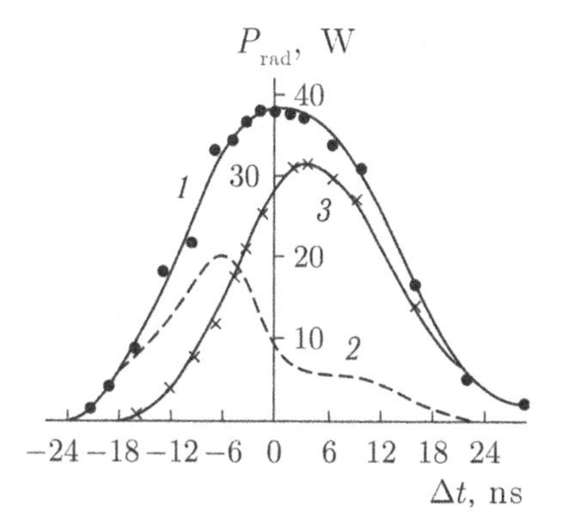

Fig. 4.24. The dependence of the average radiation power in the total (*1*), background (*2*), and diffraction (*3*) beams of the CVLS on the temporal detuning of the light signal of the MO with respect to the PA signal (in the absence of a diaphragm in the SFC) with an increase in the telescopic UR $M = 200$.

power input to it was 1.6%. The practical efficiency of the CVLS is defined as the ratio of the output power of the radiation to the power consumed by the rectifier power source. The total losses of power consumed from the rectifier power source, in the elements of the charging and discharge electrical circuits of the high-voltage modulator of the pump pulses were 40–50%.

A further decrease in the level of background radiation at the output of the CVLS can be achieved by reducing the diameter of the aperture opening in the SFC. Figure 4.25 shows the experimental dependence of the average power of the background component at the output of the SFC on the diameter of the aperture opening. From this dependence it follows that the value of the average background radiation power below 1 mW (which corresponds to a power density of less than 0.3 mW/cm^2) can be obtained with the aperture diameter of the diaphragm of less than 0.2 mm. However, even at diameters of 0.2–0.3 mm, because of the oscillations in the position of the axis of the directional pattern of the low-divergent beams and, primarily, the diffraction component of the MO at the output of the SFC, there are oscillations of energy in the pulse reaching 30–50%, and at the output of the PA – up to 10–15%. By shielding the MO beam with a pipe, it was practically possible to get rid of undesirable refractive phenomena caused by air-heat fluxes and to achieve a relatively stable operation of CVLS with a diaphragm diameter of up to 0.3 mm. The further increase in the stability of the output beam characteristics of the laser system was, in this case, mainly due to the need to increase its resistance to external mechanical influences. In the experimental setups in Section 4.2, for example, to eliminate the instabilities of the position of the axis of the beam pattern of diffraction beams caused by mechanical fluctuations in the floor of industrial premises, a massive optical table with vibration damping supports is used as the supporting structure.

The CVLS was used to study the distribution of radiation intensity (total and at separate wavelengths) in the cross section of the beam at both the input and output of the PA. To separate the individual lines, a prism was used (position 16 in Fig. 4.21). With an increase in the resonator $M = 200$, a deep dip, corresponding to the geometric shadow from the opaque exit mirror of the telescopic UR, occurred directly at the output of the MO in the intensity distribution of the radiation beam. At the input of the PA (the optical path of the light signal from MO to PA is 7 m), at the centre of this shadow, due to diffraction at the output mirror of the UR, a Poisson spot was

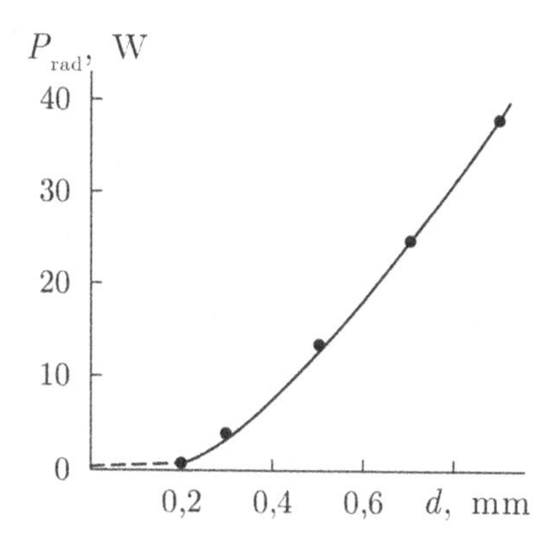

Fig. 4.25. Dependence of the average power of the background radiation beam at the output of the SFC on the diameter of the aperture of its diaphragm with increasing telescopic UR M = 200.

observed. After passing this radiation beam through the PA, the depth of the intensity dip at the centre of the output beam decreased substantially and was no more than 25–30%. The level of intensity of the Poisson spot was sufficient to introduce into saturation the active medium RM in the near-axis region of the beam. When glass menisci were used as the output mirror of the resonator (without coating), the dip in the distribution of the near-axis region of the MO was absent. At the same time, at both the input and the output of the PA, the intensity distribution was close to U-shaped (Figu. 4.26), which indicates a high uniformity in the distribution of the gain over the cross section of the active medium, as well as the high efficiency of its use.

Investigations of CVLS of the type MO–SFC–PA in the mode of operation of MO with a single convex mirror and industrial sealed-off AE Kristall GL-201 (GL-205A).

Experimental installation, methods and means of measurement. A schematic diagram of the experimental CVLS of the MO–SFC–PA type with a single convex mirror in the MO [16, 232,233] is shown in Fig. 4.27 *a*. The MO and PA (items 1 and 2) used the industrial sealed-off AE Kristall GL-201 (GL-205A – new model) (see Tables 2.4 and 2.5). The MO worked in the regime with a single convex mirror (position 7) having a radius of curvature R from 0.6 to 10 cm. Between the MO and the PA there is an SFC formed by concave spherical mirrors (positions 9 and 10) and a diaphragm (position 11). The mirror (item 9) had a radius of curvature R = 1.6 m and diameter

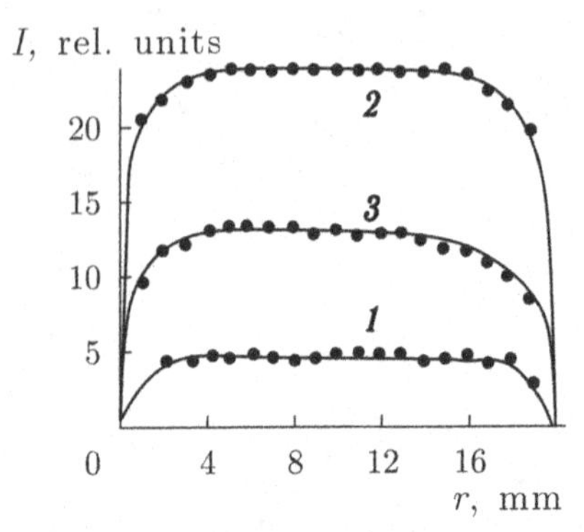

Fig. 4.26. The distribution of radiation intensity in the cross section of the beam at the input (*1*) and the output (*2*) of the PA at two wavelengths ($\lambda = 0.51$ and 0.58 μm) and at a wavelength $\lambda = 0.58$ μm (*3*), *r* is the distance along the beam diameter.

$D_3 = 72$ mm, the mirror (item 10) – respectively $R = 0.75$ m and $D_3 = 35$ mm. The radiation beam from the MO to the SFC and from the SFC to the PA was directed with the help of flat rotary mirrors (position 8) having a multilayer dielectric coating with a reflection coefficient of ~ 99%.

Heating and excitation of AE in the experimental CVLS, as in the previous case (section 4.1), provided a two-channel synchronized pulse power source containing two high-voltage rectifiers of the power source (item 3) and two high-voltage nanosecond pump pulses (position 4) based on water-cooled hydrogen thyratrons TG1-2000/35. The MO modulator was made in a direct electric circuit, the PA – according to the scheme of transformer voltage doubling with a magnetic link of the current pulse compression. The pump modulators were started from a common master pulse generator (item 5) equipped with an adjustable nanosecond delay line (item 6), which allowed the light pulses of the MO and PA to move relative to each other within ±50 ns. The working PRF was 8 kHz. The studies were carried out in the steady-state optimum temperature regime of the AE, which was provided for MO at a power supply from the 2.5 kW rectifier and the voltage at the anode of the thyratron of the modulator of 17 kV, for PA – 3.5 kW and 21 kV, respectively.

The measuring equipment (Fig. 4.27 *b*) made it possible to investigate the temporal, spatial, and energy characteristics of the radiation at the output of both the MO and the PA. The average radiation power was measured with a M136 millivoltmeter with a TI-3 laser radiation power converter connected to it (item 15). To register the radiation pulses, a FEK-14K coaxial photocell (item 16)

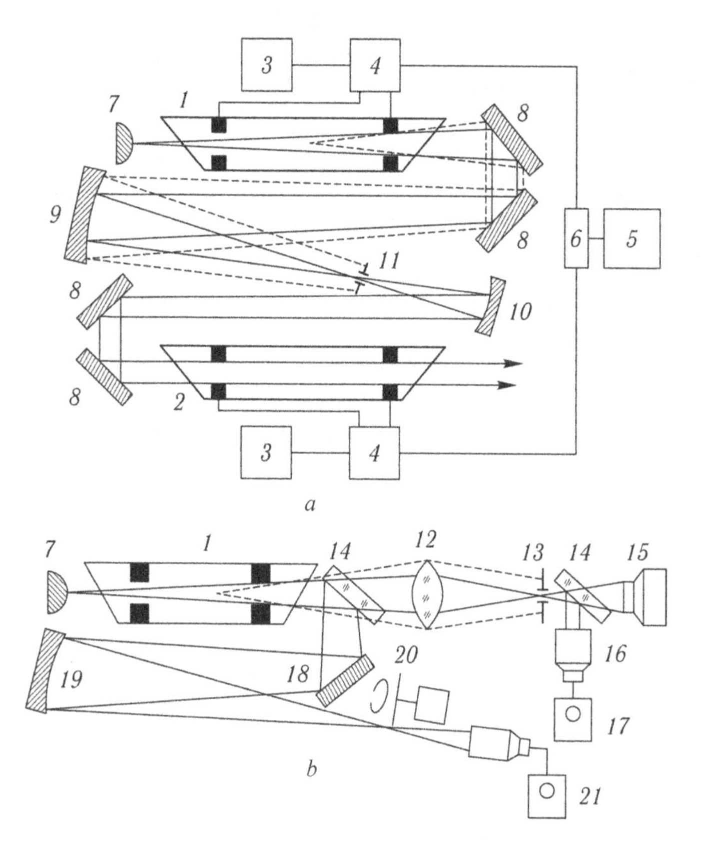

Fig. 4.27. Experimental CVLS of the MO–SFC–PA type in the mode of operation of MO with a single convex mirror and industrial sealed-off AE Kristall GL-201 (*a*) and separately MO with measuring equipment (*b*): *1* – AE MO; *2* – AE PA; *3* – high-voltage rectifiers of the power source, *4* – high-voltage modulators of nanosecond pump pulses, *5* – master pulse generator; *6* – nanosecond delay line; *7* – convex spherical mirror; *8* – flat rotary mirrors; *9*, *10* – concave spherical mirrors of the SFC, 11, *13* – selection diaphragms; *12* – focusing lens, *14* – beam splitter plate; *15* – millivoltmeter M136 with the converter of laser radiation power TI-3; *16* – photocell FEK-14 K; *17* – oscilloscope C1-75 or GDS-840S; *18* – flat rotary mirrors; *19* – focusing mirror; *20* – rotating disc; *21* – oscilloscope C8-7 A.

and an oscilloscope C1-75 or GDS-840S (item 17) were used. The intensity distributions in the focal plane of the lens (item 12) and the focusing mirror were taken using a rotating disk (item 20) with a hole diameter of 0.1 mm, a photocell (item 16) and a memory oscilloscope C8-7A (item 21) and in the radiation focusing plane, along which the geometrical (θ_{geom}) and limiting (θ) beam divergences were estimated. The focus of the radiation at the MO output was carried out by a clear lens (item 12) with a focal length F = 0.7 m or a concave spherical mirror (item 19) with a radius of curvature $R = 5$ m, at the output of the PA – a mirror with $R = 15$ m.

The results of research CVLS and their analysis. The output radiation of an MO with an AE GL-201 in the mode of operation with a single convex mirror has a two-beam structure – it contains beams of superradiance with geometric divergence $\theta_{geom} = 50$ and 18 mrad [16, 232, 233]. The geometric divergences of the beams were estimated from their diameters in the focal plane of the lens (position 12 in Fig. 4.27 *b*) with a focal length $F = 0.7$ m, which was set directly at the output of the MO. The diameter was measured by a conventional measuring ruler with a millimeter scale, and also by the distribution of the radiation intensity and was 34 and 13 mm, respectively (Fig. 4.28 *a*).

The geometric divergence of the beams can be reduced to a certain limiting value using a lens, a mirror or a collimator based on them. The smaller the radius of curvature of the convex mirror (position 7 in Fig. 4.27 *b*), the closer the divergence of the qualitative beam with $\theta_{geom} = 18$ mrad to the diffraction limit. For an experimental estimate of divergence, the output radiation was focused by a mirror (item 19) with a radius of curvature $R = 2.5$ m (Fig. 4.27 *b*) established at a distance of 1.75 m from the AE, and the intensity distribution in the focusing plane of a qualitative beam (Fig. 4.28 *b*), by which the diameter of its spot was calculated (without taking into account the diffraction wings). Next, the distance from the focusing plane to the point where the diameter of the beam is equal to the diameter of the discharge channel ($D_{chan} = 2$ cm) was measured.

And then the ratio of the diameter of the focused spot to the measured distance determines the limiting divergence of the beam. The experimental values of the divergence are in good agreement with the calculated data. Divergences close to the diffraction limit are achieved only when the radius of curvature of the convex mirror is two orders of magnitude shorter than the distance from the mirror to the output aperture of the AE and are calculated by formulas 4.5 and 2.8 (section 4.2) [16].

To record the radiation pulses and measure the power in the beams of the MO, its output radiation was focused by a lens (position 12) with a focal length $F = 0.7$ m (Fig. 4.27 *b*). A diaphragm was installed in the focusing plane of the qualitative beam (with $\theta_{geom} = 18$ mrad), and the diameter of the hole of the diaphragm corresponded to the diameter of its waist (0.3–1.5 mm). The

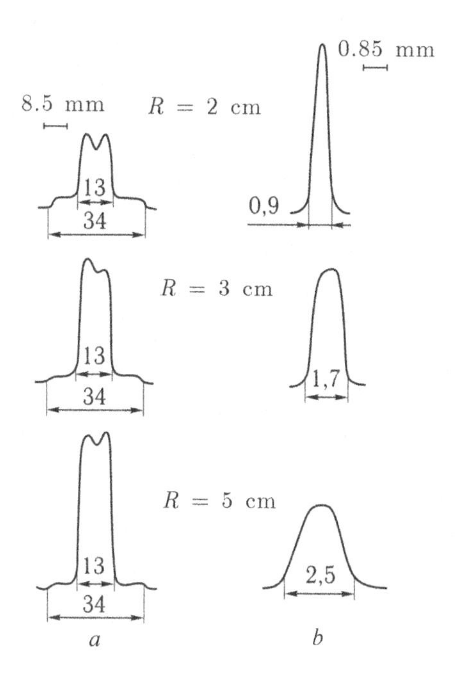

Fig. 4.28. The distribution of the intensity of the radiation of the MO of the CVLS in the mode of operation with a single convex mirror in the focal plane of the lens with $F = 0.7$ m (*a*) and in the plane of beam focusing with a mirror with $F = 2.5$ m (*b*) for different radii of curvature of the mirror. The lens was located directly on the output of the AE, the mirror – at a distance of 1.75 m.

diaphragm isolated a qualitative beam from the background of a beam of superradiance with low spatial coherence ($\theta_{geom} = 50$ mrad). For the diaphragm, depending on the kind of measurements, either a photocell (item 16) was placed on which the radiation was removed by a beam splitter plate (key 14) or a power sensor was used (item 15). Figure 4.29 *a* shows the oscillograms of the radiation pulses of the total beam (*1*) and individual superradiance beams, with θ_{geom} = 50 (*2*) and 18 mrad (*3*). It can be seen that the radiation pulse of a qualitative beam with $\theta_{geom} = 18$ mrad lags behind the background beam pulse with $\theta_{geom} = 50$ mrad for 2–3 ns. This lag is determined practically by the time of the double passage of radiation from the near end of the discharge channel of the AE to the convex mirror. At small radii of the curvature of the mirror (R = 0.6–2.0 cm), the vertices of the oscillograms of the radiation are somewhat blurred, which indicates slight oscillations of energy in the pulse. At $R = 0.6$ cm, the instability of the pulse energy did not exceed 3%.

As the radius of curvature of the mirror decreases, the radiation power in a qualitative beam with θ_{geom} = 18 mrad decreases, and in a beam with θ_{geom} = 50 mrad increases. Such a redistribution of power between the beams is associated with a decrease in the working surface of the mirror. For example, the diameter of the working surface of a mirror with a radius of curvature R = 10 cm is about 5.5 mm, with R = 0.6 cm – 0.05 mm, i.e., the working areas differ by almost four orders of magnitude. When R varied from 10 to 0.6 cm, the average radiation power in a qualitative beam decreased from 4 to 0.1 W, divergence from 0.8 to 0.15 mrad, and in a beam with low coherence (50 mrad), the power increased from 0.6 to 2.2 watts.

To the disadvantage of the MO with a single mirror, one must attribute, first, the impossibility of achieving diffraction divergence and, secondly, the relatively low power of radiation in a qualitative beam. In order to increase the radiation power in a qualitative beam, CVLS studies of the MO–SFC–PA type were carried out.

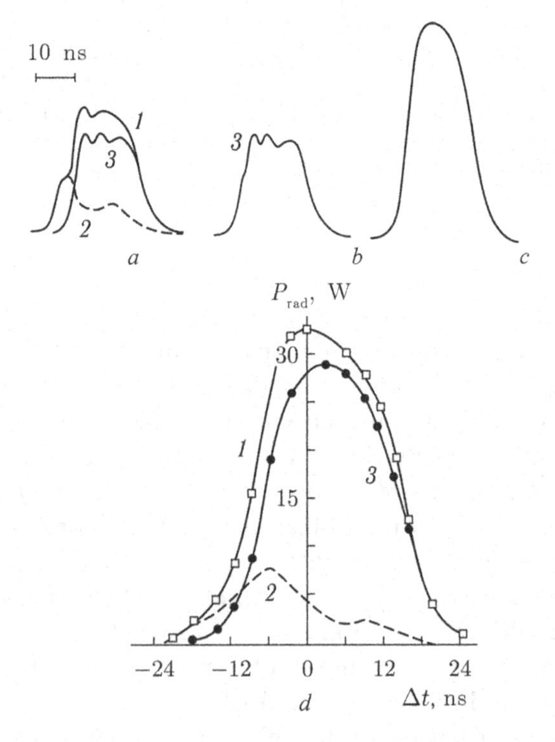

Fig. 4.29. Oscillograms of the pulses from the CVLS radiation at the output of the MO (*a*), SFC (*b*) and PA (*c*) and the dependence of the average radiation power (*d*) in the total (*1*), background (*2*) and qualitative (*3*) beams at the output of the PA on the mismatch time of the light signal of the MO relative to the PA signal (in the absence of a diaphragm in the SFC).

Using PFCs (items 9, 10 and 11 in Fig. 4.27 *a*) placed between the MO (item 1) and the PA (item 2), a qualitative MO beam was extracted from the background beam with low spatial coherence, conversion of the separated beam into a parallel beam with a minimum divergence and matching it with the aperture of the PA channel were carried out [16, 230]. The diaphragm (position 11) was set in the focusing plane of the qualitative beam of the MO. To effectively suppress the background radiation component, the diameter of the diaphragm aperture was chosen close to the diameter of the focusing spot of the qualitative beam. The output mirror of the SFC (item 10), the focus of which is aligned with the focusing plane, performs the function of the converting and matching element.

An expression for calculating the limiting divergence of the output radiation beam for a given system (the prediction of the system) and the parameters of the SFC elements is derived from the consideration of the geometric path of rays in the CVLS using formula (4.5) [16, 232, 233]:

$$\theta_{\text{lim.syst,}} = \frac{1}{2L_1(R+2l)}(RD_{\text{chan}}(\text{MO}) + D_{\text{chan(PA)}}(R+2l) - \\ -([RD_{\text{chan(MO)}} + D_{\text{chan(PA)}}(R+2l)R]^2 - 4L_1(R+2l)D^2_{\text{chan(MO)}}R/l)^{1/2}) + \\ +\theta_{\text{dif.syst.}}, \tag{4.19}$$

$$F_{\text{out}} = \frac{L_{\text{col}}}{1+\theta_{\text{lim.syst.}}/\theta_{\text{lim.(MO)}}},$$

where

$$\theta_{\text{lim.(MO)}} = \frac{D_{\text{chan(MO)}}R}{l\left[R + \frac{L_0}{l}(R+21)\right]} + \frac{2.44\lambda}{D}$$

(in accordance with formula (2.8)),

$$F_{\text{in}} = \left[\frac{1}{L_{\text{col}} - \text{F}_{\text{out}}} + \frac{1}{L_0 + \frac{lR}{R+2l}}\right]^{-1}, \quad d_{\text{hole}} \geqslant \text{F}_{\text{out}}\theta_{\text{lim.syst.}}, \tag{4.20}$$

where R is the radius of curvature of the convex mirror of the MO,

l is the distance from the convex mirror to the output aperture of the MO, $D_{chan(MO)}$ is the diameter of the output aperture (practically the discharge channel of the AE) of the MO, $D_{chan(PA)}$ is the output aperture diameter of the PA, $D = (L_0/l) \cdot D_{chan(MO)}$ is the diameter of the radiation beam at the input mirror of the collimator, L_0 is the distance from the convex mirror of the MO to the input mirror of the SFC, L_{col} is the distance between the mirrors of the collimator, L_1 is the distance from the output mirror of the SFC to the output aperture of the PA, F_{in} and F_{out} is the focal length of the input and output mirrors of the SFC, d_{hole} is the aperture opening of the SFC, $\theta_{dif.syst} = 2.44\lambda/D_{chan(PA)}$ is the diffraction divergence of the output beam of radiation of the CVLS.

Experimental studies of the CVLS characteristics were carried out at radii of curvature of a convex mirror M = 1, 2, 3, 5, 7 and 10 cm and distances l = 115 cm, L_0 = 320 cm, L_{col} = 140 cm and L_1 = 280 cm. The mirrors of the SFC had focal distances calculated by the formulas (4.19) and (4.20). For example, at R = 3 cm, the focal length of the input mirror of the collimator was F_{in} = 78 cm, and the output one – F_{out} = 36.5 cm.

Figure 4.29 *d* shows the dependence of the average radiation power at the output of the CVLS in the total, background, and qualitative (θ_{lim} = 0.35 mrad) beams on the temporal detuning of the light signals of the MO and the PA at a mirror radius R = 3 cm in the absence of a diaphragm in the SFC. The positive values of Δt correspond to the lead, while the negative values correspond to the lag of the signal MO with respect to the signal of the PA. At zero detuning, the total radiation power has a maximum value (33 W), while the background beam has about 5 W (~15 %). If the signal is delayed by more than 20 ns, this signal does not pass through the PA, i.e., the active medium of the PA, because of the high concentration of copper atoms with populated metastable levels, becomes highly absorbing. When the pulses of the MO radiation advance the pulses of the PA by more than 20 ns, the MO signal is not completely absorbed, which indicates that the metastable levels of copper atoms are partially populated at the initial stage of the development of the discharge current pulse.

As the radius of the MO mirror decreases, the percentage of power at the output of the PA in the background beam increases, which is due to the increase in MO power in the beam of ultraviolet radiation with θ_{geom} = 50 mrad. The introduction into the SFC of diaphragms with hole diameters of 0.3–1.0 mm (the diameter depends on the radius of curvature of the MO mirror), which exceeded by several times the calculated values of d_0 = 0.06–0.3 mm, led to an almost complete elimination of the influence of the background signal of the MO for the operation of the PA and obtaining a single-beam structure of the output radiation. Oscillograms of the radiation pulses recorded at the output of the SFC and PA (Fig. 4.29 *b*, *c*) also indicate the efficiency of suppression of the background signal. The intensity distribution of the output beam of the system in the far zone has the same form as the distribution of the qualitative beam of the MO (Fig. 4.28 *b*).

Figure 4.30 shows the experimental curves reflecting the relationship between the characteristics of the output radiation beam of the system and the radius of curvature *R* of the mirror of the MO. It is seen that when the *R* varies from 1 to 10 cm, the divergence of the output beam varies from 0.18 to 0.98 mrad (curve 1), and the average radiation power is from 27 to 34 W (curve 3). At the same time, the practical efficiency of the CVLS amounted to 0.45–0.57%, the individual PA – 0.75–0.9% (power take-off 26.5–31 W), the efficiency of the PA at the power input to it – 1.5–1,8 %. Curve 6

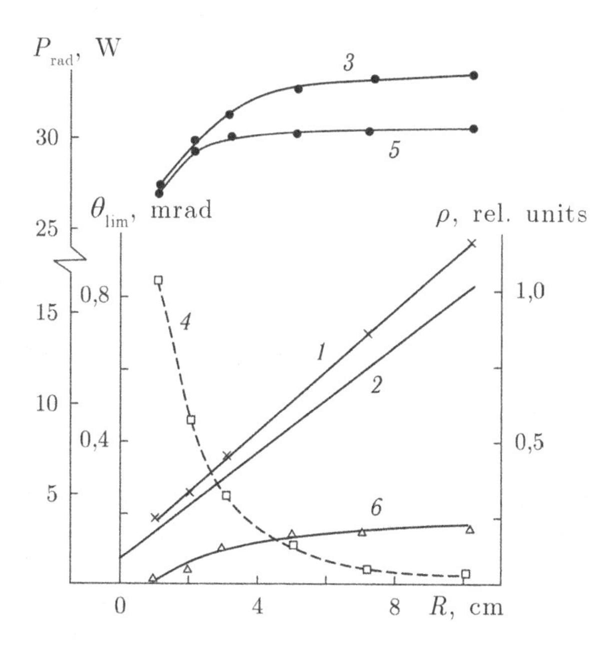

Fig. 4.30. Dependences of the experimental (*1*) and calculated (*2*) divergences of θ, the average radiation power P_{rad} (*3*) and the power density in the focused spot ρ (*4*) of the CVLS, the power pick-up with the PA (*5*) and the beam power at the input of the PA (*6*) on the radius curvature of the convex mirror of the MO.

characterizes the change in the radiation power in a qualitative beam at the input of the PA. It follows from the course of curve 5 that the regime of complete saturation of the active medium PA is practically achieved at a radius of curvature of the mirror $R = 3$ cm ($P_{rad} = 2–3$ W – see curve *6*), and the power of radiation is about 30 W. In the mode of operation of the MO with the telescopic UR, the removal of the power from the MO is 10–15% less. This is due, firstly, to the fact that the duration of the pulses from the emission of a qualitative beam from the MO with a single mirror is larger than from a MO with a telescopic UR, for the time of one double passage of radiation in the resonator (i.e. at $2L/c = 10$ ns) and, secondly, with the absence in the beam of the MO with one mirror of a dip at the centre of the intensity distribution. The instability of the pulsed energy and the axis of the radiation pattern of the output beam of radiation CVLS in the steady-state thermal regime amounted to less than 3% and 0.01 mrad/h, respectively. In the MO with the telescopic UR, these values are much larger, which indicates a higher sensitivity of the laser system with the resonator to external influences. Curve 4 expresses the dependence of the peak radiation power density of the CVLS (in relative units) on the radius of curvature of the MO mirror. For example, when focusing an output beam with an objective with a focal length $F = 5$ cm, as R varies from 10 to 1 cm, the peak power density increases from $1.5 \cdot 10^{10}$ to 3.5×10^{11} W/cm^2 (approximately 33 times). This power density makes it possible to carry out precise microprocessing of almost any metal material up to 1 mm thick.

Investigations of CVLS of the MO–SFC–PA type with telescopic UR and elongated AE Kristall GL-201E and GL-201E32 as the power amplifier. An increase in the output power of radiation in the CVLS of the type MO–SFC–PA can be achieved by increasing the volume of the AM of a separate AE and by increasing the number of AEs in the PA.

Below are the main results of the CVLS research, in which the elongated sealed-off self-heating AE Kristall model GL-201E with the volume of the active medium (AM) $V_{AM} = 350$ cm^3 is used as the PA [253]. Today, instead of AE GL-201E, a modernized new GL-201B model with the same diameter ($D_{chan} = 2$ cm) and AM volume is used (see Table 4.4). The AE GL-201E (GL-201B) was developed on the basis of industrial sealed-off AE GL-201 (GL-201A) with $V_{AM} = 250$ cm^3. The increase in V_{AM} in the AE GL-201E was achieved by extending the discharge channel by 30 cm in accordance with the GL-201 (GL-201A).

Experimental installation. The experimental CVLS of the MO–SFC–PA type with the sealed-off AE Kristall GL-201D as the PA [254] and the methods and means of measurement are similar to those presented in the sections 4.2 and 4.3. In this CVLS, as in the systems considered above, the AE in the MO is represented by the sealed-off AE, model GL-201.

In order to reduce the inductance of the discharge circuit, the AE GL-201E was placed in a conical screen made of eight copper strips measuring 1.5 × 120 cm. The screen diameter at the anode node of the AE was 12 cm and at the cathode 20 cm. The AE provided heating and excitation for a two-channel synchronized high-voltage power source with water-cooled hydrogen thyratrons in high-voltage modulators of nanosecond pump pulses and with a common trigger pulse generator. In the modulator feeding the MO, the thyratron TGI1-2000/35 was used, in the modulator supplying PA the more powerful thyratron TGI1-2500/50 was installes. At the output of the triggering pulse generator, a delay line (±50 ns) was connected to ensure synchronization of the light pulses of the MO and the PA. To increase the excitation efficiency of the AE, the high-voltage modulators were made according to the scheme with a transformer voltage doubling and a magnetic link of the current pulse compression. The compression link was a water-cooled copper tube with M1000NM ferrite rings threaded on it with the dimensions K20 × 12 × 6 mm. The number of rings in the compression MO was 80, in the PA – 150. The capacitance of the storage capacitors was 2200 pF, the sharpening capacitor for GL-201 245 pF, GL-201E 160 pF, PRF was 8.6 kHz, operating pressure of the AE GL-201 250 mm Hg, GL-201E 180 mm Hg. The optimum power consumption from the power source rectifier for the AE GL-201 was 3.3 kW at a voltage of 20 kV at the anode of the thyratron, for GL-201D the corresponding values were 5.3 kW and 23 kV. For AE GL-201E, the duration of the voltage pulses was 70 ns, and the amplitude was 28 kV; the corresponding characteristics of the current pulses were 140 ns and 0.4 kA. The MO worked both in the regime with a single convex mirror with a radius of curvature R = 0.6–10 cm, and with a telescopic UR with an increase in M = 200 [230, 231]. An SFC was installed between the MO and the PA intended to isolate the qualitative beam of the MO and to match it with the aperture of the discharge channel of the PA. The characteristics of the AE GL-201E with flat and planar (R = 3.5 m) resonators were investigated. The reflection coefficient of the blind mirrors in the resonators was 99%. A plane-parallel plate without a coating was used as an output mirror.

Characteristics of the sealed-off AE GL-201E in the regime of a generator with flat and planar resonators and in the regime with one convex mirror. In the CVL with the AE Kristall GL-201E, when working with a flat resonator, the total average output beam power was 30 W, with a plane-spherical resonator 35 W, the duration of the radiation pulses at the base was 30 ns. About 70% of the power in both cases was concentrated in the beams formed directly by the resonator, and 30% in two (always present) superradiance beams [227–233]. The geometric divergences of the superradiance beams are 40 and 15 mrad. No more than 1–2% of the power was concentrated in a beam with θ_{geom} = 40 mrad.

The divergence of the resonator radiation beam when working with a plane resonator was about 3 mrad, which is 3.5 times smaller than when working with a plane-spherical resonator, which value is 45 times larger than the diffraction limit (θ_{dif} = 0.07 mrad). Usually the beams with low divergence are produced using telescopic URs with an increase of M = 100–300. When the AE GL-201E with UR was operating, the percentage of power in the diffraction beam was no more than 10–15%. Therefore, it is advantageous to use long AEs as the PA for powerful CVLS.

In the case of a single-mirror operating mode of the CVL, the output radiation has a strictly two-beam structure: it contains superradiance beams with θ_{geom} = 40 and 15 mrad. The radiation characteristics of the second beam (θ_{geom} = 15 mrad) can be controlled within wide limits, changing the radius of curvature of the convex mirror [232, 233]. When the radius of the mirror is two orders of magnitude smaller than the distance from the mirror to the output aperture of the AE, this radiation beam has a quality close to the diffraction one, so that it can be collimated into a narrowly focused cylindrical beam, focused into a spot of small diameter with a high peak power density (~10^{11} W/cm^2), as well as isolated by means of a spatial filter from a background (aperture) beam with low coherence (40 mrad).

For the AE GL-201 Fig. 4.31 shows the dependence of the average radiation power in the total (curve *1*), background (*2*) and qualitative (*3*) beams and the calculated divergence of the qualitative beam (*4*) on the radius of curvature of the convex mirror. The calculated

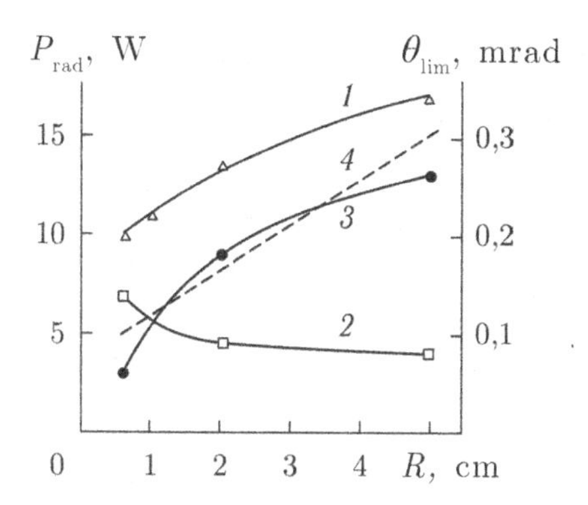

Fig. 4.31. Dependences of the average radiation power in the total (*1*), background (*2*), qualitative (*3*) beams and the calculated divergence (*4*) of the qualitative beam of CVL with the AE GL-201E on the radius of curvature of the convex mirror.

divergence agrees well with the experimental data. A change in the radius of curvature of the mirror from 0.6 to 5 cm leads to a change in the power in a qualitative beam from 3 to 13 W (curve *3*), the calculated divergence from 0.1 to 0.3 mrad. For the AE GL-201 due to its shorter length, the same divergences are achieved with smaller radii of curvature. Thus, for a GL-201E a beam with a divergence of 0.3 mrad is formed at R = 5 cm, and for the AE GL-201 – at R = 3 cm, but in the latter case, the radiation power is approximately half that in the first. With a decrease in the radius of curvature of the mirror, the efficiency of the GL-201 AE increases faster than the efficiency of GL-201: the radiation power at a divergence close to the diffraction limit (θ = 0.1–0.15 mrad) is 5–10 times greater in the first case. Therefore, in order to obtain a relatively powerful qualitative beam of radiation in the operating mode of CVL with a single mirror, it is more advantageous to use long AEs. But the maximum efficiency of the long AE is achieved when they are used as a power amplifier in the MO–SFC–PA laser systems.

Research results and analysis. The dependence of the average radiation power at the output of the CVLS with the Kristall GL-201E AE as the PA on the temporal detuning of the light pulse MO relative to the pulse of the PA with the radius of curvature of the convex mirror R = 5 cm [253] is shown in Fig. 4.32. The radiation power in the isolated (by means of the SFC) qualitative beam of the MO at the input of the PA was 5.3 W; at the output (with the PA switched off) it was 3.7 W. With the optimal adjustment of the CVLS (detuning zero), the output power of the radiation was 55 W, of which 51.3 W is the power taken from the PA. In the same conditions, the power take-off in the case of GL-201 was about 30 W. Thus, an increase in the length of the discharge channel by 1.32 times led to an increase in the energy capacity by a factor of 1.7. An additional increase in the radiation power with the AE GL-201D is associated with the improvement of the electrical matching of the high-voltage modulator of nanosecond pump pulses with an extended AE and a relative increase in the volume of the active medium due to the establishment of a more uniform temperature distribution along the discharge channel. As can be seen from Fig. 4.32, the output power of the radiation is very sensitive to the temporal detuning. For example, when the detuning time is 4 ns, the power is reduced by 10–15%. Therefore, in order to maintain the system in an optimal mode, the two-channel power source must provide stable parameters of the pump pulses and a high degree of their synchronization.

Figure 4.33 shows the dependence of the average radiation power (*1*) of the CVLS and the power take-off with the PA (*2*) on the power at the input of the PA. The power of the radiation at the input of the PA varied in the range from 0.6 to 8.5 W by changing the radius of the convex MO mirror from 0.6 to 10 cm. At the same time, the divergence of the output beam of the CVLS varied within 0.2–1 mrad. As a result, the output power of the system increased from 50 to 60 watts, and the power removed from the amplifier was from 49.5 to 54 watts. At the maximum radiation power, the efficiency of the PA based on the input power in the AE was 2.5%.

In the mode of operation of MO with the telescopic UR at M = 200 and the power at the input of the PA 8 W, the output

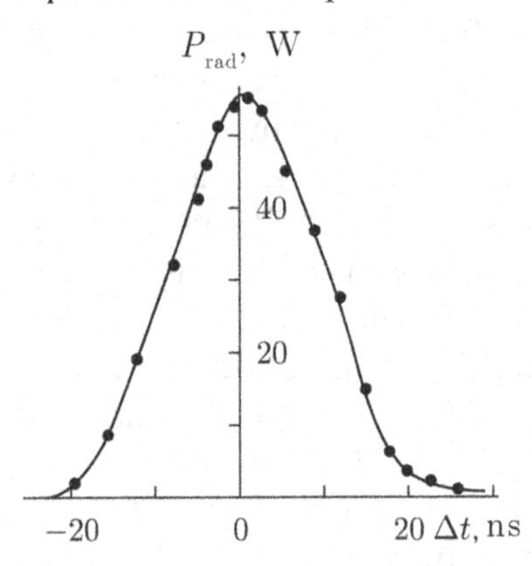

Fig. 4.32. Dependence of the average radiation power of the copper vapour laser system MO (AE GL-201)–SFC–PA (AE GL-201E) at a radius of curvature of the convex mirror of the MO R = 5 cm from the detuning time of the light signal of the MO with respect to the PA signal.

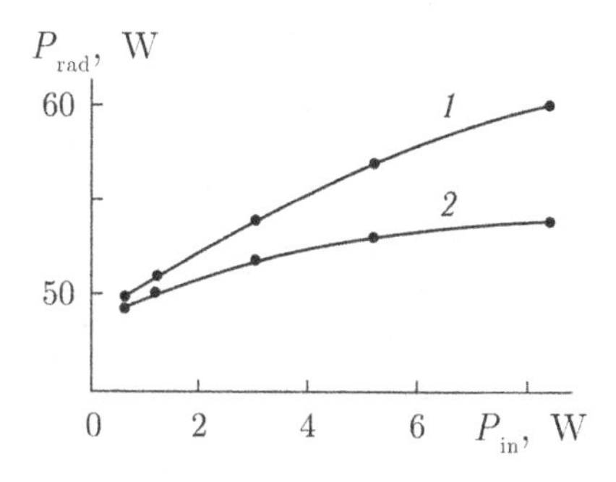

Fig. 4.33. Dependence of the average radiation power (*1*) of the CVLS of the type MO (AE GL-201)–SFC–PA (AE GL-201E) with one convex mirror in the MO and the dependence of the power take-off with the PA (*2*) on the radiation power at the input of the PA.

power of the CVLS radiation reached about 60 W (divergence 0.2 mrad), practical efficiency 0.8%. Under the same conditions, the laser system in which both the MO and the PA used the GL-201 AE had an output power of about 35 W and a practical efficiency of 0.5% [231]. From a comparison of these data it follows that the increase in the output power of the CVLS by a factor of 1.7 (60/35) is achieved by increasing the power consumption by only 10%.

Investigations of CVLS of the MO–SFC–PA type with the telescopic UR and two elongated AE GL-201E as PA. The MO of this CVLS used a telescopic UR with an increase of $M = 60$ and a sealed AE Kulon GL-204 with an average radiation power of 3 W. In the first experiment, two AEs GL-201D were used in the PA to increase the output power of the CVLS. The pumping of both AEs GL-201E was carried out from a two-channel synchronized lamp power supply IPL-10–001 with a PRF of 12.5 kHz. The power of the laser system with two AEs GL-201E as the PA increased to 70 W (the output power of the first PA was 30.5 W). The divergence of the radiation beam was 0.4 mrad, the energy in the pulse was 5.6 mJ, the pulse (peak) power 370 kW ($P = W/\tau_{pulse}$, where W is the energy in the pulse, τ_{pulse} is the pulse duration at half the height). The practical efficiency of the system was 0.93%, the efficiency of the amplification stage was 1.08%, the efficiency of the individual AE GL-201E was about twice as high – 2.15%.

When two GL-201E32 AEs were used as the PA, the output power of the laser pulse was 105 watts at 10 kHz PRF, divergence 0.3 mrad, pulse energy 10 mJ, peak power 500 kW, practical system efficiency

0.87%, while the amplification stage was 1%, the efficiency of a separate GL-201E32 AE reached 2%.

The AE GL-201E32 was pumped from thyratron power sources in which the high-voltage pulse modulators used powerful water-cooled thyratrons TGI1-2500/50 with stabilized heating of a hydrogen generator and a cathode. The voltage of the feeding three-phase network was stabilized with the help of a stabilizer of the CTC-2M10 type. The diameter and length of the discharge channel of the AE were 32 and 1230 mm, respectively. The power consumption from the power source rectifier for one AE was 5 kW, for the second – 5.5 kW.

The attained values of the output radiation power in the CVLS with two AEs GL-201E and two AEs GL-201E32 as the PA (70 and 105 W) are not limiting. A further increase in power can be obtained by antireflecting the output windows of the AE, reducing the losses in the rotary mirrors, improving the characteristics of the pump current pulses, and temporarily matching the duration of the signals of the MO and the PA. When used instead of the AE Kulon GL-204 Kristall GL-201, the radiation pulses of which have a long duration of 10 ns ($\tau_{pulse} \approx 30$ ns), the power pick-up from one AE GL-201D32 ($\tau_{pulse} \approx 35$ ns) has increased up to 70 W. To increase the duration of the MO pulse different optical delay lines can be used. Figure 4.34 shows the delay line consisting of four flat reflecting mirrors. The increase in the duration of the MO signal with the GL-204 AE by 10 ns, obtained using this line, resulted in an increase in the system output by approximately 10%.

CVLS of the MO–PA type with several amplifying AEs. It is possible to substantially increase the radiation power of the CVLS of the MO–PA type by increasing the number of amplifying AEs in the PA. However, with losses at the exit windows of AEs caused by Fresnel reflection (4% from each window face), it is possible that with a relatively small number of series-connected AEs in the PA, the addition of the next AE hardly increases the total radiation power. Such losses can be reduced by two-sided antireflecting output windows of the AE. The radiation power of such a multimodal CVLS is the sum of the powers of the individual AEs of the PA and the MO:

$$P_{CVLS} = P_0 + P_1 + P_2 + \ldots + P_{n-1} + P_n, \quad (4.21)$$

where P_0 is the radiation power from the MO, n is the number of

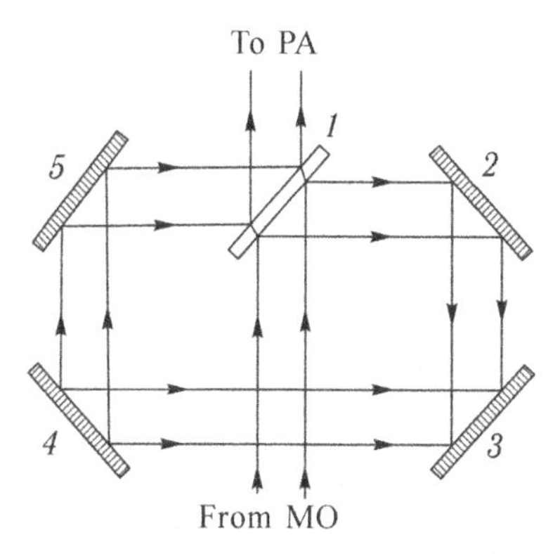

Fig. 4.34. Variant of the optical system for increasing the duration of the pulses of the radiation of the MO: *1* – beam splitter plate; *2–5* – flat mirrors.

AEs in the PA, P_i is the radiation power from the *i*-th AE, $i = 1,..., n$,

$$P_0 = P_{MO} \cdot \tau_w^{2n+1}, \tag{4.22}$$

where τ_w is the transmittance of one window of the AE, P_{MO} is the radiation power of the MO at $\tau_w = 1$,

$$P_1 = P_{PA} \cdot \tau_w^{2n-1}, \tag{4.23}$$

$$P_2 = P_{PA} \cdot \tau_w^{2n-3}, \tag{4.24}$$

$$P_{n-1} = P_{PA} \tau_w^3, \tag{4.25}$$

$$P_n = P_{PA} \tau_w^3, \tag{4.26}$$

where P_{PA} is the radiation power of one AE at $\tau_w = 1$.

Substituting expressions (4.22)–(4.26) into the formula (4.21), we have:

$$P_{CVLS} = P_{MO} \tau_w^{2n+1} + P_{PA} \cdot (\tau_w^{2n-1} + \tau_w^{2n-3} + \ldots + \tau_w^3 + \tau_w), \tag{4.27}$$

where $\tau_w^{2n-1} + \tau_w^{2n-3} + ... + \tau_w^3 + \tau_w$ – increasing geometric progression with the denominator $q = \tau_w^{-2}$. As a result, a relatively simple formula is obtained for calculating the output power of the CVLS radiation, consisting of one MO and *n* identical AEs in the PA:

$$P_{\text{CVLS}} = P_{\text{MO}}\tau_w^{2n+1} + P_{\text{PA}}\frac{\tau_w(1-\tau_w^{2n})}{1-\tau_w^2}. \tag{4.28}$$

Since $\tau_w < 1$, then for a large number of the amplifying AEs ($n \to \infty$) the expression (4.28) tends to the value

$$P_{\text{CVLS}} = P_{\text{PA}}\frac{\tau_w}{1-\tau_w^2}.$$

The need to reduce losses on the AE windows can be demonstrated by the example when there is no antireflection and $\tau_w = 0.92$. In this case, the maximum output power of the system does not exceed $6P_{\text{PA}}$:

$$P_{\text{CVLS}} = P_{\text{PA}}\frac{0.92}{1-0.92^2} \approx 6P_{\text{PA}}.$$

Figure 4.35 shows the dependences of the normalized radiation power of the laser system ($P_{\text{out}}/P_{\text{PA}}$) on the number of amplifying AEs, calculated from formula (4.28). It is seen that the greater the loss on the windows, the stronger the dependence deviates from the linear one (which corresponds to the idealized case of zero losses ($\tau_w = 1$)) and goes faster to saturation.

A new generation of industrial sealed-off self-heating AEs of the series Kristall GL-205A (Kristall LT-30Cu), GL-205B (Kristall LT-40Cu) and GL-205C (Kristall LT-50Cu) (see Table 4.4) [16, 42, 269–271] with a guaranteed lifetime of more than 1500 hours is produced with an antireflecting exit window ($\tau_w \geq 0.98$). If we set, for example, the output radiation power of the CVLS at the level of 0.8 of the power at a transmittance $\tau_w = 1$, then in the case of the antireflecting windows with $\tau_w = 0.98$, it is advisable to put no more than 12 AEs in one amplifying optical line. If the AE output windows are used without superradiance, i.e., with a transmittance $\tau_w = 0.92$, then at the indicated level (0.8) it makes sense to use no more than three AEs.

The development and application of multi-module, high-power CVLS of the MO–PA type on the basis of industrial sealed-off AEs Kristall, produced by the Istok Company, are carried out at the Kurchatov Institute (Moscow), Medical Sterilization Systems LLC (Odintsovo, Moscow Region) and the Institute of Semiconductor Physics (Novosibirsk).

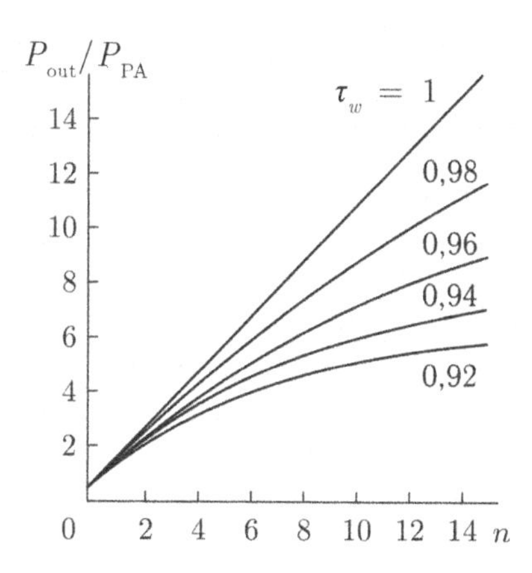

Fig. 4.35. The dependence of the normalized radiation power of the laser system MO–PA (P_{out}/P_{PA}) on the number of amplifying AEs (n).

Each amplification path of CVLS usually includes two or three AEs GL-205B (Kristall LT-40Cu) or GL-205C (Kristall LT-50Cu). The most commonly used MO is the AE GL-205A (Kristall LT-30Cu) with a telescopic UR or, less often, the AE GL-205B in the regime with one convex mirror. If, for example, three AEs GL-205Bs are used as PAs, each of which provides a power take-off of about 50 W and the AE of 205 A as the MO with a power in a qualitative beam of 10 W and with a transmittance of the antireflecting windows $\tau_w = 0.98$, then the output power of the system in accordance with the formula 4.28 is

$$P_{CVLS} = 10 \cdot 0.98^7 + 50\frac{0.98(1-0.98^6)}{1-0.98^2} = 150 \text{ W}.$$

When three AEs GL-205C are used (at a power take-off from an AE of about 60 W), the output power of the CVLS under the same conditions as for the GL-205B AE should be 178 W. However, it should be emphasized that these values of the laser radiation power can be achieved only under the condition of ideal spatial and temporal matching of all the AEs entering into it.

4.7. Investigation of the properties of the active medium of a pulsed CVL using CVLS

In this pulsed CVLS its MO used the AE Kulon of the GL-206E

model with a telescopic type of UR with an increase of $M = 180$ was used, in the PA – in one case AE GL-206E, in the other, a more powerful AE GL-206I (Table 4.3). One of the main characteristics of a laser system for a given PRF, as was shown above, is the dependence of the output radiation power on the synchronization of the pulses of the MO relative to the pulses of the PA. In Fig. 4.36 such a dependence of the radiation power on the temporal detuning of the light signal of the MO with respect to the light signal of the PA is presented for the CVLS using the same AE GL-206E in the MO and PA (PRF 13.6 kHz), in Fig. 4.37 – for a more powerful CVLS with the use of AE GL-206I in the PA (PRF 14 kHz).

With a zero temporal detuning of the signals of the MO and PA, the output power of the radiation and the efficiency are maximal. The maximum value of the average radiation power of the CVLS with the AE GL-206E in the MO and PA is 16–17 W at an efficiency of 0.4% (Fig. 4.36), with an AE in the PA – 24–25 W at an efficiency of 0.5% (Fig. 4.36). When the light signal of the MO is temporally detuned both in the direction of the delay from the PA signal and in the direction of its advance, the radiation power curve, due to the short time of existence of population inversion in the AM, falls quite sharply. At detuning within ±25 ns, the radiation power is reduced to the level of the input signal from the MO (3–4 W). The sign '–' corresponds to the delay, '+' – to advance. If the MO signal lags behind by more than 40–50 ns, this signal is completely absorbed in the AM of the PA due to a sharp increase in the concentration of copper atoms with populated metastable levels at the decay of the discharge current pulse. When the MO signal advances by 20–40 ns, its partial absorption takes place, and it reaches a minimum power value (1.8 W at $t = 30$ ns).

The appearance of a zone of weak absorption is due to the population of metastable levels of a part of copper atoms at the initial stage of the development of pump current pulses (discharges). When advancing for a time longer than 60 ns, the active medium of the PA becomes practically transparent for the signal of the MO. In this mode, the signal power of the MO at the output of the PA was ~2.5 W (the signal strength of the MO signal at the input is ~4 W). Part of the power – about 0.8 W (20%) – was lost on the windows of the AE of the PA, the other part – 0.7 W (17%) – probably absorbed by its active medium. The latter testifies that the active medium of the PA in the interpulse period at the PRF of 13.5–14 kHz is not completely restored.

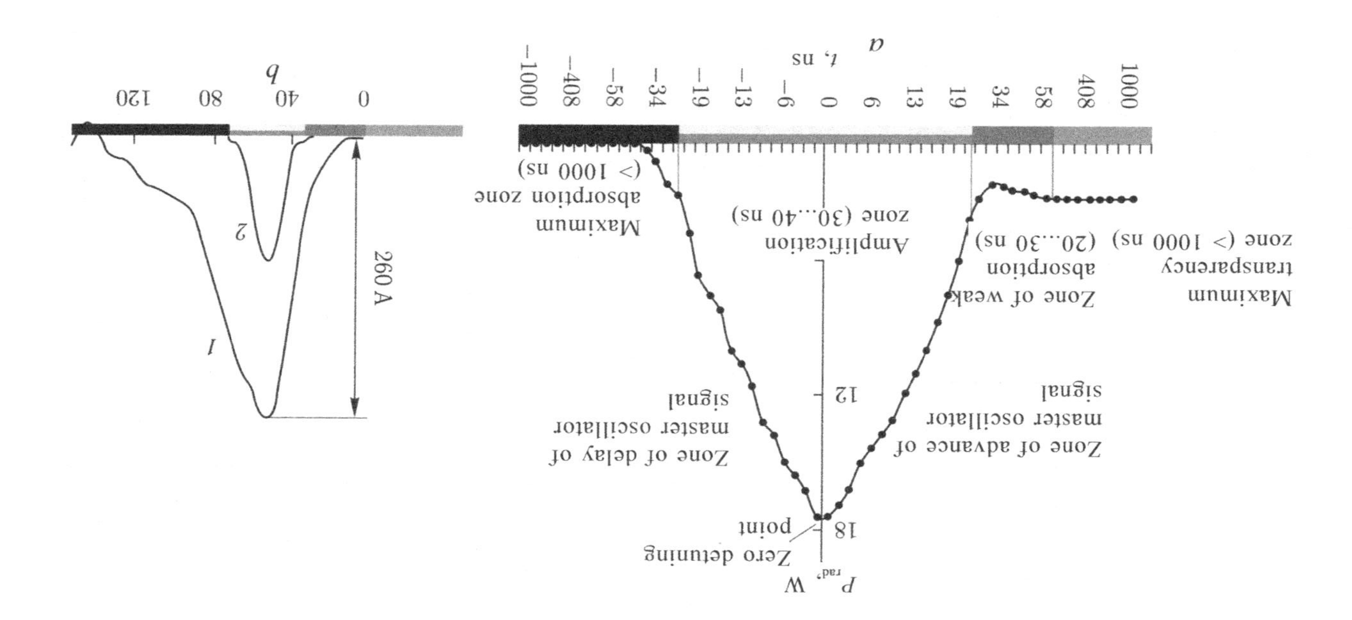

Fig. 4.36. Dependence of the average radiation power of a pulsed CVLS with industrial sealed-off AE Kulon GL-206E on the temporal detuning of the pulses of the MO radiation relative to the pulses of the PA within ±1000 ns, the PRF of 13.6 kHz.

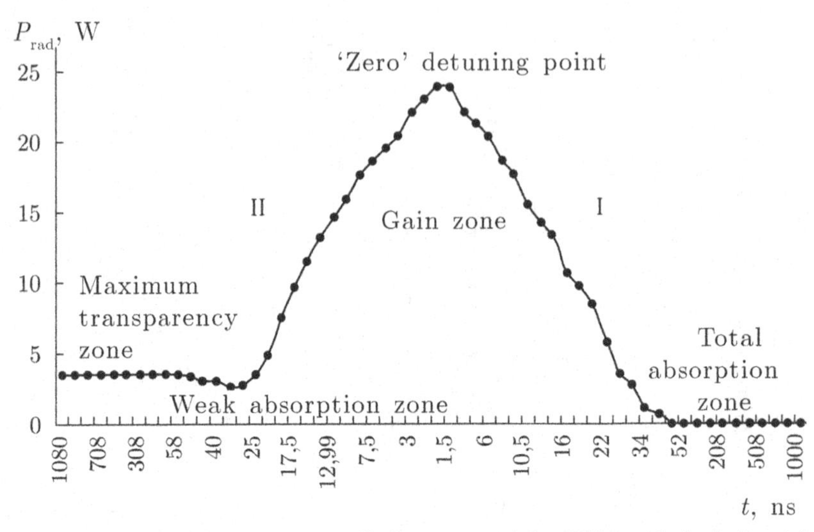

Fig. 4.37. Dependence of the average radiation power of the CVLS with the industrial sealed-off AE Kulon GL-206E (MO) and Kulon GL-206I (PA) on the time delay of the pulses of MO radiation relative to the PA pulses within ±1000 ns, PRF 14.7 kHz. The minus sign corresponds to the lag (I), the plus sign to the lead of the light signal of the MO relative to the signal of the PA (II).

Thus, the graphs clearly show four consecutive successive characteristic time zones in the AM of the pulsed CVL: a zone of weak absorption, a gain zone, a zone of total absorption, and a zone of maximum transparency. Owing to the presence of such zones in the AM of the laser in the CVLS of the MO–PA type, due to the dissynchronization of its light signals with respect to each other, it is possible to operatively control the radiation power from zero to the maximum value and the PRF according to a given law. This method of operational power control is protected by the patent for invention of the Russian Federation No. 2 264 011 'Method of excitation of radiation pulses from laser systems on self-contained transitions' [365]. The presence of a zone of total absorption in the active medium of the PA allowed us to develop methods and electronic devices (high-speed gates) for controlling the parameters of the output radiation for CVL and CVLS. They allow to create modes of the pulse and packet modulations of the output radiation according to any predetermined algorithm. The time of exposure of the light signal of the MO in the absorption zone of the PA and, correspondingly, the time it takes to find it in the amplification zone, is determined by the number of pulses N in the radiation packet per second:

$$N = f(t_0 - t_1).$$

where t_0 = 1 s, t_1 is the dwell time of the MO signal in the absorption zone of the PA, and f is the pulse repetition frequency (PRF).

This method of pulse and packet modulation of the output radiation is applied in ALTI Karavella-1 and Karavella-1M, created on the basis of CVLS: the light signal of the MO is shifted from the point of zero dissynchronization to the side of the signal from the PA for a time of 150 ns (to the zone of total absorption). The possibility of ALTI Karavella operation in modulation modes allows to significantly increase the efficiency of managing the technological processes of manufacturing parts, increase the productivity of material processing, make cuts and holes with minimum dimensions, minimum roughness and a zone of thermal impact.

4.8. Conclusions and results for chapter 4

1. The distinctive properties and peculiarities of the dynamics of output radiation formation in the CVLs with nanosecond pulse duration are considered. It is established that in the CVL in the regime of a separate generator with an optical resonator the structure of the output radiation is multibeam, and each beam has its spatial, temporal and energy characteristics. The distribution of the radiation intensity of the CVL with the multibeam structure is uneven (stepwise) and therefore is unsuitable for high-quality processing of materials, which prevented the creation of modern technological equipment on the basis of CVL.

2. A unique experimental setup and measurement techniques for qualitative studies of the characteristics of the output radiation of CVL with UR with two convex mirrors and the telescopic UR were developed, which was due to the high sensitivity of the elements and the field of these resonators and, accordingly, the radiation parameters to external mechanical, acoustic and thermal effects and air heat fluxes.

3. The conditions for qualitative investigations of the radiation of the CVL with unstable resonators have been achieved through the use of a set of design and scientific and technical solutions in an experimental installation: the use as a carrier of all optical elements and assemblies of a massive honeycomb optical table, air insulators between the table and its four supports and rubber insulators between the supports and the production floor, the placement of a heat-loaded

high-temperature laser AE in a water-cooled heat sink with a thermal sensor of the feedback, sealing the space between the mirrors of the UR and the heat sink and establishing a restrictive aperture before the blind mirror of the UR with two convex mirrors. These measures have led to the almost complete disappearance of the instabilities of the position of the axis of the beam pattern of radiation beams and pulsed energy.

4. The structure, spatial, temporal and energy characteristics of the laser radiation in the regime of a separate generator with optical systems possessing high spatial selectivity were studied: with one convex mirror, with the UR with two convex mirrors and the telescopic UR, and conditions were determined for the formation of diffraction-quality single-beam radiation in them and high stability of the parameters.

5. It has been established that the structure of the output radiation during the operation of the CVL in the mode of the generator with one convex mirror is two-beam: an incoherent beam of superradiance formed by the total geometric aperture of the discharge AE from the amplifying spontaneous 'seeds' and with a high coherence (qualitative) beam formed by the mirror and the output aperture of the channel, the second beam lags behind the first for a time $\Delta t = \frac{2l'_1}{c}$, where l'_1 is the distance from the mirror to the AM and c is the speed of light. The parameters of the second qualitative (useful) beam of superradiance can be varied within wide limits by changing the radius of curvature of the mirror (R) and the distance (l) from the mirror to the output aperture of the discharge channel with a diameter D_{chan} in wide limits.

6. Based on the laws of geometric optics, taking into account the diffraction limit of divergence, a formula is derived for calculating the divergence of a qualitative (useful) beam of superradiance when the CVL is operating in the regime of a generator with a single convex mirror, which has the following form

$$\theta = \frac{R \cdot D_{chan}}{2l \cdot (R+l)} + \frac{2.44\lambda}{D_{chan}}.$$

The smaller the radius of curvature of the mirror (R) and the diameter of the aperture of the discharge channel (D_{chan}) and the greater the distance from the mirror to the output aperture of the discharge channel (l), the closer the divergence (θ) to the diffraction limit.

At radii R one or two orders of magnitude smaller than the distance l, the beam divergence becomes close to the diffraction ($\theta = 2–4\theta_{dif}$), and the peak power density in the spot of the focused beam reaches 10^9 W/cm^2, sufficient for productive microprocessing of foil materials and cutting of solders (0.02–0.1 mm). For the treatment of thicker materials, a single-convex mirror with a single convex mirror is used as a master oscillator in the CVLS of the MO–SFC–PA type, when using the Kristall AE of the GL-205A and GL-205B models, the power of the radiation increases by more than an order of magnitude 30–60 W).

7. An optical scheme for increasing the power of a qualitative beam of radiation and an optical scheme for reducing its divergence to a diffraction limit during the operation of a linear oscillator in the regime of a generator with a single convex mirror have been developed and investigated. The power is increased by an order of magnitude and the divergence is reduced to the diffraction limit.

8. It is established that in the CVL in the generator mode with the UR with two convex mirrors it is possible to form a diffraction-type single-beam radiation ($\theta_{dif} = 2.44\lambda/D_{chan}$) when three interrelated conditions are fulfilled:

– the distance (l'_1) from the blind mirror of the UR to the near end of the AM (practically to the discharge channel of the AE) must be at least half the length of the AM ($l_{AM}/2$) and the distance (l'_2) from the output mirror to the near end of the AM and not more than half the distance traveled by the radiation during the lifetime of the inversion ($\tau_{inv} \cdot c/2$) and the sum of the length of the AM and the distance from the output mirror to the near end of the AM ($l_{AM}+l'_2$), i.e.,

$$\frac{l_{AM}}{2} + l'_2 < l'_1 < \frac{\tau_{inv} \cdot c}{2} - (l_{AM} + l'_2);$$

– the radius of curvature of the surface of the blind mirror of the UR should be 1–2 orders of magnitude smaller than the distance (l_1) to the output aperture of the discharge channel AE,

$$R_{1(blind)} < l_1 / 10 - 100 = 1 - 3 \text{ cm};$$

– the output mirror of the UR should be maximally close to the exit window of the AE and have a radius of curvature of the surface greater than that of the blind mirror ($R_{2(out)} > R_{1(blind)}$). In this case, the feedback from the AM begins earlier from the exit mirror of the resonator, which is in addition to the first and second conditions for suppressing the parasitic (background) superradiance beam formed from the blind mirror with a divergence greater than the diffraction, but close to it.

9. A new UR design with two convex mirrors was developed and investigated for the CVL in which an enlarged convex–concave meniscus (a positive lens) with a mirrotr spot reflecting laser radiation 1–1.5 mm in size at the centre of its convex surface is used as an exit mirror, facing the AE. The focus of the meniscus is combined with the focus of a blind convex mirror, which is a condition for collimating the generated divergent beam of radiation into a cylindrical (flat) beam with a minimum divergence. The divergence of the radiation beam is determined by the formula

$$\theta = \frac{D_{chan} \cdot R_1 \cdot R_2}{4 \cdot (R_1 + 2f_2 + 2L)(R_2 + 2l_{S_2})(1 + f_1)} + \frac{2.44\lambda}{D_{chan}},$$

derived on the basis of the laws of geometric optics, taking into account the diffraction limit by divergence. R_1 and R_2 are the radii of curvature of the blind and output mirrors of the UR, D_{chan} is the diameter of the aperture of the discharge channel of the AE, L is the length of the UR, l is the distance from the mirror from R_1 to the output aperture of the discharge channel, l_{S_2} is the distance from the mirror from R_2 to the output aperture, f_1 and f_2 are the distances from the mirror to the images in them of the aperture.

10. In the CVL in the generator mode with the UR with two convex mirrors, due to large losses in the UR, the power of the diffraction beam is no more than 10% of the total radiation power, which limits its capabilities in terms of the thickness of the microprocessing of materials. For example, in the CVL with the 15 W AE of the Kulon model GL-206D, the power of the diffraction beam is about 1 W, with 20 W of the AE model GL-206I – 1.5 W and only thin-film coatings and foil materials are processed efficiently (5–50 μm).

11. The most significant application of CVL in the mode with UR with two convex mirrors is the use of it as an MO in the CVLS of the MO–SFC–PA type. The average radiation power and, correspondingly, the peak power density of the diffraction beam in the CVLS with the use of industrial sealed-off AEs as the PA is increased by 1–2

orders of magnitude: from the AE of the Kulon series of the GL206E and GL206I models to 15–25 W and 10^{11}–10^{12} W/cm^2, with the AE Kristall the models GL-205A, GL-205B and GL-205C – up to 30–100 W and 10^{12}–10^{13} W/cm^2, which significantly increases the productivity of microprocessing and the thickness of the processed materials to 1–2 mm.

12. It is established that the structure of the output radiation in the operation of CVLs in the generator mode with the telescopic UR is multibeam: two incoherent superradiance beams (always formed) and several resonator beams, the number of which is limited by the time of existence of the population inversion (τ = 30–40 ns) and the gain in the resonator (M). At the resonator gains $M \geq 60$ two resonator beams are formed, the second of which is diffractive and is the imposition of two or three diffracted beams following one another. At $15 \leq M < 60$ three resonator beams are formed, the third of which is diffractive, for $M < 15$ – also three beams, and the fourth diffraction one, due to the limited time of existence of the inversion, does not have time to form. The first resonator beam lags behind the second superradiance beam and the resonator beams lag behind each other for a time of one double pass through the radiation in the resonator ($\Delta t = 2L/c$, where L is the resonator length and c is the speed of light), all the beams partially overlapping in space and in time, competing with each other on power in the process of its formation.

13. Based on the laws of geometric optics, taking into account the diffraction limit of divergence, a formula is derived for calculating the divergence of resonator radiation beams during the operation of CVL in the generator mode with the telescopic UR, which has the following form:

$$\theta = \frac{D_{\text{chan}}}{(ML + l)M^{n-1}} + \frac{2.44\lambda}{D_{\text{chan}}}.$$

With the increase in the resonator (M) and in the number of passes (n) of radiation in the resonator, the beam divergence decreases and at M = 100–300 it becomes close to diffraction.

14. In the CVL in the generator mode with the telescopic UR with the gains M = 100–300, the power in the diffraction beam reaches a maximum value (up to 50% of the total output radiation power), and the peak power density in the spot of focused radiation is up to 10^{11} W/cm^2, sufficient for productive microprocessing of metallic

materials up to 0.3 mm thick and a large range of non-metallic thicknesses up to 0.7 mm. Qualitative microprocessing is provided by using a spatial filter–collimator (SFC) outputting the diffractive component from the multibeam radiation structure at the output of this CVL.

15. The use of CVL in the generator mode with the telescopic UR as a MO in the CVLS of the MO–SFC–PA type was the basis for the creation of a modern industrial ALTI Karavella-1 with a radiation power of 10–15 W for precision microprocessing of metallic materials up to 0.5 mm thick and non-metallic up to 1–1.2 mm of electronic components.

The use of CVLs in the mode of working with the UR with two convex mirrors as the MO in the CVLS of the MO–SFC–PA type was the basis for creating the most powerful industrial ALTI Karavella-1M (20–25 W) for precision microprocessing of metallic materials up to 1 mm thick and non-metallic materials up to 1.5–2 mm for electronic components.

16. The CVL in the mode of operation with the telescopic UR and with the SFC at its output was the basis for the creation of compact low-power modern industrial ALTIs Karavella-2 and Karavella-2M (5–8 W) for precision microprocessing of metallic materials up to 0.3 mm and non-metallic materials to 0.5–0.7 mm for electronic components.

17. The spatial, temporal, and energy characteristics of the radiation of high-power CVLSs operating according to the MO–SFC–PA scheme with industrial sealed-off self-heating AEs of the Kristall series were studied in the MO modes with a telescopic type of UR and with one convex mirror, and conditions were determined to ensure efficient operation of the CVLS.

18. The maximum efficiency (2–3%), the radiation power (30–110 W) and the peak power density (up to 10^{13} W/cm^2) are provided in the CVLS operating under the MO–PA scheme due to:

– the introduction of a spatial filter–collimator (SFC) between the MO and PA, which provides the separation of the diffraction component from the MO multibeam radiation and its subsequent spatial matching with the aperture of the active medium of the PA;

– synchronization (alignment) in time of the light signals of the MO and the PA with accuracy not worse than ±2 ns;

– to ensure the level of power density of radiation at the input of the PA is not less than 0.5–1 W/cm^2;

– to ensure the stability of the position of the beam axis axis of the radiation beam $\Delta\theta \leq \theta_{dif}/10^2$–$10^3$.

At maximum peak power densities (10^{12}–10^{13} W/cm^2), the maximum productivity of microprocessing of materials up to 2–3 mm thick is ensured.

19. In the CVLS of the MO-SFC-PA type with industrial sealed-off AE Kristall models GL-201 (new model GL-205A) the average power of the radiation in a diffraction-quality beam is 34–35 W, practical efficiency ≅0.5% (efficiency of the PA ≅1.6%) at PRF 8 kHz, which is 4-5 times greater than the power of a separate MO. Using a sealed-off AE GL-201E (GL-201B) as a PA with a 1.3-fold extended discharge channel led to an increase in the radiation power of the CVLS to 60 W with a practical efficiency of 0.8% and an efficiency of 2.5%. In the case of the low-power AE Kulon GL-204 (2 W) as the MO and two AEs GL-201E (GL-201B), the power of radiation was more than 70 W with divergence 0.4 mrad, practical efficiency ≅0.93%, and the efficiency of the PA ≅2.1% at a PRF of 12.5 kHz, two experimental AEs GL-201E32 also as the PA – 105 W with a divergence of 0.3 mrad, 0.87% and 2%, respectively, at PRF 10 kHz.

20. The average radiation power of the CVLS (P_{CVLS}) consisting of n identical AEs as the PA, depending on the transmittance of the output windows, is determined by the formula

$$P_{CVLS} = \mathrm{P}_{MO}\tau_w^{2n+1} + P_{PA}\frac{\tau_w(1-\tau_w^{2n})}{1-\tau_w^2},$$

where P_{MO} is the average output power of the MO radiation beam at the input of the first AE of the PA, R_{PA} is the average radiation power of one AE of the PA, and τ_w is the transmittance of one AE window.

P_{CVLS} with non-clarified windows of AE in the PA can not exceed the power of six AEs in the PA ($P_{CVLS} \leq 6P_{PA}$).

21. The CVLS is an ideal device not only for determining the amplifying properties of a single AE, but also for studying its thermal gas lens (TGL), estimating the gas temperature and the state of the active medium before and after the disappearance of populations inversion.

22. On the basis of studies of the dependence of the radiation power of the CVLS on the time lag of the light signal MO relative to the signal of the PA, it is established that the AM of the pulsed CVL with respect to its own radiation has four characteristic time zones consecutively following each other and repeatinf from pulse

to pulse: weak absorption with a duration of 20–30 ns (occurs at the initial stage of development of the pump current pulse), amplification lasting 30–40 ns (arises on the steep leading edge of the current pulse), total absorption of duration 1000 ns (occurs at the section the current pulse and follows a pulse) and maximum transparency duration 1000 ns (before a new current pulse).

These properties of the AM of the pulsed CVL became the basis for the creation of new methods and electronic devices for operational power control and PRF of the radiation in the created industrial ALTIs Karavella-1, Karavella-1M, Karavella-2 and Karavella-2M for precision microprocessing of materials.

23. From the point of view of practical application of pulsed CVL and on its basis CVLS as a part of technological equipment for productive and qualitative microprocessing of electronic component materials, two of its optical operating modes are promising: with the UR with two convex mirrors and the telescopic UR. Only these UR and the use of SFC make it possible to form single-beam radiation with a diffraction (minimum) divergence and a high level of average power (10–100 W), so that the maximum peak power density (10^{11}–10^{13} W/cm^2) is reached in the focused spot of radiation (5–20 μm) .

The main conclusions and results investigations of pulsed CVLs and CVLS with optical resonators and systems are also valid for pulsed lasers with other metal vapours, as well as for gas and solid-state lasers with a short duration of population inversion.

5

Industrial copper vapour lasers and copper vapour laser systems based on the new generation of sealed-off active elements and new optical systems

The creation of modern industrial technological copper vapour lasers (CVL) and more powerful copper vapour laser systems (CVLS) with high reliability, efficiency and quality of radiation and improved mass and size indicators remains today one of the urgent tasks. The need for the development of CVL and CVLS is due to the wide possibilities of their application in science, technology and medicine. One of the promising and developing technological applications is the microprocessing of components for electronics.

5.1. The first generation of industrial CVLs

Industrial CVL Kriostat. The pulsed CVL Kriostat (LGI-201 according to Technical Instructions(TI)) is the world's first industrial copper vapour laser, which was developed in the USSR in the Istok Enterprise in 1974–1975. The Kriostat CVL is created on the basis of a sealed-off self-heating AE TLG-5 [222] with a diameter of the discharge channel of 12 mm and a high-voltage pulse power source of the IP-18 model.

The Kriostat laser is water-cooled, with a water consumption of about 3 l/min. In the high-voltage modulator IP-18, which generates current pulses for pumping AE (active element), a hydrogen thyratron of the TGI1-2000/35 type is used as the pulse switch [146]. The modulator is made in a direct electric circuit in which the thyratron TGI1-2000/35, AE TLG-5 and a storage capacitor with C_{stor} = 2200 pF form a single discharge circuit. A sharpening capacitor connected in parallel with the AE has a capacitance C_{sh} = 100 pF. The average power and divergence of the radiation of the Kriostat CVL were 3–6 W and 2–3 mrad, the efficiency 0.2%, the guaranteed operating time is 200 hours with the optimum 10 kHz PRF (pulse repetition frequency) and the power consumption from the three-phase 2.5 kW network. The design features and drawbacks of this laser are discussed in detail in [16, 26].

In the period until 1990, about 100 units of industrial CVL Kriostat were produced at the Istok Enterprise. Most of the CVL was completed with a wavelength-tunable laser on dye solutions of the brand LZhI-504 (λ = 530–900 nm) for a wide range of spectroscopic studies.

But because of low efficiency, low guaranteed operating time and poor quality of radiation, the application of the first industrial CVL in the future has become inappropriate.

The first powerful industrial CVL Kurs. The industrial CVL Kurs (LGI-202 according to TI) was developed in 1990–1991. The

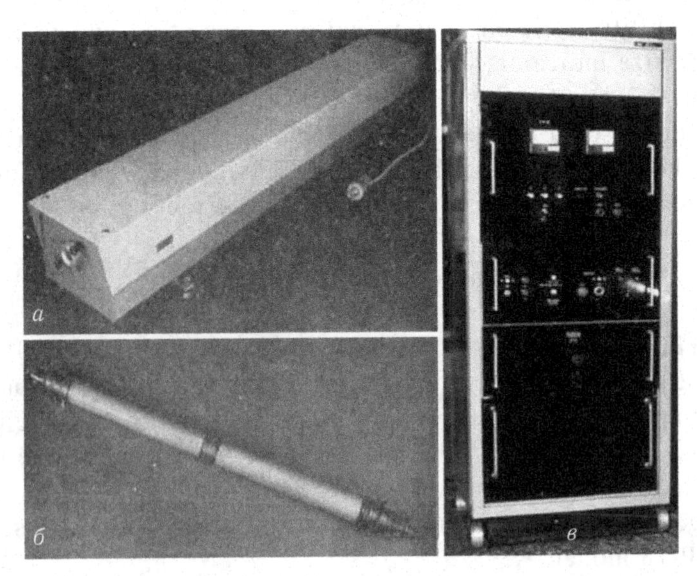

Fig. 5.1. The appearance of the radiator (*a*), the sealed-off self-heating AE TLG-5 (*b*) included in it and the high-voltage power supply IP-18 (*c*) of the first industrial pulsed CVL Kriostat.

main attention in development was paid to improving the energy parameters, reliability and quality of radiation. The Kurs laser, shown in Fig. 5.3, consists of a modernized power source IP-18 and a Klen cylindrical radiator (ILGI-202) with the first powerful sealed-off AE Kristall (GL-201). The power consumed by the laser from the three-phase network is not more than 3.5 kW, cooling by water.

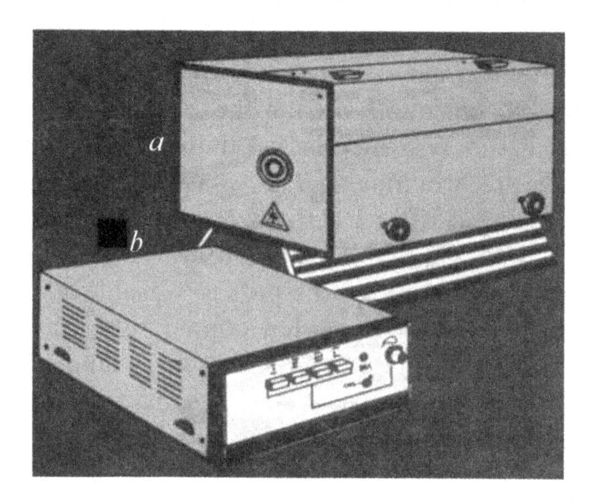

Fig. 5.2. Liquid pulse laser LZhI-504: *a* – radiator, *b* – flow unit. Pumping of LZhI-504 is produced by the radiation of the Kriostat CVL.

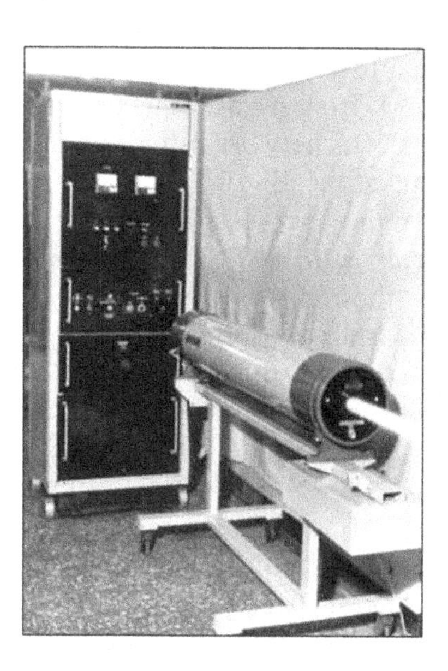

Fig. 5.3. The appearance of the Kurs industrial CVL (LGI-202).

The hydrotract of the laser passes successively through the IP-18 power source and the ILGI-202 radiator. The water flow for effective cooling of the thyratron TГGI1-2000/35 of the IP-18 modulator and the radiator housing is 4–5 l/min.

In the Kurs CVL (LGI-202), the power source IP-18 was used from the Kriostat series laser (LGI-201). At the initial stage of the development of the laser, the source of the IP-18 was not subjected to structural and circuit changes (its pump modulator was made in a direct circuit). The average output power of the LGI-202 laser with the optimum power consumption from the 2.5–2.6 kW rectifier and 10 kHz PRF was 12–14 W, which is 2–3 times higher than the power and efficiency of the previous LGI-201 laser. To ensure effective pumping of the sealed-off AE GL-201 in the IP-18 modulator, a circuit with a capacitive voltage doubling and a magnetic link for compressing the pump current pulses was built in [16]. The compression link was a water-cooled copper tube as a conductor with 120 ferrite rings of the M2000NM grade stamped on it with dimensions K20 × 12 × 6 mm. The storage capacitor with a PRF of 8–9 kHz was in the form of two series-connected capacitors with a capacitance of C_{stor} = 1500 pF, with a PRF of 10–12 kHz – with C_{stor} = 1000 pF. The optimum power consumption from the IP-18 rectifier with the PRF of 10–12 kHz was 3.3–3.0 kW, and at values below 10 kHz it increased to 3.6–3.7 kW. When the water pressure at the input of the system is 1.4–1.5 atm a nominal flow of water (4–5 l/min) is provided to cool the laser. To protect the supply three-phase electrical network from high-frequency radio interference that occurs in the pulsed CVL, network filters of the FSP-10 brand are connected at the input of the IP-18.

The nanosecond pump pulses generated in the modulator of the IP-18 power supply with 8–12 kHz PRF are transmitted via a high-voltage cable to the radiator's AE GL-201 for its heating and excitation. Figure 5.4 shows oscillograms of voltage and current pulses of the AE GL-201 with the execution of a high-voltage pump modulator in a direct circuit and a voltage doubling scheme with a 10 kHz PRF. The high-voltage pulse cable is designed for an average power of up to 5 kW and does not emit interference into the surrounding space. It passed a long (more than 2000 h) test when working with voltage pulses having an amplitude of 20–25 kV and a duration of 90–120 ns. Such a cable consists of a high-voltage wire PVMR-10–2.5 ms-12,5, three insulating tubes TV-40 (A) with a diameter of 14, 16 and 20 mm and two metal braids

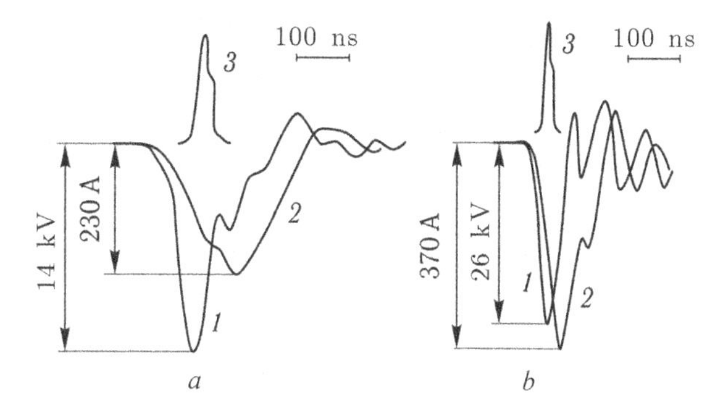

Fig. 5.4. The oscillograms of the pulses of voltage (*1*), current (*2*), and radiation (*3*) of the Kurs laser (LGI-202) with the the IP-18 modulator produced in the direct electric circuit (*a*) and with the voltage doubling circuit (*b*).

PML16-24. The strand of the high-voltage wire is copper–silver, its cross section is 2.5 mm^2, the insulation with a diameter of 12.5 mm is made of silicone material. Assembling the high-voltage cable is done in the following sequence: first, an insulating tube with an inner diameter of 14 mm is put on the high-voltage wire, then a tube with a diameter of 16 mm and a braid, then a tube with a diameter of 20 mm and again a braid. The first (internal) cable braiding is used as a reverse coaxial current lead, the outer one – as a screen grid. Tubes with diameters of 14 and 16 mm are designed to strengthen the insulation between the high-voltage wire and the inner braid, a tube with a diameter of 20 mm – to insulate the braids from each other. To prevent the formation of corona discharge at the ends of the cable, they are filled with a high-voltage sealant of the VGO-1 type. At the same time, the metal tube body of the radiator performs the function of the reverse current conductor. The cable temperature in normal operating conditions does not exceed 35°C.

Figure 5.5 shows the dependence of the average power of the laser radiation on the capacitance of the sharpening capacitor.

The maximum power is achieved at a capacitance C_{sh} = 340 pF. In the absence of a sharpening capacitor, the radiation power is approximately 17% less.

The Klen radiator has an original design, which is shown in Fig. 5.6.

The dimensions of the radiator: length 1510 mm, width (diameter) 212 mm, height 280 mm, weight 40 kg.

The main elements of the radiator are the sealed AE GL-201, the pipe water-cooled casing, end caps, mirrors of the optical resonator

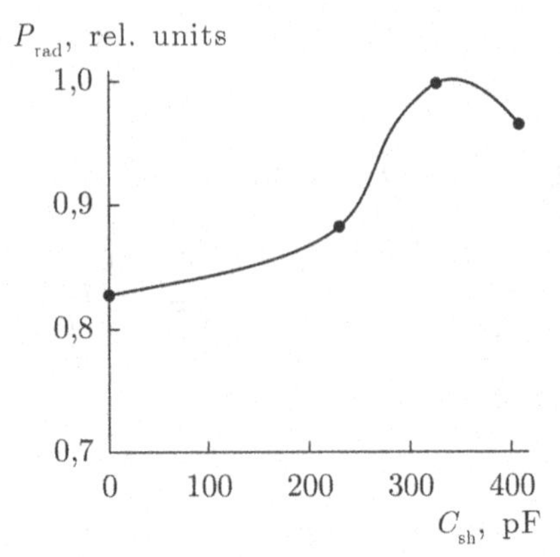

Fig. 5.5. Dependence of the average radiation power of the Kurs CVL (LGI-202) on the capacitance of the sharpening capacitor.

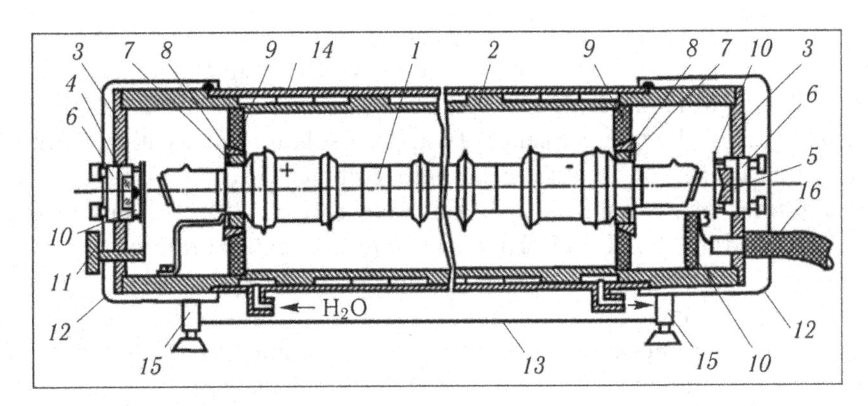

Fig. 5.6. The design of the Klen radiator (ILGI-202): *1* – sealed-off AE Kristall GL-201; *2* – the tubular case; *3* – end caps; *4* and *5* – mirrors of the optical resonator; *6* – the mechanism of adjustment of mirrors; *7* – metal half-rings; *8* – insulating rings; *10* – heat shields; *11* – mechanical shutter; *12*, *13*, *14* – decorative covers; *15* – supports; *16* – high-voltage cable.

and their adjustment mechanisms, AE fastening elements, thermal screens, mechanical shutter, decorative casing, adjustable supports.

The construction of the bearing shell is protected by patent for invention No. 1 813 307 RF [315] and is presented separately in Fig. 5.7.

The body was a pipe (item 1) made of AMts aluminium alloy, along its outer surface annular chambers with a depth of 5 mm were cut to circulate the cooling liquid (water), on the inner surface from the ends there were provided seats with a diameter of 184 mm for the fastening elements of the AE GL-201. The internal diameter of the body is 180 mm, the wall thickness is 12 mm, the length is

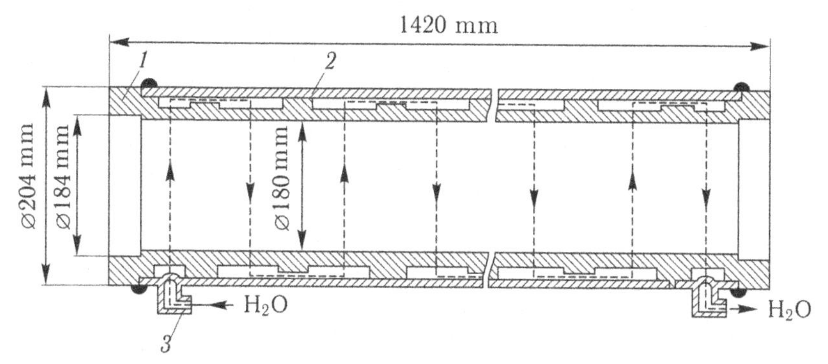

Fig. 5.7. The construction of the load-carrying case of the Klen radiator: *1* – pipe; *2* – casing; *3* and *4* – cooling system connections.

1420 mm. Sealing of annular chambers of cooling is provided with the help of a cylindrical casing (item 2) 2 mm thick also made of AMts alloy. The casing (item 2) is put on the pipe (item 1), tightly tightened with clamps and sealed by welding along the longitudinal joint and along the ring ends with the pipe (argon-arc welding). The choice of AMts alloy as the material of the housing is due to its good weldability and corrosion resistance, high thermal and electrical conductivity, and low specific gravity.

To pass water from one chamber to another and create a uniform temperature field across the dividing chambers, the upper and lower grooves are successively intersected in the cross section of the casing, the grooves in the upper partitions being the upper ones. With this design, the inlet and outlet connections are located at the bottom of the casing, which is also important for the aesthetic appearance of the device. It should be noted that the sequential method of cooling prevents the formation of sediments in the chambers of the body and their contamination.

An important point in the design of the radiator is the choice of the gap between the supporting tube body and the AE GL-201, which is under high pulse voltage. The gap, in turn, determines the inner diameter of the pipe, and the diameter – the mass of the radiator. In the nominal power supply mode of the AE, the power consumption from the rectifier of the power supply unit IP-18 at the PRF of 8–12 kHz is 3.5–3 kW, in the forced warm-up mode it is 4–3.5 kW. At the same time, the amplitude of the pulse voltage on the high-voltage electrode reaches 20–25 kV with a pulse duration of 100–120 ns. The experimentally found dependence of the breakdown voltage on the gap size (Fig. 5.8) showed that a gap of 14 mm corresponds to a value of 25 kV.

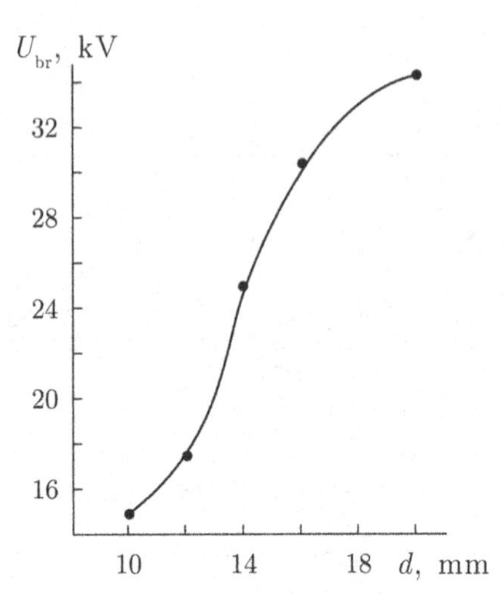

Fig. 5.8. Dependence of the breakdown voltage on the gap between the high-voltage electrode AE GL-201 and the bearing tube body of the Klen radiator at 8 kHz PRF.

To ensure a safety margin for the breakdown voltage, a 20 mm gap is chosen, which corresponds to a tube with an internal diameter of 180 mm. In addition, the outer diameter of the electrode assemblies is reduced from 140 to 134 mm in the AE. Thanks to these measures, a reliable structural reserve for breakdown voltage (up to 35 kV) was created.

The tests of a real sample of a radiator with an internal pipe diameter of 180 mm and an AE diameter of 134 mm confirmed the conclusions about the high electrical strength of this design.

Before installing the GL-201 AE (see Fig. 5.6) into the supporting tube casing (item 2), two 90.5 mm diameter electrode assemblies are threaded and screwed together with two half-rings (item 7) of the Al alloy. On the inner surface of each half ring there are two protrusions to provide electrical contact with the electrodes of the AE and to form an air-insulated layer 1.7 mm thick between the half-rings and the electrodes. AE with half-rings is installed inside the tubular body on fluoroplastic rings (key 9) with a transverse cut and a conical inner surface. For rigid fixation of AE in relation to the body, metallic rings (item 8) with a conical outer surface are provided. These rings, by means of the screws passing through them, screwed into the half rings, move along the axis along the surface of the half-rings, squeezing out the fluoroplastic insulators with a cut (item 9) to the wall of the housing. In the insulator located above the high-voltage electrode (cathode) of the AE, there are annular grooves to eliminate accidental surface breakdowns.

The optimal mode of operation in the AE is achieved when the power consumed by it is ~1.8 kW. The temperature of the envelope of the AE is about 30°C. The maximum operating temperature of fluoroplastic insulators should not exceed 220°C, since the release of harmful fluorine-containing substances begins. In the optimum mode, due to the above design features of the water-cooled half-rings, the maximum temperature of the insulator does not exceed 180°C.

The alignment mechanisms (item 6, Fig. 5.6) of the mirrors of the optical resonator (positions 4 and 5) are installed in the end aluminum flanges (item 3) with screws. To align the centre of the mirrors with the optical axis of the AE, it is possible to move the alignment mechanisms in the radial direction. The tuning of the resonator, consisting in the alignment of the mirror axes, is carried out with the adjustment screws of the alignment mechanisms, and their rigid fixation with the help of locking nuts and screws. Thin-walled screens (item 10) made of an aluminium alloy are designed to protect the alignment mechanisms from thermal radiation from the ends of the hot AE. Details of adjusting mechanisms are made of aluminum alloy, the main part of them are parts used in the Kareliya radiator (ILGI-201). Long-term operation of the Kareliya and Klen devices showed that the chosen design of the alignment mechanisms has high practical qualities.

A mechanical shutter with an aluminium curtain (item 11) designed to protect against laser radiation has two fixed working positions: 'open' and 'closed'. The shutter is conveniently integrated into the end flange (item 3) at the output of the radiation. The transfer of the shutter from one position to another is carried out by means of a handle with a catch located on the front (output) panel of the radiator.

The length of the optical resonator in the radiator (see Fig. 5.6.) is determined by the length of the tube body and is 1360 mm (length of the AE GL-201 1300 mm). The diameter of the mirrors of the resonator is 35 mm, the diameter of the laser beam is 20 mm. The energy, spatial and temporal characteristics of the radiation of the sealed-off AE GL-201 with different types of optical resonator have been extensively studied in [227–233] (see Chapter 4). By means of a compromise solution, based on the main application fields (laser pumping on dye solutions and transfer of radiation by means of a light guide to an object, etc.), a telescopic unstable resonator (UR) was chosen for the LGI-202 laser with a gain $M = 5$. The radius of curvature of the 'blind' concave mirror (item 5) is 3.5 m and of the output convex mirror (item 4) is 0.7 m. The output mirror having a diameter of 4 mm is glued with the help of TK-1 glue onto a plane-parallel, antireflecting thin plate of optical glass K-8 with a diameter of 35 mm. The coating of mirrors is multilayer dielectric, the reflection coefficient is ~99%. The angle between the optical axes of the plate and the mirror attached to it is 4–5°, which is necessary to eliminate the inverse parasitic connection between the plate and the active medium of the AE. When the output plate is installed perpendicular to the optical axis of the AE (i.e. turns by about 4°), a mirror with a radius of curvature $R = 0.7$ m ceases to work, and a plane-spherical resonator mode is realized. If the blind mirror is removed from the resonator, then the operating mode with a single convex mirror with $R = 0.7$ m is realized.

Table 5.1 shows the total output radiation power ($P_{\Sigma rad}$), power and divergence of the qualitative beam ($P_{rad.qual}$ and θ_{qual}) of the CVL with optical resonators of various types under the same conditions of pumping (excitation) of the AE.

At the same time, the high-voltage modulator of the power supply unit IP-18 was made according to the scheme of capacitive voltage doubling with a link of magnetic compression of the pump current pulses. The PRF was 10 kHz, the power consumption from the rectifier IP-18 3.2 kW. From the data given in the table it is seen that the maximum radiation power in a qualitative beam is achieved when using a telescopic UR with a gain of $M = 5$ (16.3 W). The plane and plane spherical resonators are inferior to the telescopic resonators with $M = 5$: they have lower power and greater divergence of the radiation beam. The single-mirror mode is also inferior in the same parameters, but the intensity distribution in the focusing

Tab. 5.1. The total power, power and divergence of the diffraction radiation beam of the Kurs CVL (LGI-202) with different types of optical resonators

Parameter / Resonator type	ΣP_{rad}, W	$P_{rad. qual}$, W	θ_{qual}, mrad
Plane-parallel	20.5	14	3.7
Plane-spherical with $R = 3.5$ m	23.2	16.1	9.3
Single-mirror with $R = 0.7$ m	11.4	11	3.8
Telescopic UR: $M = 5$ $M = 300$	22.6 20	16.3 8	2.5 0.12

plane is more uniform. When a resonator with $M = 5$ is replaced by a resonator with $M = 300$, the divergence of the beam decreases by an order of magnitude (from 2.5 to 0.12 mrad) and the power by a factor of two (from 16.3 to 8 W). At $M = 300$, the peak power density in the focusing spot reaches 10^{11} –10^{12} W/cm^2, which makes it possible to process virtually any material.

It should be emphasized that the design of the Klen radiator ensures a high stability of the position of the axis of the radiation pattern and the pulse energy. These qualities are achieved due to the fact that the supporting structure and alignment mechanisms with end flanges are made of the same material, the cooling system provides a uniform temperature field along the axis and along the cross section of the housing and the resonator space is closed.

The readiness time of the Kurs laser (LGI-202) is determined by the rate of the self-heating AE GL-201 output to the stationary thermal mode, when the output power of the radiation is set at a certain level. Figure 5.9 shows the dependence of the average power of the laser radiation on the warm-up time at the optimum power consumption consumed by the IP-18 power supply rectifier in the case of a direct modulator of the pump modulator and in the case of a voltage doubling scheme with a magnetic unit for compressing the pump current pulses at 10 kHz. With a direct circuit, the power consumption was 2.7 kW, with the circuit doubling the voltage 3.3 kW. In the first case (curve *1*), the generation started in 36 min, and the time to reach a stable level of radiation power, circuit (*1*) and a voltage doubling scheme with a magnetic pulse compression link

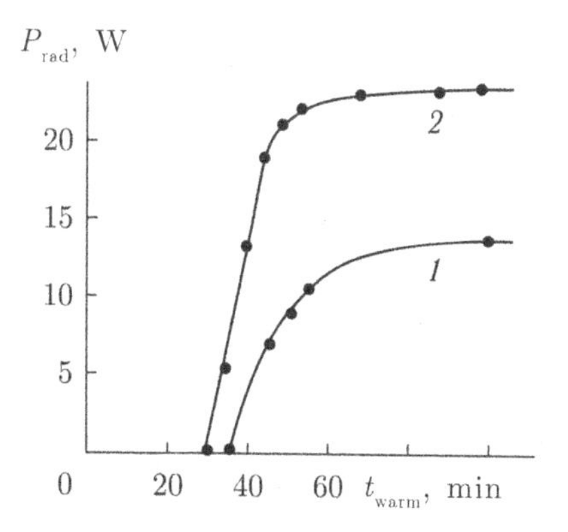

Fig. 5.9. Dependence of the average radiation power of the Kurs CVL (LGI-202) with a 10 kHz PRF on the warm-up time: the high-voltage modulator IP-18 is made in a direct electric circuit (*1*) and according to the voltage doubling scheme with the magnetic compression of the current pulses (*2*).

(*2*) equal to 13.5 W was ~80 min, in the second case (curve *2*) the corresponding values are ~30 min, 23 W and ~60 min.

In the forced warm-up mode, exceeding the nominal power consumption by 15%, the laser readiness time is reduced by approximately 10%.

Tests of the Kurs CVL (LGI-202) in order to determine the operating time to failure were carried out in a cyclic mode for two electrical circuits for the execution of a high-voltage pump modulator IP-18. The failure criterion is the reduction of the radiation power by 20% in relation to the nominal value, as well as the laser shutdown, which requires repair to restore it. Routine operations to replace certain items with a low warranty period services and cleaning of the water cooling system, conducted during operation, were not associated with failures. The tests were carried out on three Kurs CVLs.

The dependence of the average power of laser radiation on the operating time under the direct design of the pump modulator and the circuit for doubling the voltage with an 8-hour cyclic regime is shown in Fig. 5.10. The laser cut-off time between cycles was 1 hour. With a direct circuit, the duration of the laser test with a 10 kHz PRF was about 1850 h, with a voltage doubling scheme with a 9 and 10 kHz PRF 1000 h for each frequency. In the first case, the radiation power decreased for 1000 hours of operation from 12 to 11 W (curve *1*) at the optimum power consumed by the rectifier, P_{rect} = 2.6 kW, in the second from 24 to 22 kW (curve *2*) at P_{rect} = 3, 25 kW and PRI of 9 kHz and from 22.5 to 20.5 W (curve 3) with Pvpr = 3 kW and 10 kHz PRI.

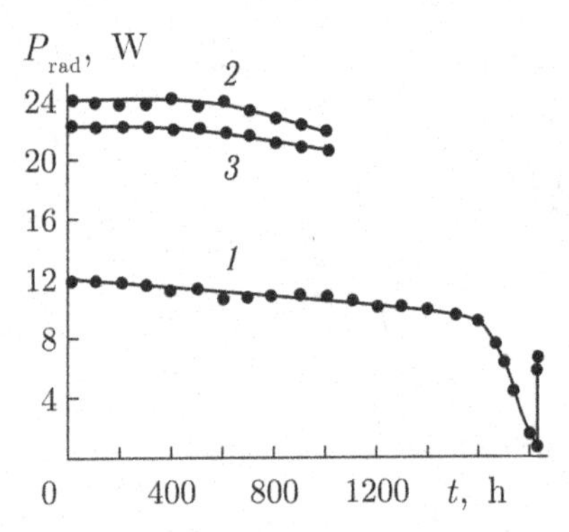

Fig. 5.10. Dependence of the average output power of the laser LGI-202 on the time of use: the modulator of the power supply IP-18 is made in a direct circuit, 10 kHz (1) and double voltage, 9 kHz (2) and 10 kHz (3).

Thus, for 1000-hour operating time, the laser radiation power is reduced by about 10%. The reduction in power is due to the dusting of the exit windows and the appearance of dendritic formations in tantalum copper generators. During the tests in each laser, the thyratron TGI1-2000/35 was changed twice (after 400–600 h) and the cooling system was cleaned three times (basically the thyratron TGI1-2000/35 cooling nozzles were clogged). With a direct design of the pump modulator in the test time interval from 1600 to 1850 h, there was a sharp decrease in the radiation power from 9.5 to 0.7 W. The replacement in the high-voltage modulator of the PI-18 pumping of the thyratron TGI1-2000/35 with a new one resulted in the restoration of the power to the level of 7 W. The main reasons for the decrease in the radiation power during this period were the increase in power losses in the thyratron and the depletion of copper in the outer AE generators.

The main parameters of the Kurs CVL (LGI-202) and, for comparison, the Kriostat CVL (LGI-201) are presented in Table 5.2.

From the data given in Table 5.2, it follows that the energy parameters of the LGI-202 laser when the modulator is executed in accordance with the voltage doubling scheme are almost twice as high as when executed in the direct circuit. Therefore, the main variant was the execution of a laser modulator with a voltage doubling and with a magnetic link for the compression of current pulses. In comparison with the first serial copper vapour laser, LGI-201, the LGI-202 laser has 4–5 times higher the average and peak power and energy in the pulse, three times higher efficiency, higher reliability and radiation quality. A total of about 20 CVLs LGI-202 were produced.

Parts of the elements and individual blocks of the modernized IP-18 of the LGI-202 laser are now morally obsolete and a power source has been created on a modern element base with improved mass and size parameters. The LGI-202M laser with a new power source has a power consumption of up to 4 kW and its mass is four times less than the mass of the IP-18 (100 kg). The appearance of the LGI-202M laser is shown in Fig. 5.11.

5.2. A new generation of industrial CVLs of the Kulon series

The basis for the development of a new generation of industrial pulsed CVL of the Kulon series was laid in the period 1998–2003. A new generation of CVL Kulon and on their basis lasers on gold

Table 5.2. The main parameters of pulsed CVL of models LGI-202 and LGI-201

Parameter name	LGİ-202		LGI-201 with direct circuit
	With a direct scheme	With a circuit for doubling the voltage	
Pulse repetition frequency, kHz	8–12	8–12	8–12
Diameter of the radiation beam, mm	20	20	12
Divergence of a qualitative radiation beam, mrad	2.5 (M=5);	0.12 (M=300)	3
Average power of total radiation, W	12–14	20–24	3–6
The duration of the radiation pulse (at half-height), ns	~20	~15	~20
Energy in the radiation pulse, mJ	1–1.75	1.7–3	0.25–0.75
Peak radiation power, kW	50–87.5	110–200	12.5–37.5
Power consumed from the rectifier, kW	2.4–2.7	3–3.7	2.2–2.5
Power consumed from a three-phase network, kW	≤3.5	≤4.5	□3.2
Practical efficiency (from the power source rectifier),%	≤0.52	≤0.8	□0.27
Laser efficiency (from the network),%	≤0.4	≤0.53	□0.2
Operating life to failure, h	1000	1000	200
Water consumption, l/min	4	5	3
Weight of radiator, kg	~40	~40	~40
Laser weight, kg	~400	~400	~400

vapours (GVL), a mixture of copper and gold vapour (CGVL), in combination with lasers on solutions of organic dyes (DVL) and with frequency doubling on non-linear crystals (NC) has been developed with close scientific and technical cooperation with the VELIT Company (Istra) on the basis of serial sealed-off self-heating AE Kulon (GL-206 according to TI) [16, 26, 41, 42, 141-145, 157-183]. The following industrial lasers belong to the new generation: CVL Kulon 01 with an average output power ≥10 W, GVL Kulon 02

Fig. 5.11. Appearance of CVL model LGI-202M.

with the power ≥1.5 W, CGVL Kulon 03 with power ≥4.5 W and CVL Kulon 04 power ≥15 W, with the modulation of the radiation of the CVL Kulon 05 and Kulon 06 with the power ≥10 W and ≥15 W, respectively (YuVEI.433,713.001 TU). The lasers of the Kulon series are distinguished by high reliability, efficiency and stability of parameters, and small weight and size parameters. Despite the results, the work to further improve their design and excitation conditions continues. The laser design is protected by 8 patents [160–163, 316, 317, 319, 321], one of which [161] is for the excitation method. Domestic and foreign analogs are not available.

Industrial CLVs Kulon-01, GVL Kulon 02, CGVL Kulon 03 and CVL Kulon 04. Figure 5.12 represents the appearance of industrial impulse CVL Kulon-01, GVL Kulon 02, CGVL Kulon 03 and CVL Kulon-04 and separately, on the right, its radiator and serial sealed self-heating AE Kulon (see Chapter 2) [16, 41, 42]. The first was designed CVL Kulon-01, radiation power ≥ 10 watts. On its basis, other industrial lasers for metal vapors were created and investigated: GVL Kulon 02 with a power of ≥ 1.5 W, CGVL Kulon 03 with power ≥ 4.5 W and CVL Kulon 04, power ≥ 15 W [16, 157, 158, 160, 162-164, 176, 178, 181, 183]. The appearance, design and composition of industrial pulsed CVL, GVL and CGVL are identical. The CVL Kulon-01 uses the sealed-off AE Kulon model GL-206D with a diameter of the aperture of the discharge channel 14 mm, in the GVL Kulon 02 and CGVL Kulon-3 – sealed-off AE of the same construction and geometric dimensions as GL-206E and differ only in the composition of the active substance (gold or copper and gold). The CVL Kulon-4 uses the sealed-off AE Kulon, model GL-206I,

Fig. 5.12. Industrial CVLs Kulon 01, GVL Kulon 02, CVLS Kulon 03 and CVL Kulon 04 and their radiator (right) and sealed-off AE.

the design of which is similar to GL-206D, but longer by 130 mm, which increased the radiation power by 1.4 times.

The industrial pulsed CVLs, GVLs and CGVLs of the Kulon series include a radiator with a flat resonator or with a telescopic UR and a high-voltage switching power supply providing pumping, automatic output of the AE to the optimum operating temperature regime and its stable operation [158, 164, 176, 181, 183]. The emitter is a three-rod optomechanical block with the sealed-off AE of the Kulon series installed on its axis, along the ends there are mirrors of the resonator in the mechanisms of their alignment (Fig. 5.12). Round the AE, coaxially, there is cylindrical heat shield with a removable top cover and a temperature sensor. A high voltage impulse voltage of negative polarity is fed from the power source to the cathode of the AE, and the anode of the AE is grounded. The radiator is equipped with a forced air cooling system with two exhaust fans mounted on the rear end face of the laser. Structurally, the Kulon laser is made in the form of a monoblock, which has a horizontal layout (Fig. 5.13, top view). The monoblock measures 1250 × 335 × 190 mm and weighs no more than 45 kg.

The design of the monoblock and the technical parameters of the high-voltage pulse PI allow for the installation and pumping of the four most powerful AE Kulon – GL-206E (Kulon LT-10Cu), GL-206G (Kulon LT-1.5Au) without additional laser modifications. GL-206J (Kulon LT-Au-Cu) and GL-206I (Kulon LT-15Cu) [164]. It should be noted that the design of the radiator in this laser makes

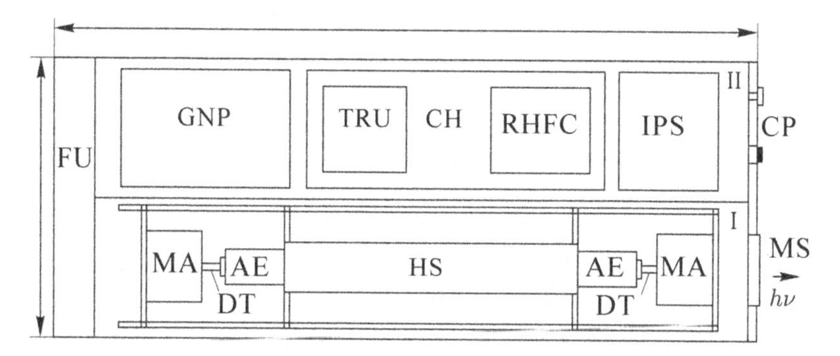

Fig. 5.13. The layout of industrial CVL Kulon-01, GVL Kulon 02, CGVL Kulon 03 and CVL Kulon 04. *I* – radiator: AE – active element; HS – heat shield of the AE; MA – mechanisms of alignment of mirrors of an optical resonator; MS – mechanical shutter; DT – dustproof dielectric tubes; II – pulsed power supply: IPS – input power supply; RHFC – block of the rectifier and high-frequency converter; TRU – transformer-rectifying unit; CH – charger; GNP – high-voltage generator of nanosecond pulses; CP – control panel; FU – fan unit.

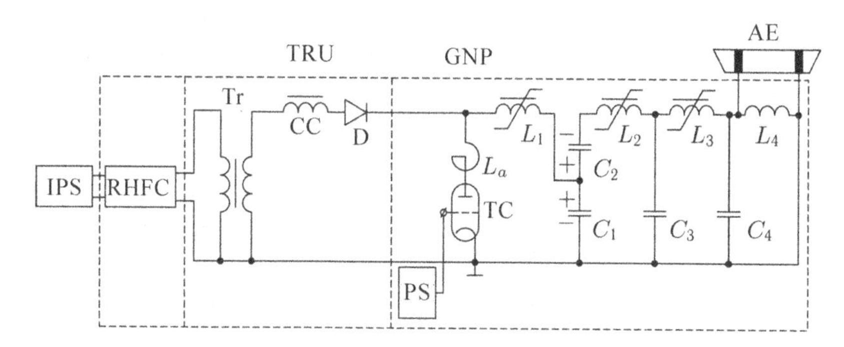

Fig. 5.14. The basic electrical circuit of the high-voltage power source of nanosecond impulses of industrial CVLs Kulon 01, GVL Kulon 02, CGVL Kulon 03 and CVL Kulon 04: IPS – input power unit; RHFC – block of the rectifier and high-frequency converter; TRU – transformer rectifier unit: Tr – high-voltage transformer, Dr – charge choke, D – high-voltage diodes; GNP – high-voltage generator of nanosecond pulses (thyratron–magnetic pulse shaper); TC – thyratron commutator; PS – pulse submodulator.

it possible to use less powerful models of the AE Kulon series [16, 41, 42].

A schematic electrical diagram of the high-voltage power of nanosecond pulses of CVL, GVL and CGVL of the Kulon is shown in Fig. 5.14.

The pulse power source of industrial CVLs, GVLs and CVGLs contains the input power unit (IPU), the rectifier and resonant converter unit, the transformer rectifier block (TRB) and the high-voltage generator of nanosecond pulses (GNP). The IPS contains starting commutation equipment, a control and indication device, a protective shutdown button, power supplies for supplying auxiliary voltages. The operation of the block is carried out as follows. The automatic switch Power-220 is switched on on the front panel of the power source. At the same time, the fan unit (FU) starts to operate, which during the operation of CVL cools the power source channels and the AE emitter. Then the electromechanical switch turns clockwise to the ON position and the auxiliary power supplies and the power supply of the pulsed thyratron are turned on. After a 7-minute warm-up of the thyratron, pulsed pumping pulses are provided to the AE, which are necessary for preheating the AE. After 6 minutes, the amplitude of the current pulses, in the automatic mode, increases to the nominal value. During the next 60 minutes, the AE is smoothly warmed up to a stationary operating temperature and the maximum laser radiation power is established. To turn off the laser, turn the electromechanical switch in the Off position and switch off the Power-220 switch after at least 30 minutes.

The RRCU and TRB form an adjustable high-voltage power supply charger. RRCU consists of a power rectifier, a battery of filter capacitors, a control board and two JGBT modules. TRB consists of a high-voltage pulse transformer (Tr), a charge choke (Ch) and a high-voltage rectifier in the form of a series of pulse diodes (D). These two units are designed to provide resonant mono-pulse charging of a capacitor storage of a nanosecond pulse generator (NPG) and to stabilize the charging voltage level to within ±2%, to protect the generator elements against short circuit currents and overvoltages at idle running. By changing the level of the charging voltage, the output power of the power source is regulated in accordance with the requirements for the energy consumption of the laser.

The essence of the method for regulating the charging voltage on a capacitor storage device of NPG consists in the following [163]. The capacitive accumulator is mono-pulse charged from a low-voltage power supply (power rectifier) through a single-contact quasiresonant transistor converter (JGBT-modules) during the flow of only a part of the direct half-wave of the charging current to the required voltage on the capacitive storage. Then the accumulator is discharged at a constant frequency of the current pulses through the thyratron commutator of the NP|G to the gas-discharge AE at the time determined by the time of passage through the zero value of the same half-wave of the direct charge current, the duration of which is equal to the half-period of the natural frequency of the charging circuit. As a result, a current pulse is generated in the AE with the required amplitude, front, and duration. The pulse frequency is determined by the mode of operation necessary for the laser. In operation, the core of the high-voltage transformer is not saturated, so the maximum frequency of the pump pulses is not limited to the operating mode of the device elements. By changing the setting level of the reference voltage on the control board of JGBT-modules, it is possible to change the level of the charging voltage of the capacitive accumulator from the maximum value practically to zero.

The pulse submodulator (PS) is designed to form control grid voltages that transfer a pulsed thyratron (TGI) to a conducting state [162]. A feature of PS when a hydrogen generator of a tetrode structure is applied to a TMPS is the introduction of feedback circuits in an electrical circuit to provide indirect control of charge concentrations in its cathode-current area and control of the level of voltages on NPG elements and the entire power source, which increases the reliability of both the thyratron commutator and the laser as a whole. The device also has a voltage stabilization board for the hydrogen generator of the thyratron. It is necessary for the formation of a constant stabilized voltage (6.3 V) applied to the heater of the thyratron hydrogen generator.

The NPG, the basic electrical circuit of which is shown in Fig. 5.14, is made in the form of a thyratron-magnetic current pulse shaper (TMPS) with a capacitive voltage doubling and two magnetic compression links. The NPG contains a high voltage pulsed thyratron commutator (TGI), a pulsed submodulator (PS) that generates firing pulses on a thyratron grid, three storage capacitors (C_1, C_2, C_3), three nonlinear inductors (L_1, L_2, L_3), inductance (L_4) and a sharpening capacitor (C_4) connected in parallel to the AE, a common device bus and an anode reactor (L_a). The output of the high voltage battery charger is connected to the terminal of the anode reactor L_a and the output of the first inductor L_1, the second terminal of which is connected to the common point of two series-connected storage capacitors C_1 and C_2. The second outlet of the reactor L_a is connected to the anode of the thyratron switch TGI, the output of the first storage capacitor C_1 to the common bus of the device, and the output of the second storage capacitor C_2 through the second nonlinear inductance L_2 to the output of the third storage capacitor C_3, which through the third nonlinear inductance L_3– a sharpening capacitor C_4 and an AE electrode with a parallel inductance L_4. The second

AE electrode through the common bus is connected to the second terminal of the third storage capacitor C_3, the cathode of the thyratron commutator, and the second terminal of the battery charger TBB. This technical solution of the electrical circuit makes it possible to reduce switching losses in the thyratron switch of the TGI and increase the efficiency of the pulsed laser.

NPG works as follows. Before the formation of the operating pump pumping pulse, the storage capacitors C_1 and C_2 are charged from the memory through the first nonlinear inductance L_1 and inductance L_4 to a certain amplitude value of the voltage with the polarity indicated in the diagram (Fig. 5.14). After switching on the thyratron switch of the TGI from the PS at the first moment of time, all the voltage on the capacitor C_1 (equal to the output of the charger) is applied by the inductance L_1. The nonlinear inductance L_1 is a toroidal structure with an annular core of a ferromagnetic material, which passes into a saturated state after a certain time interval determined by the material and the core section, the number of turns of the core winding and the voltage on the storage capacitor C_1. In this case, the value of the nonlinear inductance L_1 decreases sharply and further charge transfer of the capacitor C_1 is determined by the characteristic impedance of the circuit formed by the capacitance C_1, the inductance L_1 in the saturated state, the anode reactor L_a, and the thyratron commutator. The charge of the capacitor C_1 occurs to an amplitude close to the voltage of the high-voltage memory and the second non-linear inductance L_2 is under doubled potential of the series-connected capacitors C_1 and C_2. The inductance core L_2 is saturated. The second L_2 and third L_3 inductances are also made on annular cores of a ferromagnetic material. The inductance of L_2 is chosen so that its saturation occurs at the instant of time corresponding to the complete recharging of capacitor C_1 to the opposite sign. At the time of saturation of the core of the nonlinear inductance L_2 the storage capacitors C_1 and C_2 are discharged to the third storage capacitor C_3 with a current amplitude determined by the characteristic impedance of the circuit formed by the capacitors C_1, C_2 and C_3 and also by the nonlinear inductance L_2 in the saturated state. That is, the electrical energy of the series-connected capacitors C_1 and C_2 is converted into the electrical energy of the capacitor C_3, but in a time much shorter than the time of the direct anode current pulse through the thyratron commutator. After charging the third storage capacitor C_3 to an amplitude approximately equal to twice the voltage of the memory, the core of the third nonlinear inductance

L_3 saturates. The parameters of this inductor are chosen so that its saturation occurs at the time of the full charge of capacitor C_3. As a result, the sharpening capacitance C_4 of the AE with the parallel inductance L_4 is energized at capacitance C_3 and a current pulse of the required amplitude and duration is formed in the AE. Thus, in the magnetic links of compression, there is a consecutive transfer of energy from the storage capacitors to the AE, while simultaneously compressing the pulses in time. The magnetization reversal of the ferromagnetic cores of nonlinear inductances L_1, L_2 and L_3 in the reverse direction occurs as a result of the charge current of the storage capacitors C_1 and C_2 from the memory and does not require special reversal reversal circuits. The aggravating capacitance C_4 provides the front of the pump current pulse. Inductance L_4 serves to charge capacitor C_2 and to short-circuit the discharge gap of the AE in the interimpulse period, which creates the necessary conditions for the dispersal of the metastable levels of copper atoms.

In the NPG with the electrical circuit considered, at the initial stage of switching on, the current of the thyratron commutator does not exceed the value of the magnetization current of the nonlinear inductance L_1, and after saturation of its core, the rate of increase and the amplitude of the commutator current is limited by the inductance of the anode reactor L_a. As a result of a decrease in both the slew rate and the pulse amplitude of the anode current, the switching losses in the thyratron decrease and the resource of its work increases. But on the other hand, the introduction of the anode reactor L_a dramatically changes the parameters of the discharge circuit and degrades the lasing characteristics due to the strong influence of the rate of increase in the pump current pulses on the excitation of the active medium. Therefore, the decrease in the amplitude and the increase in the duration of the leading edge of the current pulses in the GNI are compensated by the introduction of the storage capacitor C_3 and the inductance L_3 into the electrical circuit, which ensures correction of the initial pulse and the formation of the required pump pulse. In addition, the increase in the generation power and the laser power parameters is due to a greater degree to the matching of the output parameters of the STI with the characteristics of the gas-discharge channel. The ratio of the capacitances of the three storage capacitors C_1, C_2 and C_3 is 1: 0.95: 0.4 and the nonlinear inductances L_2 and L_3 in the saturated state of 10: 1 provide such matching of the generator and load parameters, at which the AE generation power increases.

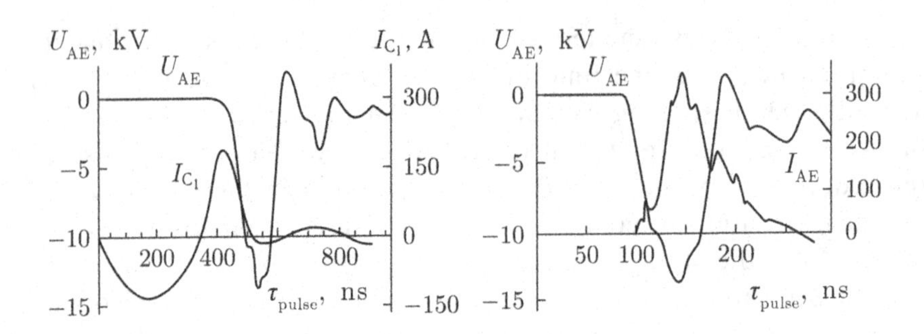

Fig. 5.15. The oscillograms of the current pulse of the first storage capacitor (I_{C1}), voltage pulses (U_{AE}) and discharge current (I_{AE}) of the AE GL-206D of the Kulon 01 CVL.

A common thyratron, for example, TGI1-1000/25 manufactured by the enterprise Contact (Saratov), and a thyratron of the tetrahedral type TGI2-1000/25K type can be used as a thyratron commutator in NPG. The optimized capacitance of storage capacitor C_2 is equal to 1000 pF, the sharpening capacitor C_4 is 110 pF. As capacitors C_1, C_2 C_3 and C_4, low-inductance ceramic capacitors KVI-3 are used. The inductance of L_1 in the saturated state is approximately 15 μH, the inductance of L_2 is 8 μH and L_3 is 2.2 μH. The inductance of AE with the reverse current conductor is 0.8 μH, the inductance of L_4 is 80 μH.

The results of tests of Kulon industrial lasers. When testing industrial lasers Kulon, the voltage at the thyratron anode was recorded, the shape and amplitude of the current pulses of the first storage capacitor C_1 and the voltage and pump voltage pulses through the AE, the power consumed from the rectifier of the PI power, and the average power of the radiation. The practical efficiency of a laser was calculated. Figure 5.15 shows the oscillograms of the current pulse of the first storage capacitor (I_{C1}), voltage pulses (U_{AE}) and pumping current (I_{AE}) of the AE CVL Kulon 01.

The oscillograms of the pulses of the voltage and discharge current of the AE GL-206E (LT-10Cu Kulon) and Kulon 01 (Fig. 5.15, *b*) show that, compared to the single-ended version, the current amplitude through the thyratron is reduced to 120 A (i.e. 2.5 times), and its duration, on the contrary, was increased to 350 ns (2.5 times) [157, 158, 160, 164]. In this case, the decrease in the amplitude and the increase in the duration of the anode current pulses through the thyratron commutator due to the introduction of the anode reactor L_a and initially impairing the parameters of the pump pulses are

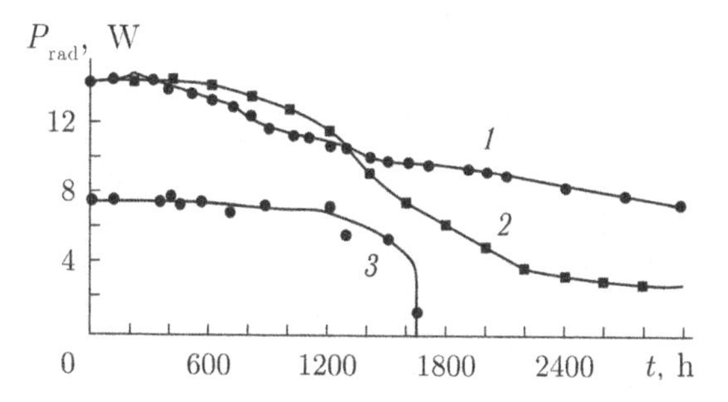

Fig. 5.16. Dependence of the average radiation power of industrial CVL Kulon 01 with sealed-off AE GL-206E No. 127 183 (curve *1*) and No. 127 155 (curve *2*) and sealed-off AE GL-206D No. 127 149 (curve *3*).

compensated by introducing into the TMPS a second component of the magnetic compression of the pulses. Thus, at the same time, a significant reduction in energy losses in the thyratron commutator is achieved and the pump efficiency remains at the same level. The decrease in losses in the thyratron was confirmed during the experiment by a decrease in the temperature of the thyratron anode from 210 to 105°C, while the contribution of the filament energy of the thyratron cathode to the anode temperature was about 35°C. The circuitry solutions adopted in the development of the industrial laser made it possible to reduce not only the switching losses in the thyratron, but also to increase the reliability of the laser elements as a whole and to ensure the stability of the output radiation parameters.

The industrial CVL Kulon 01 with sealed-off self-heating AE model GL-206D (No. 127 183) with an increased content of active copper substance was subject to long-term resource tests. CVL tests were carried out with an 8-hour cyclic mode at 15 kHz PRF and an initial average radiation power of 14.5 watts. During 3000 hours, the average power of radiation decreased from 14.5 to 7.5 W. As a thyratron commutator in the TMFI laser, a thyratron of the TGI2-1000/25K tetrahedral design was used. During the life tests, which lasted up to 3600 hours, no failure was detected in the laser operation.

The dependence of the average radiation power for this CVL Kulon 01 with AE No. 127 183 on the operating time is shown in Fig. 5.16 (curve 1). Here, for comparison, analogous dependences with AE GL-206E No. 127 155 (curve *2*) and GL-206D (Kulon LT-5Cu) No. 127 149 (curve 3) obtained during bench tests are also shown.

Table 5.3. Radiation power and efficiency of industrial metal vapour lasers of the Kulon series

Laser model	Average radiation power, W	Practical efficiency,%	Percentage of power at individual wavelengths		
			0.5106 μm	0.5782 μm	0.6278 μm
Kulon 01	15	0.7	50	50	—
Kulon 02	1.8	0.25	—	—	100
Kulon 03	7.4	0.3	53	38	9
Kulon 04	20	0.9	50	50	—

A comparison of the graphs shows that a more uniform (near-linear) decrease in the radiation power took place with AE GL-206E No. 127 183 (curve 1). This is due, firstly, to the fact that the charger of the power source provides a stable level of charging voltage on the storage capacitors of TMPS and, secondly, a large reserve of active copper substance in the extreme (end) copper vapour generators.

When the industrial laser Kulon was optimized, the parameters of the pump pulses were established and the main output characteristics with the most powerful models of AE on copper, gold, and gold vapour mixtures, Table 5.3.

The standard tests of the industrial CVL Kulon 01 showed that with the AE GL-206E (LT-10Cu Kulon), the maximum amplitude of the discharge current generated by the TMPS was fixed at 480 A, at which the radiation passes into the yellow region (supernumerary overheating mode active medium AE). With an optimum current amplitude of 320 A, the average radiation power of the CVL in the stationary thermal regime was 17 W, the practical efficiency was 0.7%. In the optimized mode, the amplitude of the discharge current, the average radiation power and the practical efficiency of the GVL Kulon 02 in the steady-state thermal regime were 300 A, 1.8 W, and 0.25%, CGVL Kulon 03 – 320 A, 7.4 W and 0.3% and CVL Kulon 04 – 350 A, 20 W and 0.9%.

Table 5.4 shows the range of values of the main parameters for lasers of the Kulon series. The average radiation power of 1.5 W corresponds to the CVL Kulon 01 with sealed-off AE GL-206B (Kulon LT-1.5Cu), power 20 W – CVL Kulon 04 with the most powerful sealed-off AE GL-206I (Kulon LT-20Cu) (see chapter 2).

The main advantages of the developed industrial pulsed metal vapour lasers of the Kulon series are air cooling, a long service

Table 5.4. The main parameters of industrial impulse lasers on metal vapours of the Kulon series

Parameter	Value
Wavelengths of radiation, μm	0,5106; 0,5782; 0,6278
Diameter of the radiation beam, mm	14
Average radiation power, W	1,5–20
Pulse repetition frequency, kHz	14 ± 1
Duration of radiation pulses (at half-height), ns	10 ± 2
Divergence of radiation, mrad with a flat resonator with telescopic UR (M = 200)	4 0,2
Power consumed from a single-phase network, kW	0,7–2,5
Readiness time, min	40–60
Overall dimensions, mm	1260 × 355 × 195
Weight, kg	≤46
Average operating time to failure, h	>2000
Technical service life, h	≥5*

Note. Laser cooling with forced air. The ambient temperature is not more than +35 °C. The operating mode is automatic.
*Based on the results of long-term tests of the first industrial Kulon lasers, the technical service life is more than 10 years.

life (over 2000 hours) and a service life (more than 10 years), relatively high efficiency, high stability and reproducibility of the output radiation parameters, low weight and size parameters (0.55 kg/dm^3) – can be widely used in science, technology and medicine. Industrial lasers Kulon have neither domestic nor foreign analogues.

Industrial CVL Kulon 05 and Kulon 06 with high-speed pulse modulation. The development of industrial LUM Kulon 05 and Kulon 06 with high-speed pulse modulation of radiation, with high reliability and compact design, was carried out jointly with LLC SPE VELIT (Istra, Moscow Region) [159–163, 170–171 , 180, 181]. Appearance of the CVL Kulon 05 is shown in Fig. 5.17, CVL Kulon 06 – in Fig. 5.18

Fig. 5.17. Appearance of industrial CVL Kulon 05 with high-speed pulsed modulation.

Fig. 5.18. Appearance of industrial CVL Kulon 06 with high-speed pulsed modulati^^

The design of the LLP Kulon 05 is single-module with overall dimensions of 1250 × 530 × 190 mm and weight not exceeding 60 kg, CVL Kulon 06 – two-module: power source with overall dimensions of 1250 × 395 × 185 mm and weighing not more than 50 kg and a radiator with dimensions of 1000 × 200 × 230 mm and a mass of 20 kg. The radiator consists of sealed-off AE and a cylindrical water-cooled metal heat sink. The AE is located inside the heat sink along its axis. These CVLs use a sealed-off self-heating AE of the model GL-206E (LT-10Cu Kulon) with an average radiation power ≥10 W or GL-206I (Kulon LT-15Cu) power ≥15 W produced by the Istok Company. An auxiliary low-power generator of nanosecond pulses is provided in the power supply to generate additional current pulses in the AE, allowing controlled modulation of the radiation. Therefore,

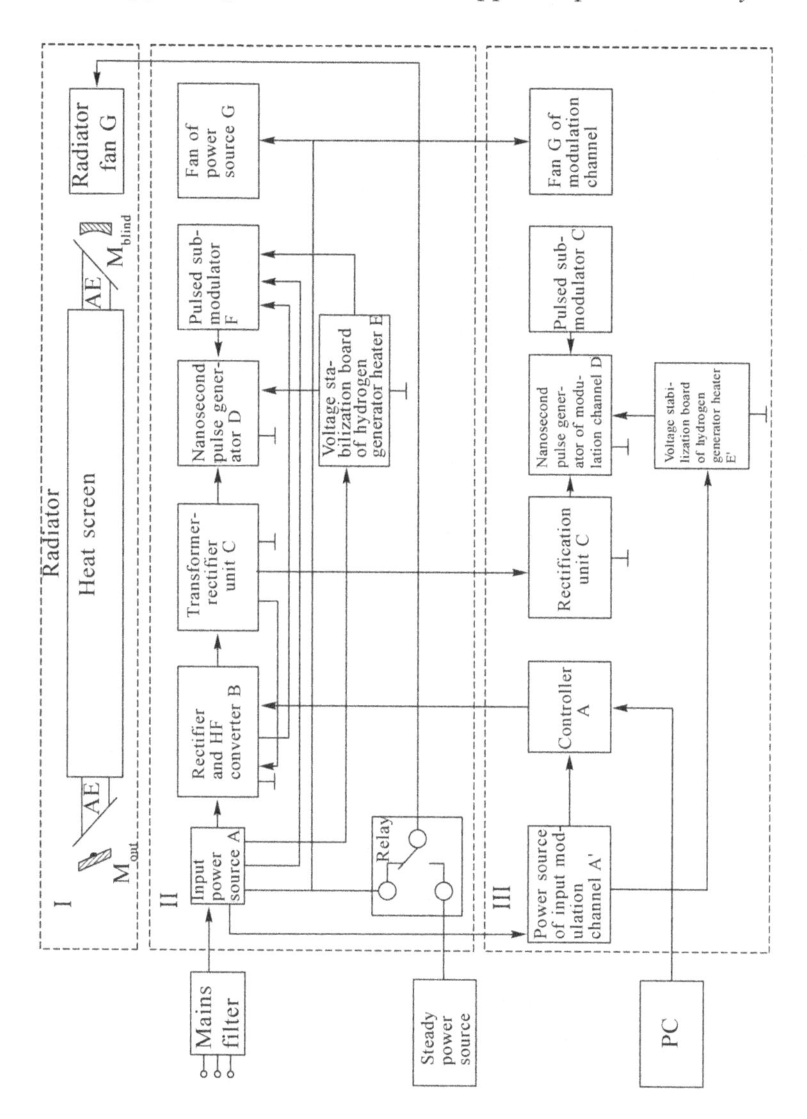

Fig. 5.19. Structural block diagram of industrial CVLs Kulon 05 and Kulon 06: *I* – radiator; *II* – power source of nanosecond pulses of the main pumping channel of the AE; *III* – power source of nanosecond pulses of the modulation channel.

a single-module CVL Kulon 05 with an auxiliary generator is wider than the basic CVL Kulon 01 by 19.5 cm.

Cooling of the LP Kulon 05 – air, forced, in the LP Kulon 06 cooling of the power source – air, forced, and the radiator – water. These CVLs operate in an automatic mode and they control radiation from a personal computer according to a given algorithm.

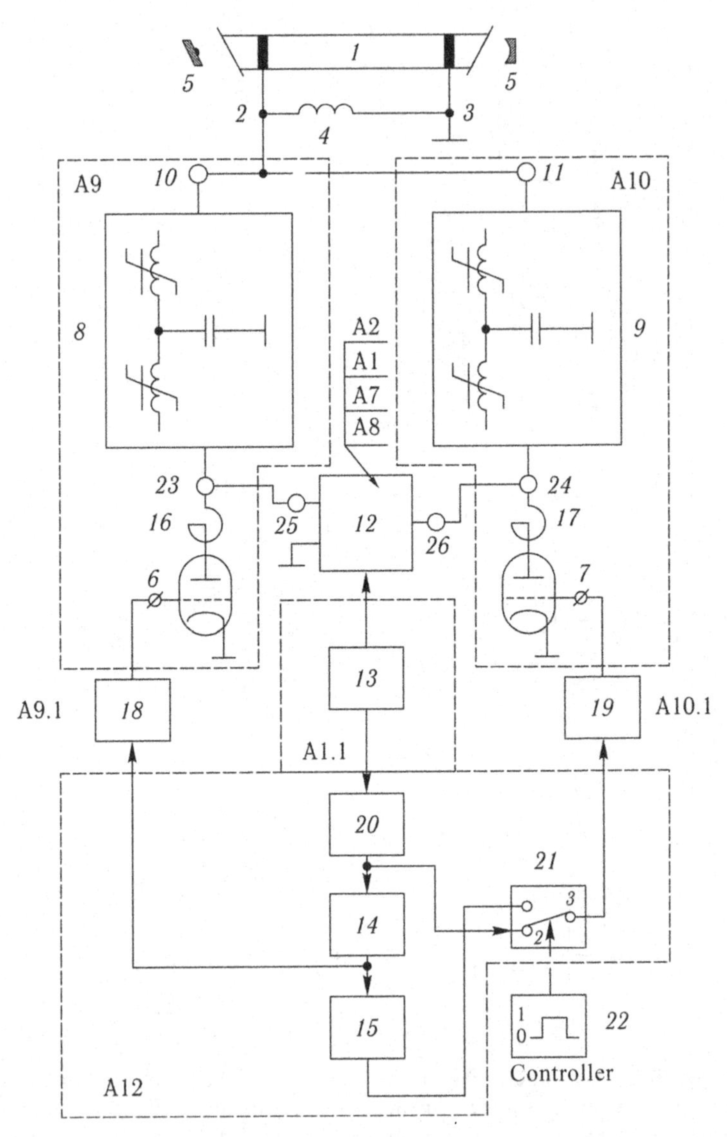

Fig. 5.20. Functional electrical diagram of industrial CVL Kulon 05 and Kulon 06: *1* – AE; *2, 3* – AE electrodes; *4* – shunt inductance; *5* – mirrors of the optical resonator, *6, 7* – high-voltage thyratron commutators; *8, 9* – magnetic compression links with capacitive energy storage devices (nonlinear artificial shaping lines); *10, 11* – terminals of capacitive energy storage devices, *12* – adjustable high-voltage power supply (transformer-rectifier block (TRB)); *13* – trigger pulse generator; *14, 15* – adjustable delay lines, *16, 17* – anode reactors; *18, 19* – impulse submodulators (ignition board of thyratron); *20* – unregulated delay line, *21* – electronic relay; *22* – controller; *23, 24* – conclusions of the anode reactors; *25, 26* – terminals of the regulated high-voltage power supply, A1 – charger,

Figure 5.19 shows the structural block diagram, Fig. 5.20 the functional electrical diagram of the industrial CVL Kulon 05 and Kulon 06.

The power of industrial lasers is provided by a single-phase alternating current network with a voltage of 220 ± 10% V, frequency 50 ± 0.5 Hz. The power consumption of the CVL from the grid does not exceed 3 kW. The power source in the CVL provides automatic output to the operating mode and maintenance of the set value of the input power in the AE during long-term operation in both continuous and cyclic modes, as well as electronic protection over the air and the external network.

The basic electrical circuit of the high-voltage generator of nanosecond pulses from the AE (A9) pumping channel and the low-power additional generator of the nanosecond modulation channel (A10) of the CVL Kulon 05 and Kulon 06 in terms of the composition of the elements and the physical principle of operation are identical with each other and with the high-voltage pulse generator of a single-channel base CVL Kulon 01. The presence of the additional modulation channel (A10) and the channel synchronization device (A12) in the CVLs Kulon 05 and Kulon 06 allow to perform power control according to a predetermined algorithm, including monopulse and packet radiation modulation, which significantly expands the possibilities of their practical application.

The physical essence of the CVL operation with radiation modulation with sealed-off self-heating AEs is as follows. In the operation of CVL with an additional pulse and constant energy input in the AE under conditions of stabilized plasma parameters, changing the delay between pulses of the main pump current (excitation) and additional low-power current pulses, it is possible to control the energy and space–time characteristics of radiation (generation) and wide limits, as well as chromaticity.

In order to provide the generation mode in the laser, an additional low-power current pulse is formed after the main excitation pulse, and an additional pulse is formed in front of the main pulse to provide the mode of extinction of the generation. The energy of additional current pulses (formed by a low-power generator) should be sufficient only for the population of metastable (lower) laser levels of the active substance – copper atoms. An additional impulse in this case determines only the processes of populating metastable levels, but does not affect the processes of their relaxation in the plasma in the interimpulse period, i.e., an additional pulse must be near the main, exciting, current pulse. Naturally, the effective control of the output energy characteristics of the laser is provided to the greatest extent when the time difference between the additional current pulse and the main pulse (the excitation pulse) is less than the lifetime of the metastable levels. With the experimental optimization of the CVL, the indicated time detuning was no more than 1 μs. In addition, from the point of view of stabilizing the parameters of the active medium plasma, this is the optimum operating mode of CVL, when the power consumed from the electric network during generation (with a lagging additional pulse) is equal to the power consumed by the laser in the case of extinction of the generation (with an advanced additional pulse). This mode is achieved by adjusting the phase and the amplitude of the additional current pulse. Figure 5.21 shows the

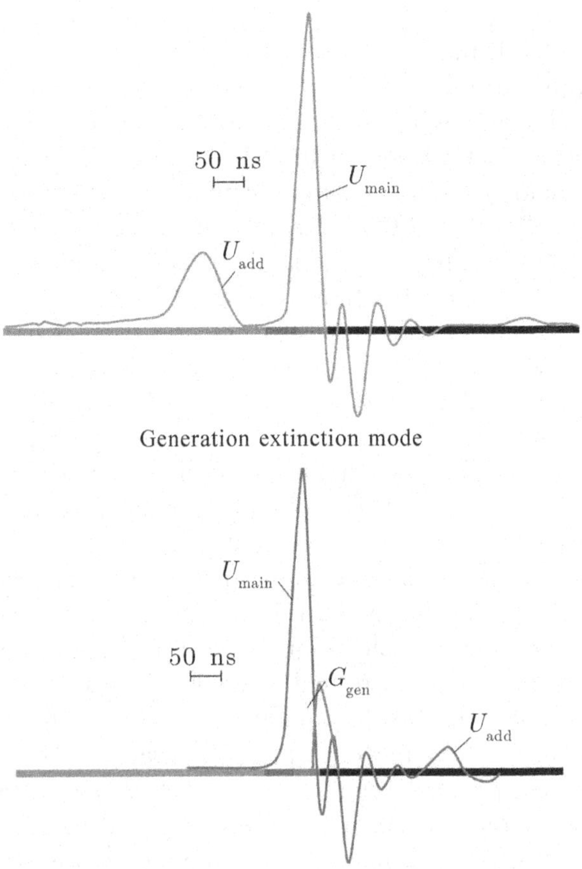

Fig. 5.21. The oscillograms of the main (U_{main}) and additional (U_{add}) pulses of the pump voltage of the CVLs Kulon 05 and Kulon 06 in the modes of extiction (upper) and in the generation mode (lower).

oscillograms of the additional and main pulses of the pump voltage (excitation) at the electrodes of the AE CVL in the quenching modes (upper oscillogram) and generation (lower), indicating the color time zones of the AM (see section 4.7).

The results of experimental tests of industrial CVLs Kulon 05 and Kulon 06 on the basis of the above controlled two-channel PI of nanosecond pulses have shown that with optimized parameters of the pump pumping pulses, the maximum values for both the radiation power and efficiency are achieved, modulation of radiation and practically trouble-free operation of the developed two-channel PI. In the usual regular pulsed-periodic frequency operation of the CVL Kulon 05 and Kulon 06, i.e., in the absence of radiation modulation, at the PRI of 12–15 kHz, the achieved values of the average radiation power with the sealed self-heating AE GL-206E (Kulon LT-10Cu) were 14–16 W, with AE GL-206I (Kulon LT-15Cu) – 18–20 W at the efficiency of the rectifier of the power source 1.1–1.3%.

The pulsed switch in the high-voltage generator of the main channel is the hydrogen thyratron TIG2-1000/25K, in the generator of the auxiliary channel TGI1-500/16. The maximum power consumption from the rectifier of the power source in the operating mode is not more than 1.5 kW, the maximum operating voltage at the anode of the thyratron of the main channel is 12 kV. The non-linear forming line of the main channel is made up of a two-link, and the non-linear forming line of the auxiliary channel is single-ended (positions 8 and 9). The inductances of the anode reactors of the main and auxiliary channels are 15 μG. An adjustable high-voltage power supply (item 12) is a constant voltage source on uncontrolled diodes and a controlled single-ended resonant converter, at the output of which a high-voltage transformer-rectifier unit is installed. The charging time of capacitive energy storage devices of nonlinear artificial forming lines is 30 μs. Adjustable delay lines (keys 14 and 15) provide a range of time control of pulses up to 1300 ns. The range of adjustment of the excitation current amplitude of the main channel up to 400 A for a duration of its front is no more than 50 ns, the voltage amplitude is up to 20 kV. The amplitude of the auxiliary channel current pulse is up to 40 A, the voltage amplitude is up to 5 kV. The duration of the excitation pulse of the main channel along the base is about 120 ns, the auxiliary pulse duration is up to 250 ns.

With these optimized parameters of the main and additional pump pulses (excitation) in the CVLs Kulon 05 and Kulon 06, the maximum power of radiation and the efficiency were reached at the working pulse repetition frequencies. The change in the response time of adjustable delay lines (items 14 and 15) and the formation, according to a predetermined law, of single pulses at the output of the controller (item 22) allows the high-speed pulse modulation of laser radiation to be accurate to one pulse, to implement any of their sequence, to set certain values of the impulse energy, etc. And more importantly, the design of the CVL provides for the possibility of controlling the operating modes from an external personnel computer.

These advantages make it possible to use pulsed CVLs and GVLs and CGVLs based on them, as well as other lasers based on self-terminating transitions, practically in all fields of science, engineering and medicine. For example, one of the advanced applications is the technology of microprocessing of metallic materials.

The use of the CVL Kulon 05 and Kulon 06 as part of automated laser technological installations (ALTI) such as Karavella-2 and Karavella-2M allows precise microprocessing of almost any metal and a wide range of non-metallic materials. The CVLs of the Kulon series are also effective for precision marking and deep engraving of parts and drawing images in volumes of transparent media (glass, quartz, sapphire).

5.3. Two-channel Karelia CVLS with high quality of radiation

The Karelia two-channel CVL was developed in 1986. This CVLS, works according to the MO–SFC–PA scheme. The aim of the development was to create a CVL with an average radiation power in a beam of diffraction quality of at least 20 W. At the beginning of the development, a large volume of theoretical and experimental studies of the energy, spatial and temporal characteristics of the laser radiation was carried out, most of which is presented in the Chapters 4 and 5, and the first industrial sealed-off AE Kristall GL-201 with a total average output power at effective pumping of 20 W and guaranteed up to 1000 h [227–232] was constructed. By this time, it became clear that, on the basis of the totality of its properties, te CVL is an almost ideal tool not only for pumping wavelength-tunable lasers on dye solutions used in technological complexes for isotope separation, but also for precise microprocessing of a number of materials used, for example, for electronic products.

The Karelia CVL includes a two-channel Karelia radiator (designation according to TU–ILGI-201) and a two-channel

Fig. 5.22. Appearance of Karelia CVL with two thyratron power sources IP-18 (on the left) and a two-channel lamp source Plaz (under the radiator).

Fig. 5.23. Appearance of Karelia CVL with two-channel lamp power supply IPL-10–001 under the radiator and the measuring chamber (right).

synchronized power supply based on two thyratrons IP-18 or a two-channel lamp-type Plaz or IPL-10–001. The radiator and power sources have independent water cooling systems. The power sources also use forced air cooling. Figure 5.22 shows the appearance of the Karelia CVL with two synchronized thyratron sources IP-18 and a two-channel lamp source Plaz under the radiator, and Fig. 5.23 – with a two-channel lamp source of the type IPL-10–001.

Figure 5.24 shows the appearance of the two-channel radiator ILGI-201 (Karelia) with closed (*a*) and open covers (*b*) in the operating mode, and Fig. 5.25 – its construction. The design of the two-channel ILGI-201 radiator corresponds to a laser system operating according to the MO–SFC–PA scheme and is most effective

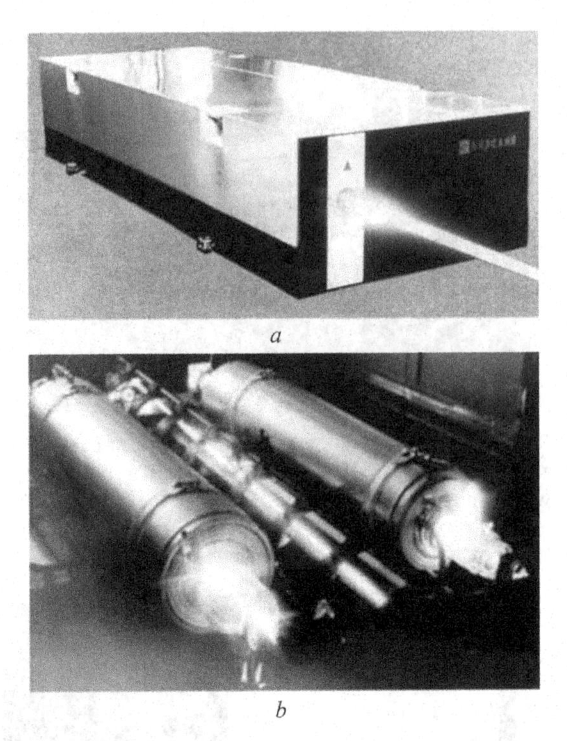

a

b

Fig. 5.24. The appearance of the two-channel radiator Karelia (ILGI-201) with closed (*a*) and open covers (*b*) in the operating mode.

for producing high-power and high-quality radiation beams [227–233].

The sealed-off self-heating GL-201 AE is used as an AE in the MO and PA of the Karelia radiator (positions 1 and 2 in Fig. 5.25). The AEs are installed in cylindrical double-wall water-cooled steel heat sinks (items 3 and 4) with an internal diameter equal to 200 mm. The AEs are attached to the heat sinks through water-cooled steel half-rings (item 5), mounted directly on its electrode assemblies, and fluoroplastic rings-isolators (6), located between the half-rings and the heat sink. The heat sinks are not only a load-bearing structure for the AE, but also perform the function of a reverse current conductor. A telescopic type UR or one convex mirror is used in the MO for the formation of a qualitative beam, .

In the design of the radiator Karelia there are two possibilities for the execution of the MO resonator and, accordingly, two versions of the SFC. Figure 5.26 *a* shows the optical scheme of the radiator in the mode of operation of a laser with the telescopic UR, Fig. 5.26 *b* – with one convex mirror.

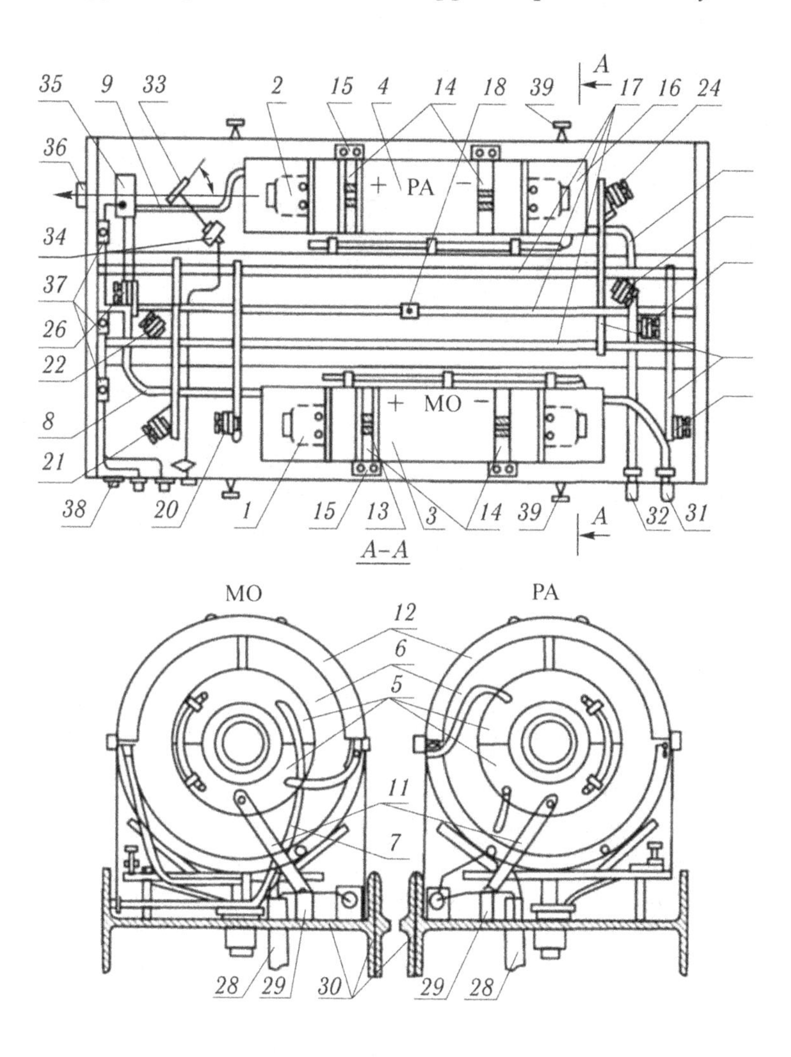

Fig. 5.25. The design of the radiator Karelia (ILGI-201): *1, 2* – sealed-off AE GL-201 MO and PA; *3, 4* – water-cooled cylindrical heat sinks; *5* – water-cooled metal half-rings; *6* – fluoroplastic ring insulators; *7, 8* – input and output hoses for cooling the heat sink of the MO; *9, 10* – inlet and outlet hoses for cooling the heat sink of the PA; 11 – anode tires; *12* – water-cooled end heaters; *13* – fluoroplastic tape; *14* – clamps for fixing the MO and PA; *15* – mechanisms for aligning MO and PA; *16* – protective screens; *17* – SFC; *18–26* – alignment mechanisms: diaphragms (*18*) SFC; Blind and output mirrors of the optical resonator (*19* and *20*); rotary mirrors (*21–24*); input and output mirrors SFC (*25* and *26*); *27* – brackets; *28* – high-voltage power cables; *29* – fluoroplastic insulators; *30* – the base of the armature of the radiator; *31, 32* – input and output connection of radiator cooling; *33* – optical coupler; *34* – converter of laser radiation power TI-3; *35, 36* – electromechanical and mechanical optical gates; *37* – interlocks; *38* – grounding bolt; *39* – radiator handles.

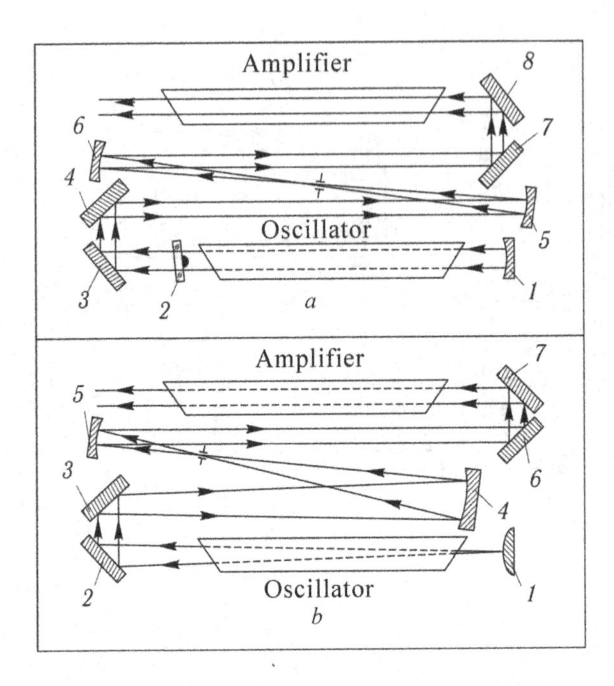

Fig. 5.26. Optical circuits of the Karelia two-channel radiator in the mode of operation of MO with the telescopic UR (*a*) and with one convex mirror (*b*): *a* = R = 275 (*1*); 1.5 (*2*) and 160 cm (*5* and *6*) and $R = \infty$ (*3*, *4*, *7*, *8*); *b* = R = 3 (or *5*) (*1*): 160 (*4*) and 75 cm (*5*), $R = \infty$ (2, 3, 6 7); D_3 = 3.5 (1, 5–7) and 7.2 cm (2–4).

The geometric dimensions of the mirrors (the radius of curvature R and the diameter of the mirror D_3) are indicated in the caption to Fig. 5.26. The increase in the telescopic UR is $M = 180$. When using such an UR, two narrowly focused beams are formed – with θ_{geom} = 0.15 mrad and θ_{dif} = 0.07 mrad. A convex mirror with a single-mirror version has a radius of curvature of R = 3 or 5 cm. In this case, a beam with a divergence θ_{lim} = 0.3 or 0.5 mrad is formed, but with a more even distribution of the radiation intensity in the far zone and a high stability of the position of the axis of the radiation pattern beam and the magnitude of the pulsed energy [227–233].

The SFC of the emitter is designed to suppress the incoherent background components of the output radiation of the MO and to spatially match the extracted qualitative beam with the aperture of the discharge channel of the PA [227, 228, 231, 232]. The SFC diaphragm, which cuts off the background components of the MO radiation, is established in the waist of the qualitative beam. The use of a mirror collimator, and not a lens collimator, precludes the appearance of parasitic feedbacks between the optical elements of the SFC and MF MO achromatic aberrations. To reduce astigmatism, the SFC was located parallel to the MO and PA in such a way that the angles of the entrance and exit of the beam were minimal. Each circuit of the resonator design corresponds to its optical elements of the SFC having certain geometric dimensions and radii of curvature. All elements of the optical system of the radiator are structurally integrated into one unit, which can be collected separately, pre-configured and then installed in the radiator. The load-bearing structure of this assembly is a three-rod frame made of aluminium alloy, on which transverse strips with mechanisms for aligning the mirrors of the resonator, collimator and rotary mirrors are fixed. The sealed-off AEs of the MO and PA are structurally connected with PFCs and are not required to be replaced when the optical system is adjusted.

In order to prevent the occurrence of achromatic aberrations in a two-wave beam and to reduce the laboriousness of optical-mechanical processing, all optical elements are mirrors (Fig. 5.26). The exception is the beam-splitting plane-parallel glass plate (item 33 in Fig. 5.25) and a glass substrate with an output convex mirror pasted on it (position 2 in Fig. 5.25, *a*) in the telescopic UR. The plate and the substrate are made of K8 optical glass and are antireflective (reflection coefficient less than 1% in the yellow–green region of the spectrum). The substrates for the mirrors are also made of K8 glass. The traditional technological process of optical–mechanical processing makes it possible to manufacture optical parts of the required shape with a high quality of the surface. The reflective coating of mirrors is a multilayer dielectric film based on zinc sulphide and magnesium fluoride.

Not all optical elements of the laser work under the same conditions, therefore their guaranteed operating time is different. This applies primarily to mirrors located on the side of the high-voltage cathode nodes of the AE. In a high-voltage electric field, suspended particles in the air acquire an electric charge and are deposited, mostly on nearby mirrors. Already after 500 hours of operating the radiator under ordinary laboratory conditions, a diffusely reflecting dust layer is formed on the surface of the mirrors, which significantly reduces the output power and the radiation quality. To reduce the effect of dust, it is necessary to seal the space between the exit windows of the AE and the mirrors. Note that the main security measure when working with mirrors is the isolation of their surface from direct physical contact with any solid and liquid bodies. Cleaning the surface can be done using medical cotton wool soaked in pure acetone or alcohol, and a squirrel art brush.

The TI-3 power converter is used as a receiver for the laser power indicator, which is characterized by high operating parameters. The operating range of TI-3–0.1 is 100 W, the time of establishment of the stationary mode is 50 s. The radiation to the input of the receiver TI-3 (item 34 in Fig. 5.25) is fed from the beam splitter plate (item 33), installed at the output of the PA. Thermal EMF, which appeared in the TI-3 power converter and is proportional to the power of the incident radiation, either goes directly to the indicator device (for example, the M-135 millivoltmeter) or serves as a signal for introducing feedback into the power supply in order to maintain the radiation power at a given level.

To cover the radiation beam at the output of the radiator, an electromechanical shutter with a traction electromagnet EU 2201H.248 (item 35 in Fig. 5.25) is installed as a shutter, having the following parameters: operating voltage and current 24 V, current 0.5 A, average speed beam overlap of 0.25 m/s, a spreading time of 5 ms. Overlapping of the beam in the shutter is done with the help of a dumb flat mirror fixed in it with a reflection coefficient ≥99% (mirror diameter 35 mm).

The electromechanical gate is inertial and in cannot be used in many cases of CVL applications, for example, in technologies where operational control of the characteristics is required. For the purpose of increasing the speed, an electronic radiation power control circuit was created. The principle of operation of this scheme is based on the partial or total absorption of MO radiation in the active medium of the PA, which is achieved by changing the delay time of the optical signal of the MO with respect to the signal of the PA (Chapter 4, Section 4.3) [231, 232]. The delay time for complete absorption depends on the excitation conditions of the active MO and PA media and is usually at least 25–30 ns. Thus, when the radiator is pumped from the thyratron power supplies IP-18, the delay time of the MO signal in relation to the signal of the PA is not less than 40 ns, with a lamp source of the Plaz or IPL-10–001 type, not less than 25 ns.

For full overlapping of the radiation beam, for safety reasons, a mechanical shutter–damper (item 36 in Fig. 5.25) located on the front panel of the radiator is used.

The basis of the load-bearing structure of the radiator is a welded frame made of steel pipes of rectangular cross-section with dimensions of 60 × 30 × 3 mm. On the base of this frame, three long I-bars with a section of 105 × 240 mm from D16T aluminium alloy are fixed. They contain all the main elements of the radiator: MO, PA, SFC, the output beam splitter, insulators for fastening high-voltage power cables, inlet and outlet chokes of the cooling system, power indicator receiver, electromechanical and mechanical gates, electrical interlocks and covers.

The covers of the radiator not only determine its shape and aesthetic appearance, but also fulfill the function of protecting the ether from radio interference. The covers are made of 1.2 mm thick steel and overlap by 15 mm at the joints.

The emitter is mounted on four supports, which allow to adjust the height of the output beam within ±10 mm and the orientation of its direction in space with accuracy not worse than 0.5 mrad.

The MO and the PA with cylindrical heat sinks are attached to the supporting structure by means of horizontal and vertical supports, which can be adjusted horizontally and vertically. When installing and replacing the AE in the MO and the PA, the adjustable supports allow, if necessary, to combine the geometric axes of the AE with the axis of the optical system of the radiator.

The mirrors of the MO and the SFC, as well as the rotary mirrors are fixed in the unified alignment mechanisms, which allow to adjust the optical elements with a sufficiently high accuracy: 0.01 rad per one revolution of the adjusting screw (19–26 in Fig. 5.25).

The power emitted in the radiator, depending on the pumping conditions of the AE GL-201, is in the range 3.6–4 kW – 1.8–2 kW from each AE. To exclude overheating of the AE and radiator components and associated negative consequences (decrease in longevity, deterioration in the stability of energy parameters and the position of the axis of the radiation pattern, increase in the cladding temperature), a forced water cooling system is used. Water carries away heat from the heaters of the AE, its end sections and the electromechanical shutter. The water flow is about 5 l/min with the pressure at the input of the system up to 2 atm. The temperature difference between the input and output of the cooling system is approximately 12°C.

To protect maintenance personnel from electric shock and laser radiation, blocking devices, warning signs and inscriptions, absorbing coatings are provided in the design. Electrical interlocks (item 37 in Fig. 5.25) are installed for each of the three removable covers of the radiator. On these same covers there are signs of electrical danger. On the faceplate of the radiator, above its output aperture, a laser hazard mark is affixed, and the aperture (outlet) itself can be overlapped by a mechanical shutter if necessary (item 36). The radiator has a protective earthing bolt with a ground sign on the labeled label (item 38 in Fig. 5.25), connections for water supply and drainage (drain and water labels). All the individual parts and components of the radiator under the lining are painted black – to reduce the level of scattered and re-reflected laser radiation.

The energy parameters of the radiator are determined to a great extent by the type of power source. The high longevity and reproducibility of the parameters of the AE GL-201 and the reliable radiator ILGI-201 created on their basis made it possible to objectively compare the parameters of the two-channel thyratron and lamp power supplies.

Two samples of the Karelia CVL with thyratron power sources IP-18 were manufactured and tested. In the first sample, the high-voltage pump modulator of each power source was made in a direct electric circuit, in the second – to improve the excitation efficiency of the AE – according to the scheme of voltage doubling and magnetic compression of current pulses. The triggering of the hydrogen thyratrons TGI1-2000/35 of the IP-18 modulators was started from the common master pulse generator (MPG) located in one of the power supplies. In another power supply instead of MPG the MO–PA channel synchronization block and the voltage stabilization of hydrogen thyratrons were used. The synchronization unit is structurally a cylindrical wire (copper) rheostat, to the middle mobile terminal of which is connected the output from MPG, and to the terminal terminals – the grids of the thyratrons. The synchronization unit operates as a delay line, which allows shifting the current pulses of the MO and PA channels relative to each other within 50 ns. Initially, pulsed boosting autotransformers were used in the two-channel thyratron power supply with a voltage doubling scheme which were combined together with the magnetic links of the pulse compression into a separate unit. Then, a more convenient circuit with a capacitive voltage doubling was applied. In the latter case, the elements for increasing the voltage and magnetic compression of the current pulses were compactly installed directly in the modulators of the power supplies. A schematic electrical diagram of the two-channel thyratron power supply based on two IP-18s with a direct circuit for the execution of pump modulators is shown in Fig. 5.27 *a*, its external appearance is shown in Fig.

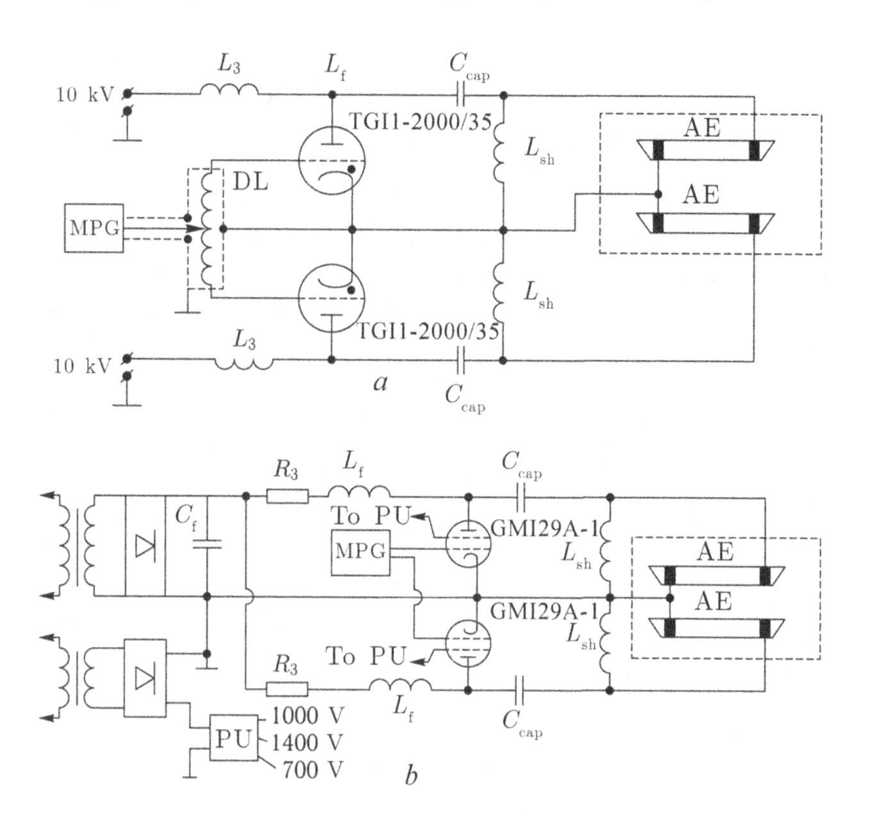

Fig. 5.27. Schematic diagram of two-channel thyratron (*a*) and lamp (*b*) power source: MPG – master pulse oscillator, DL – delay line, PU – power supply unit.

5.22. The block for synchronization and stabilization of the glow voltage of the thyratrons is installed in the left power source (in the block with the white panel) instead of the MPG block. The common master pulse oscillator is located in the right power source, above which there is a block with an autotransformer of voltage doubling and a magnetic link of the impulse compression. Both IP-18 power supplies are connected to the network via a voltage regulator of the STS2M10 type, which allows to increase the stability of the output parameters of the laser.

Two types of a two-channel lamp power source – Plaz and IPL-10-001 on the basis of modulating water-cooled vacuum lamps GMI-29 A-1 – were investigated. Their appearance is shown in Figs. 5.22 and 5.23. Under the Karelia radiator in Fig. 5.23 there is the IPL-10–001 modulator block, on the right under the measuring chamber there is a current source for supplying the modulators, to the left – the control cabinet of the source. The principal electrical circuits of these power supplies are practically the same (Fig. 5.27 *b*). In systems for stabilizing the power of laser radiation, there are differences. In IPL-10-001, part of the laser radiation converted by the TI-3 sensor into an electrical signal is fed to the matching system, and if there is a deviation in the reference signal, a corresponding signal is sent to the control grids of the GMI-29 A-1 lamps on both channels to maintain the specified the level of average radiation power. In the Plaz, the average current in the modulator of each channel is maintained at a given level. Output parameters of the Karelia radiator with these lamp sources are approximately the same. The Plaz has a higher anode voltage and as a result less loss of power on the lamps and a lower water flow. When using the lamp power sources, the power consumption of the AE is higher, and therefore the operating conditions of its cathode and discharge channel are heavier than when using thyratron power supplies.

Table 5.5 presents the main results of the Kareliya CVL research with two-channel thyratron power sources based on two IP-18 and lamp sources IPL-10–001 and Plaz. The Karelia CVL operates according to the MO–SFC–PA scheme.

The operating modes of power supplies (voltages, currents, switched powers, PRF) are optimized based on the conditions of maximum radiation power. With the use of the lamp power sources, pump pulses with higher PRF, steeper fronts and high temporal stability are formed. The latter circumstance is important for providing synchronous operation of the MO–PA system. The instability of pulse synchronization in the lamp sources does not exceed 0.5 ns, in the thyratron pulses it is 4 times higher.

As can be seen from this table, when using a thyratron power supply based on two IP-18s with a direct electrical circuit of the modulators, the average radiation power at 9 kHz PRF was about 20 W, the efficiency of the radiator was 0.67%, the practical efficiency (from the rectifier of the power source) 0.4% and laser efficiency (from the network) 0.3%. When using the voltage doubling scheme and magnetic compression of the current pulses, the corresponding values are 33 W, 0.92%, 0.5% and 0.4%. In the case of the lamp power sources, higher energy characteristics are achieved, since the current pulses generated by them have a shorter duration (~90 ns) and a steeper front (~40 ns). Due to the short duration of the pump pulses, the power consumed by the radiator is higher and is 3.9 kW (in the case of thyratron power supplies 3 and 3.6 kW for the direct circuit and for the voltage doubling circuit, respectively).

Table 5.5. Parameters of the Karelia two-channel CVLs with thyratron and lamp power sources

Parameter		Thyratron power supply based on two IP -18		lamp power source of type Plaz
		with direct scheme	with the voltage doubling circuit	
Power supply	Volume, m^3	1.4	1.55	1.8
	Weight, kg	720	750	1050
Rectifier of power source	Voltage, kV	6.5	6.5	21
	Average current, A	0.38	0.5	0.15
	Power, kW	2.5	3.25	3.15
The modulator of the power source	Anodic voltage, kV	15	25	21
	Current amplitude, A	300	380	330
	Duration of current pulses on base, ns	300	180	90
	Duration of front current pulses, ns	120	60	40
	PRF, kHz	9	9	12
	Efficiency, %	60	55	62
Radiator	Dimensions, mm	1900 × 820 × 380		
	Volume, m^3	0.6		
	Weight, kg	205		
	Power consumption, kW	3	3.6	3.9
	Average radiation power , W	19 to 21 (M = 60)	32 to 34 (M= 80)	38 to 40 (M= 180)
	Efficiency ,%	0.67	0.92	1
Practical efficiency (from the rectifier),%		0.4	0.5	0.6
Power onsumption from the network, kW		<6.5	<8	10
Laser efficiency (from the network),%		0.3	0.4	0.4
Working hours to failure, h		500	500	1000
Flow rate, l / min		5	6	8

The mean time to failure of a laser with a thyratron power supply is ~500 hours, with a lamp power supply 1000 hours, since the lifetime of the thyratrons TIG1-2000/35 was 400–600 h, and of the GMI-29 A-1 lamps are more than 1000 h. Figure 5.28 shows the dependence of the average radiation power of the Karelia CVL with a thyratron power source and its MO with telescopic UR with an increase M = 180 of the operating time. For 1000-hour operating

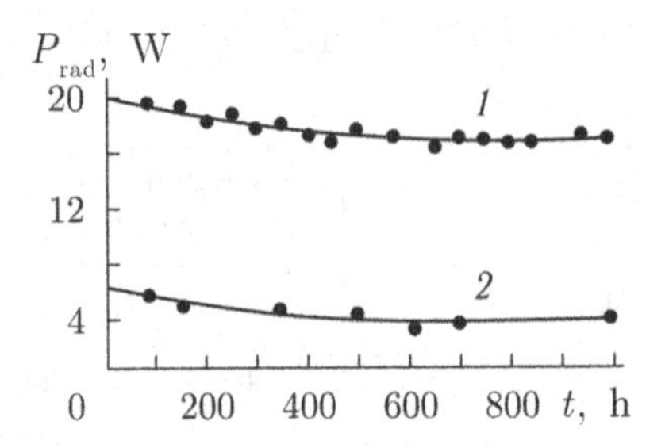

Fig. 5.28. Dependence of the average radiation power of the two-channel CVL Karelia with a thyratron power supply based on two IP-18 (*1*) and its MO with telescopic UR at $M = 180$ (*2*) on the operating time.

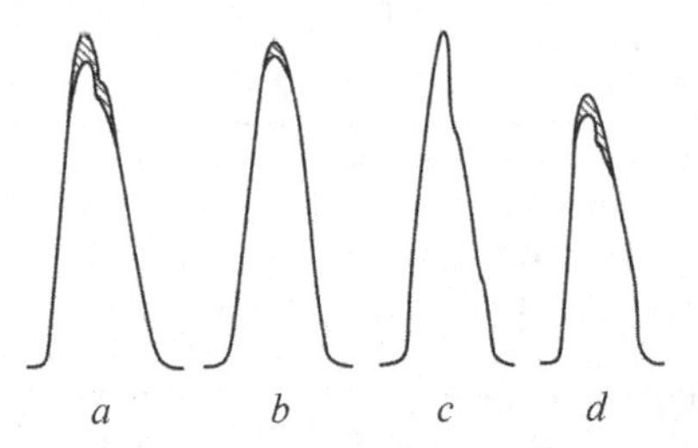

Fig. 5.29. Oscillograms of the radiation pulses of the Karelia two-channel CVL: with a thyratron power supply based on two IP-18s, made in accordance with the voltage doubling scheme (MO with telescopic UR, $M = 180$; *b* and *c* in MO mode with one mirror, $R = 3$ cm and 63 cm (respectively), and with a two-channel lamp source IPL-10–001 (*d* – MO with telescopic UR, $M = 180$).

time, the power of radiation decreased from 20 W to 16.5 W (by 17%). Replacement of thyratrons in one power source IP-18 was made in 450 hours, in the second – after 550 hours.

For technological applications, an important characteristic of the laser is the time to reach a stable thermal regime when the position of the axis of the radiation pattern is established and the energy of the radiation pulses becomes constant. Figure 5.29 shows the oscillograms of the pulses of the radiation of the Karelia CVL during the operation of the MO with the telescopic NR ($M = 180$, $\theta = 0.15$ mrad) and with one convex mirror ($R = 3$ and 63 cm, $\theta_{geom} = 0.3$ and 3.6 mrad) in steady-state thermal conditions. As can be seen from the oscillograms, the instability of the pulsed energy during the operation of MO with UR is about 5%, and in the single-mirror mode is much less (about 2%), since in the latter case the radiation beam is formed in the MO in a single pass.

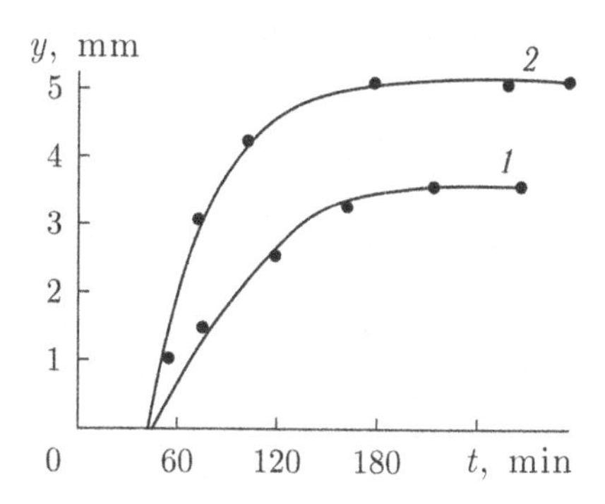

Fig. 5.30. Dependence of the displacement of the position of the axis of the radiation pattern of the output beam of the Karelia two-channel CVL with a thyratron power source on the time of heating (without screens and additional heat sinks): with MO and a single-mirror (R = 3 cm) – curve *1* and MO with the telescopic UR (M = 180) – curve *2*. The measurements were made at a distance of 9 m from the last turning mirror before entering the PA.

In the process of heating the CVL, the output power of the radiation is first established at a certain level, and then, after a certain period of time, also the axis of the radiation pattern. When warming up, the shift in the position of the axis of the radiation pattern is due to the slow establishment of temperature in the elements of the radiator structure. The AE radiator, operating in the optimal mode, dissipates about 3.6–4 kW of power into heat. Basically, this power is absorbed by heat sinks installed around the AE. However, part of the power (~0.5 kW) scattered through the ends of the AE falls into the interior of the radiator and causes the heating of its elements. Figure 5.30 shows the dependence of the shift in the axis of the beam pattern of the laser beam on the heating time for the single-mirror MO and with the telescopic UR. The characteristics are taken at a distance of 9 m from the last rotary mirror, which is installed in front of the entrance to the PA. The time to establish the stationary position of the axis of the radiation pattern, characterizing, in turn, the time of establishment of the thermal regime in the radiator was ~180 min.

The large deflection of the beam axis in the second case is due to the presence of two mirrors in the UR. The greatest instability of the position of the axis is observed in the initial warm-up period – during the first 120 min. The magnitude of the deflection of the directivity of the axis during this period with a single-mirror MO amounted to 0.2 mrad, while for the MO with the UR 0.3 mrad. In order to reduce the displacement of the position of the axis of the radiation pattern over the end sections of the AE, additional water-cooled heat sinks were installed and around them cylindrical screens with an opening in the end walls for the exit of the radiation beam. Such a design solution allowed to reduce the output of thermal radiation into the radiator space, as well as to reduce the thermal refraction of the beam in the resonator field. In this case, the temperature of the mechanisms for aligning the mirrors of the resonator decreased from 60°C to 30°C, and the displacement of the position of the axis decreased to 0.15–0.18 mrad, that is, almost 2 times. Figure 5.31 shows the synchronous shifts in the spatial displacement of the beam and the output power of thc radiation with the Plaz lamp power supply and the heat sinks in the end sections of the AE as a function of the laser warm-up time. The heating was carried out for 50 min with a 14 kHz PRF with a 12% excess of the power consumption in relation to the rated power, and then with a 12 kHz PRF in the nominal mode. The time of emission of the radiation power to the stationary level was 60 min, the time to establish the position of the axis of the radiation pattern was 80 min, that is, the process of establishing a stable regime was 2.5 times faster than in the design without additional protective elements. The instability of the position of the axis at a warming-up time from 40 to 100 min did not exceed

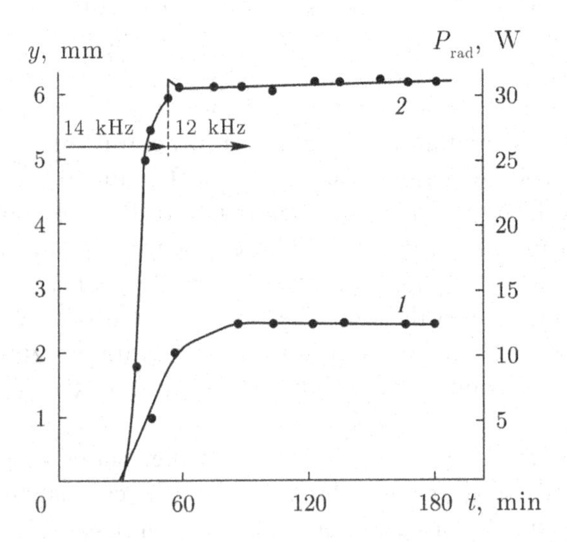

Fig. 5.31. Dependences of the displacement of the axis of the beam pattern of the output beam (*1*) and the radiation power (*2*) of the Karelia two-channel C~VL with a lamp power source and additional heat sinks and screens at the ends of the AE on the time of heating.

0.2 mrad, from 60 to 120 min 0.1 mrad and in the range of 90–150 min was about 0.01 mrad.

Six two-channel Karelia CVLs were produced: two with a thyratron power supply based on two IP-18s, one with a Plaz lamp source and three with a lamp source IPL-10–001. Two CVLs with he thyratron IP-18 have been operating practically up to the present day in the structure of Karavella-type process units for the production of precision parts used in electronics products. One CVL with the IPL-10–001 power source was used as part of a process unit making photomasks of printed circuit boards on glass with a metal coating. The remaining lasers were used for a number of scientific and experimental studies. In addition, 10 two-channel Karelia radiators were additionally produced for the acquisition of technological units for isotope separation.

5.4. Two-channel lamp-pumped laser CVLS Kulon-15

The industrial two-channel lamp CVLS Kulon-15 was developed in the Istok Company (Fryazino, Moscow region) with the participation in construction by the Altek Company (Moscow) with an average radiation power in a beam of diffraction quality up to 20 W at PRF (15 ± 1) kHz. CVLS is intended for the acquisition of technological installations that carry out productive precision microprocessing of thin sheet (up to 1 mm) metal and non-metallic materials, primarily for electronic products [16]. The laser operates in an automatic mode and is controlled by a personal computer. The advantages of the developed CVLS are ensured by the sealed-off construction of the applied industrial AE model GL-206E (LT-10 Cu Kulon) with a guaranteed lifetime of 1500 h and a service life of up to 3000 h and high stability of the output radiation parameters and the use of the Plaz-15 highly efficient two-channel synchronized lamp power source. The appearance of the CVLS Kulon-15 is shown in Fig. 5.32, and the main parameters are given in Table 5.6.

The operating mode and the optical scheme of the emitter CVL Kulon-15 are similar to the radiator Karelia. The radiator used two AE GL-206D (LT-10Cu Kulon), operating according to the effective scheme of MO-SFC-PA; the average radiation power of each AE is not less than 10 W. In the MO, a telescopic NM with $M = 200$ is used, which forms a beam of diffraction quality with a divergence of ~0.2 mrad. The AE was installed in coaxial metal heat sinks with a total water flow of 5–7 l/min. Pumping of the AE Kulon

Fig. 5.32. The appearance of the two-channel CVLS Kulon-15.

Table 5.6. The main parameters of the two-channel pulse CVLS Kulon-15

Parameter	Value
Radiation wavelength, nm	510.6; 578.2
Radiation beam diameter, mm	14
Average radiation power, not less than, W	≥20
Pulse repetition frequency, kHz	15 ± 1
Duration of radiation pulses (at half-height), ns	10 ± 2
Divergence of radiation with telescopic UR (M = 200), mrad	0,2–0,3
Power consumed from a three-phase network, kW	≤5
Readiness time, min	≤60
Channel synchronization stability, ns	± 0,5
Water consumption for radiator cooling, l/min	5–7
Overall dimensions, mm of radiator power cource	1520 × 630 × 380 770 × 600 × 1200
Weight, kg of radiator power source	≤150 ≤140
Average life to failure, h	> 1500
Technical operating life, years	≥ 5

Note. Cooling of the power source is forced air. The ambient temperature is not more than 25 °C. The operating mode is automatic.

LT-10Cu is carried out with a two-channel high-voltage pulse IP Plaz-15 with synchronization accuracy of the channels within 0.5 ns. Such synchronization ensures high stability of the output radiation characteristics (power change is not more than 2%). The switches in the source are vacuum modulating lamps GMI-32-B with air cooling.

The two-channel power source Plaz-15 consists of a control unit, a high-voltage unit, a nanosecond pulse generator and a control computer. The units are installed and fixed in a single stand. The control unit contains a microprocessor board, nanosecond vacuum lamp drivers and serves as a source of voltage for the second grids of lamps, drivers, incandescent lamp cathodes and cooling fans. The high-voltage unit is designed for converting an alternating three-phase mains voltage to a constant stabilized voltage with an amplitude of up to 20 kV, supplying the anodes of lamps GMI-29-B of a block of a nanosecond pulse generator. The operating voltage on the lamps is 18 kV. The two-channel block of the nanosecond pulse generator generates high-voltage nanosecond (τ_{pulse} < 70 ns) pulses of the emitter channels – MO and PA. It should be emphasized that the lamp power source, unlike the thyratron power supply, effectively works at the PRF up to hundreds of kilohertz, which may be important for qualitative microprocessing (the duration of the radiation pulses decreases).

On a laboratory bench with a lamp source, a more powerful sealed-off AE GL-206I (LT-15Cu Kulon) was tested with a change in the PRF to 110 kHz. With an average gain of 10 kHz, the average radiation power was 24 W, at 30 KHz 27 W, at 50 KHz 20 W, at 90 KHz 10 W and at 110 KHz 7 W. These are sufficiently high radiation powers for such PICs.

An important advantage of the Kulon-15 CVL is the possibility of operative control of the time detuning of the signals of MO and PA relative to each other. In this case, forcing the active medium of the CM to consistently operate either on amplification or on the absorption of the signal M, it is possible to change the LSI emission values within the operating frequency by any preassigned law from zero to the maximum value and, accordingly, to form the monopulse and batch modes of modulation of the output radiation. These laser operating modes allow maximum acceleration of the selection of optimal technological processes for precision processing of materials.

It should be noted that the development of industrial CVLs operating according to the MO–PA scheme should go along the way of increasing the radiation power in a beam of diffraction quality up to 40–100 W and higher, when high-performance precision processing of materials up to 2–4 mm thick is possible. It can also be hoped that industrial pulsed CVLs and on their basis technological installations in the near future will occupy a worthy place in the laser market.

5.5. Three-channel CVLS Karelia-M

The two-channel CVLS Karelia was used in 1990 as a basis to construct a modernized CVLS Karelia-M with two AEs GL-201D with a discharge channel diameter of 20 mm as the PA. The MO is a low-power AE Kulon of the grade GL-204 with a diameter of the discharge channel of 12 mm. Active elements of GL-201D were installed in water-cooled steel heat sinks, GL-204 – in aluminium heat sink, to the ends of which are attached the mechanisms of alignment of mirrors of telescopic UR. MO with SFC are established between the PA. The power to the MO was supplied from the thyratron source IP-18, the PA supply from the two-channel lamp source IPL-10–001. Synchronization of the system was performed from the generator of ignition pulses of the power supply IPL-10–001 at the optimal consumed power of AE with a PRF of 12.5 kHz (GL-204–1 kW, GL-201E–4.3 kW), the output power was 70 W (θ_{lim} = 0.4 mrad) with a practical efficiency of 0.92%. When using two AE GL-201D32 emitters with a 32-mm-diameter discharge channel with a 10-kHz PIC at the Karelia-M radiator, the output radiation power was 105 W (θ_{pred} = 0.3 mrad) with a practical efficiency of 1%. The AE GL-201E32 power was supplied from laboratory thyratron power supplies. The power consumption from the power source rectifier of one AE was 5 kW, the second AE 5.5 kW.

Figure 5.33 shows the general view of the radiator Karelia-M in the operating mode without covers.

In connection with the poor financing of development in the period 1990–2010 for the creation of modern equipment for separation of isotopes and production of highly pure substances by the laser selective technology for the needs of nuclear power and medicine and the general decline of the domestic industry, the demand for promising high-power copper vapour laser systems (CVLS) of the MO–PA type dropped noticeably.

But at the present time interest in CVLD of the MO–PA type with increased power and necessarily high quality of radiation starts to come alive, which is caused by high efficiency of using yellow-green pulsed radiation of CVL in modern technological equipment

Fig. 5.33. General view of the radiator Karelia-1M with two PA with removed covers

for enriching and obtaining pure isotopes and efficient precision microprocessing of metal and non-metallic materials. Therefore, the development of modern high-power, high-reliability and quality CVLS radiation, working in an automated mode, at this stage has become an urgent task. There is an even greater interest in the creation of industrial CVLs of small and medium radiation power levels for a wide range of spectroscopic studies, analysis of the composition of substances, for applications in medicine and other fields.

5.6. Powerful CVLS

The most powerful CVLS of the MO–PA type was developed in the USA. The average radiation power in the CVLS was brought to 72 kW, and the complex of lasers pumped by organic dyes tunable to wavelengths up to 24 kW (Section 1.3 in Chapter 1) [23. 200]. The main application of such a powerful combined laser system is obtaining highly enriched isotopes and pure substances for the needs of nuclear power. The qualitative development of laser technology over the past decade allows us to put on a new generation of this class of technological equipment. The prospective fields of application of CVLS with a small power level (10–100 W) include processing materials, as well as scientific and practical research.

In Russia, powerful high-performance CVLS of the MO–PA type in combination with lasers on dye solutions (DSL), as well as with non-linear crystals (NC), developed in the Kurchatov Institute,

Fig. 5.34. Laser technological complex for isotope separation based on three-channel CVLS and three-channel DSL.

Fig. 5.35. Two amplifying channels of CVL based on sealed-off self-heated AE GL-205C with average radiation power on each channel of 120 W.

Medical Sterilization Systems (Meditsinskie Sterilizatsionnye Sistemy) and the Institute of Semiconductor Physics (Section 1.2 in Chapter 1). In the CVLS, the industrial sealed self-heating AE series Kristall produced by the Istok Company: GL-205A, GL-205B and GL-205C with an average radiation power of 30, 40 and 50 W, respectively, is used as an AE in the MO and the PA [16, 41, 42]. In the quality of the MO in the CVLS, the AE of the GL-205A or

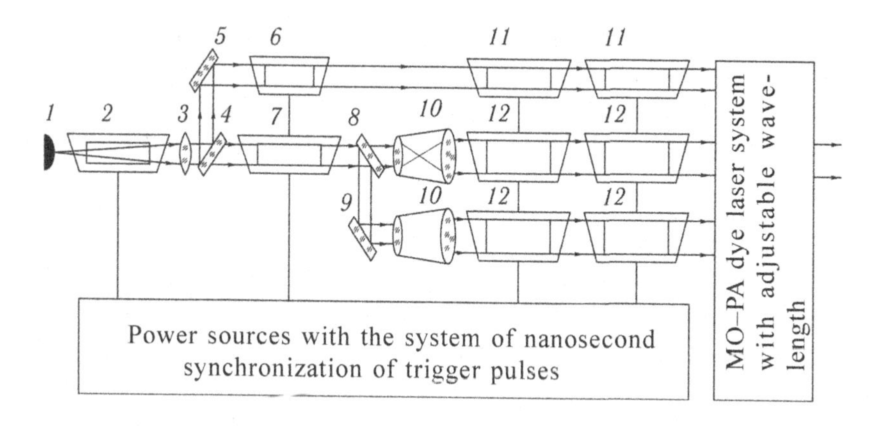

Fig. 5.36. Structural diagram of a three-channel copper vapour laser system with the MO-PA scheme, part of the technological complex for isotope separation: *1* – convex spherical mirror with a radius of curvature R = 5 cm; *2* – MO (GL-205B); *3* – clarified collimating lens; *4* and *8* – beam-splitting mirrors; *5* and *9* – rotary mirrors; *6* – PRA (GL-205A); *7* – PRA (GL-205B); *10* – expanding telescope (k = 1.6); *11* – PA (GL-205B); *12* – PA (GL-205C).

GL-205B models with an unstable resonator of the telescopic type or with one convex mirror are usually used, as the PA – GL-205B and GL-205C.

The most powerful of this class is the domestic laser technology complex developed and functioning in the Kurchatov Institute, Fig. 5.34. Figure 5.35 separately shows two amplifying channels of CVLS of this complex with an average radiation power of 120 W for each channel.

The CVLS of this complex is three-channel, the structural diagram of which is presented in Fig. 5.36.

The power source (PS) in the CVLS is multimodular and multichannel with nanosecond channel synchronization. High-voltage modulators of nanosecond pump pulses are designed and optimized in an efficient capacitive voltage doubling circuit and with two magnetic compression links, in which high-power pulsed hydrogen thyratrons of the type TGI-5000/50 are used.

The MO in the CVLS is the sealed-off AE GL-205B (item 2) used in the mode of operation with a single convex mirror with a radius of curvature R = 5 cm (position 1). Using a positive lens (item 1), the focus of which is aligned with the focus of the convex mirror (item 1), the radiation beam diverging, with high coherence, is converted into a cylindrical beam with a minimum divergence (θ = 0.5 mrad). The average beam power with a divergence θ = 0.5

mrad is 8–10 W. This beam is divided by a semitransparent mirror (item 4) into two beams of equal power. One beam is directed to the preamplifier (PRA) with the AE GL-205B (item 7), and the other beam is directed through the rotary mirror (position 5) to the second PRA (AE GL-205 A) (item 6). From the second PRA (item 6) the output beam of radiation already with a power of about 20 W is applied to the channel of the PA with two AE GL-205B and amplified to the level of 60–70 W. From the first PRA (item 7) the output beam of radiation with the increased power up to 40–45 W by a semitransparent mirror (item 8) is divided into two beams of equal power. These beams are enlarged from 20 to 32 mm (diameter of the discharge channel AE GL-205C) by means of telescopes (item 10) and are directed to two channels of PA with two AE GL-205C each (item 12) and amplified to 120–130 W. In this case, the total output power of the CVLS with three PA channels is not less than 300 W ((60 + 120 + 120) W). Then three output beams of radiation of the CVLS are directed to a three-channel tunable laser in solutions (LRK), also working according to the MO–PA scheme. In DSL, the coefficient of optical conversion of the yellow-green radiation spectrum of the CVLS (λ = 510 and 578 nm) into the near-IR radiation reaches up to 50%. Due to the use of highly selective resonators in the MO of the DSL, a narrow IR line width of 50–60 MHz is achieved. The results of practical research on the isotope separation of these groups are mainly presented at the All-Russian (international) scientific conferences ‘Physico-chemical processes in the selection of atoms and molecules’.

5.7. Conclusions and results for chapter 5

1. The analysis of the design, reliability, efficiency and quality of radiation of the first industrial single-channel CVL of the Kriostat and Kurs type and two-channel CVLS of the Karelia type was carried out. The total output of these lasers between 1975 and 1995 was about 140 units.

2. A new generation of industrial small-sized and compact pulsed CVLs with increased reliability and efficiency was developed and investigated, and on its basis CVL and CGVL: CVL Kulon-01 and Kulon 04 with an average radiation power ≥0 watts and ≥15 W, GVL Kulon 02 with power ≥1.5 W and CGVL Kulon 03 with power ≥4.5 W with PRF 12–15 kHz on the basis of serial sealed-off AE

models GL-206E, GL-206G and GL-206I (together with VELIT, Istra, MO). There are no foreign and domestic analogues.

3. A new generation of industrial small-sized CVLs with high-speed control of output radiation power was developed and investigated according to a given algorithm, including packet and monopulse modulation: Kulon 05 with an average radiation power ≥10 W and Kulon 06, power ≥15 W with PRF 12–15 kHz on the basis of serial sealed-off self-heating AE models GL-206E and GL-206I. The pulsed CVL Kulon 05 has a single-module design with air cooling, Kulon 06 is a two-module (power source and radiator) with water cooling of the radiator, has a compact design, maximum reliability and efficiency, controllability of the output parameters from an external PC (developed together with VELIT Company, Istra, MO). There are no foreign and domestic analogues.

4. The physical essence of the work of the CVL Kulon 05 and Kulon 06 with the operational control of the output radiation power, including packet and monopulse modulation, is to create conditions in the active medium for full or partial extinction of generation and maximum generation due to the formation of an additional low-power pulse discharge current in the AE. When a low-power current pulse is established in the direction of advance from the main excitation pulse, a regime of total or partial extinction of the generation is ensured (due to the population of metastable levels of copper atoms), and in the case of lagging – the regime of full generation, which was the basis for operative control by the electron method of the energy and space-time characteristics of radiation in a wide range.

5. The Kulon-15 two-channel lamp CVLS with high-speed pulse modulation of radiation and control from a PC operating according to the MO–SFC–PA scheme with an average radiation power in a beam of diffraction quality of more than 20 W and an efficiency of 0.4% at PRF (15 ± 1) kHz and channel synchronization not worse than ±0.5 ns was developed and costructed. The MO and PA use the industrial sealed-off AE models GL-206E.

6. The Karelia two-channel CVLS, operating according to the MO–SFC–PA effective scheme is based on two synchronized upgraded power supplies IP-18 and the two-channel lamp type IPL-10–001 or Plaz type with channel synchronization not worse than ±1 ns with an average radiation power in a beam of diffraction quality of 30–40 W and an efficiency factor of 0.4% with a PRF of 9–12 kHz. In the radiator, self-heating AE GL-201 (new model GL-205 A) was used as an AE in the MO and the PA with the use of telescopic

UR in MO with an increase of $M = 200$ or one convex mirror with a radius of curvature $R = 3$ cm.

7. The Karelia-M three-channel C~VLS of high power, operating according to the MO–SFC–PA scheme, was created and investigated on the basis of the two-channel CVLS Karelia using the low-power (3 W) AE Kulon model GL-204 (new model GL-205B) with a diameter of the discharge channel of 12 mm, and in the PA of two powerful AE Kristall model GL-201D (new model GL-205B) or GL-201D32 (new model GL-205 V) with a diameter of the discharge channel, respectively 20 mm and 32 mm. With two AE GL-201Ds in the PA at 12.7 kHz PRC, the average radiation power of the CVLS was 70 W (θ_{lim} = 0.4 mrad) and an efficiency of 0.92%, with two AE GL-201D32 at 10 kHz–105 W θ_{lim} = 0.3 mrad) and 1%, respectively.

8. The most powerful multimodular three-channel domestic CVLS for pumping DVL tunable along the wavelengths, operating according to the MO–PA scheme and using a preamplifier (PRA), was developed and is functioning as part of the isotope separation technology complex in the Kurchatov Institute . In CVLS, industrial AE series Kristall: GL-205A, GL-205B and GL-205C with an average radiation power of 30, 40 and 50 W are used as AE. The total average output power of the three-channel CVLS is not less than 300 W – (60 + 120 + 120) W. The 300 W yellow–green radiation of the CVLS in a tunable DSL with 50% optical efficiency is converted to near-IR radiation. Due to the use of a highly selective optical resonator in the DSL, the tunable IR radiation has a narrow line width of 50–60 MHz, which ensures efficient separation of isotopes in the technological complex.

9. The total output production of the CVL of the Kulon series was about 25 units, intended for the acquisition of modern technological and medical equipment and scientific research in the field of analysis of the composition of substances and nanotechnology, powerful CVLS – 15 units for technological applications.

6

Modern automated laser technological installation Karavella (ALTI)

6.1. Requirements for pulsed CVL and CVLS in modern technological equipment

The analysis of foreign studies on precision cutting and microdrilling of materials by pulsed radiation of CVL and CVLS, the first domestic studies on the experimental Karelia CVLS and the experimental laser technological installation of the Karavella experimental laser technological installation with an average radiation power of 20–25 W has shown the wide possibilities of using these lasers for microprocessing metal and a wide range of non-metallic materials up to 1–2 mm thick (see Chapter 2). It also follows from the analysis that for productive and qualitative microprocessing, the peak power density in a focused spot of radiation (waist) with a diameter of 10–30 μm should be 10^9–10^{12} W/cm^2 when the material is processed mainly in the evaporation mode with microexplosions followed by a vapour expansion and superheated liquid. To achieve such levels of power density, minimum heat-affected zone (HAZ) ($\leq$5–10 microns) and roughness of the cut surface ($\leq$1–2 microns), the quality of the output beam of radiation generated by optical systems should be as high as possible. Ideally, in order to ensure a high quality of radiation, the structure of the output beam of the CVL radiation (Chapter 4) should be single-beam and have a diffraction divergence ($\theta_{diff} = 2.44\lambda/D_{chan}$) with a pulse duration τ_i = 20–40 ns. The energy in the pulse should have small values W = 0.1–1 mJ at pulse repetition frequencies f = 10–20 kHz. The instability of the position of the beam axis of the radiation beam must be three orders of magnitude smaller than the diffraction limit of divergence – $\Delta\theta \leq \theta_{difr}/10^3$. The spot of the focused radiation beam should have a circular shape with a clear boundary and an intensity distribution close to Gaussian.

The main results of research and development carried out within the framework of this work on the development of industrial sealed-off AEs on copper vapour (Chapter 3), optical systems for the formation of single-beam radiation of diffraction quality with stable parameters (Chapter 4), methods for operational control of radiation power 4.7) and on their basis of pulsed CVLs operating in the regime of a separate generator with the use of SFC and more powerful CVLS working according to the MO–SFC–PA scheme (Chapter 5) became the basis for the development of a series of modern industrial automated laser technological installation (ALTI): at the first stage of Karavella-1 with an average radiation power of 10–15 W and Karavella-1M with a power of 20–25 W, in the second stage Karavella-2 and Karavella-2M with a power 6–8 W, designed for productive and high-quality precision microprocessing of electronic component materials and, in particular, microwave products [364–375, 383, 389–391, 394–400].

6.2. Industrial ALTI Karavella-1 and Karavella-1M on the basis of two-channel CVLS

The first industrial ALTI Karavella-1 was developed in the period 2001–2003, the second such class ALTI Karavella-1M – 2008–2012 on the basis of two-channel pulse CVLS Kulon-10 and Kulon-20 respectively and precision three-coordinate XYZ tables with PC control. At ALTI Karavella-1, within the framework of technological R & D project Kursor (2004–2006), the technology of contour cutting and flashing holes for the main materials used in vacuum technology was developed. The ALTI Karavella-1 became the first high-performance and compact installation of a new generation with an average radiation power in a 10–15 W diffraction quality beam and 13–14 kHz PRF intended for productive and high-quality precision microprocessing of metallic materials with a thickness of 0.05–0.5 mm and non-metallic materials up to 1 mm for products of microwave equipment and other electronic components [312,

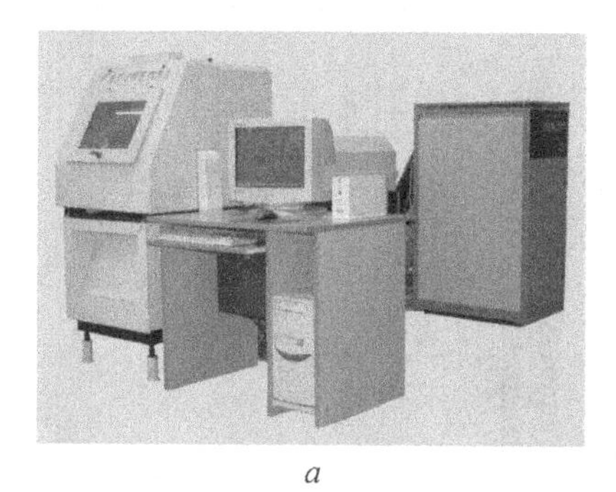

a

b

Fig. 6.1. First industrial ALTI Karabella-1 based on two-channel CVL Kulon-10 working by the MO–SFC–PA scheme for precision microprocessing of metallic materials 0.05–0.15 mm thick and non-metallic materials up to 1 mm thick: a) with closed covers; b) with open covers.

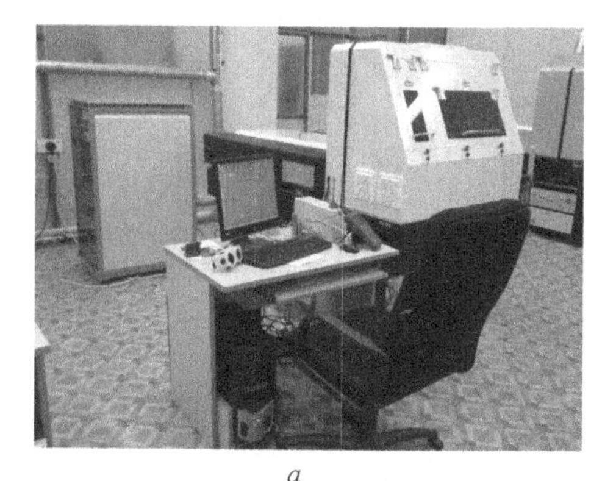

a

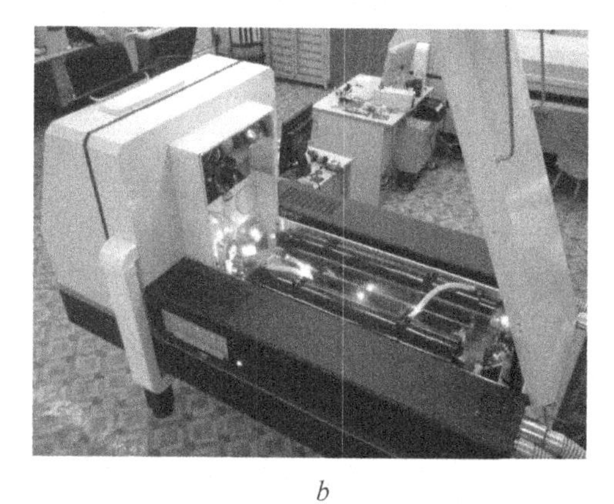

b

Fig. 6.2. Industrial ALTI Karavella-1M based on two-channel CVL Kulon-20 working by the MO–SPC-PA scheme for precision microprocessing of metallic materials 0.5–1 mm thick and non-metallic materials up to 1.5...2 mm thick: a) with closed covers; b) with open covers.

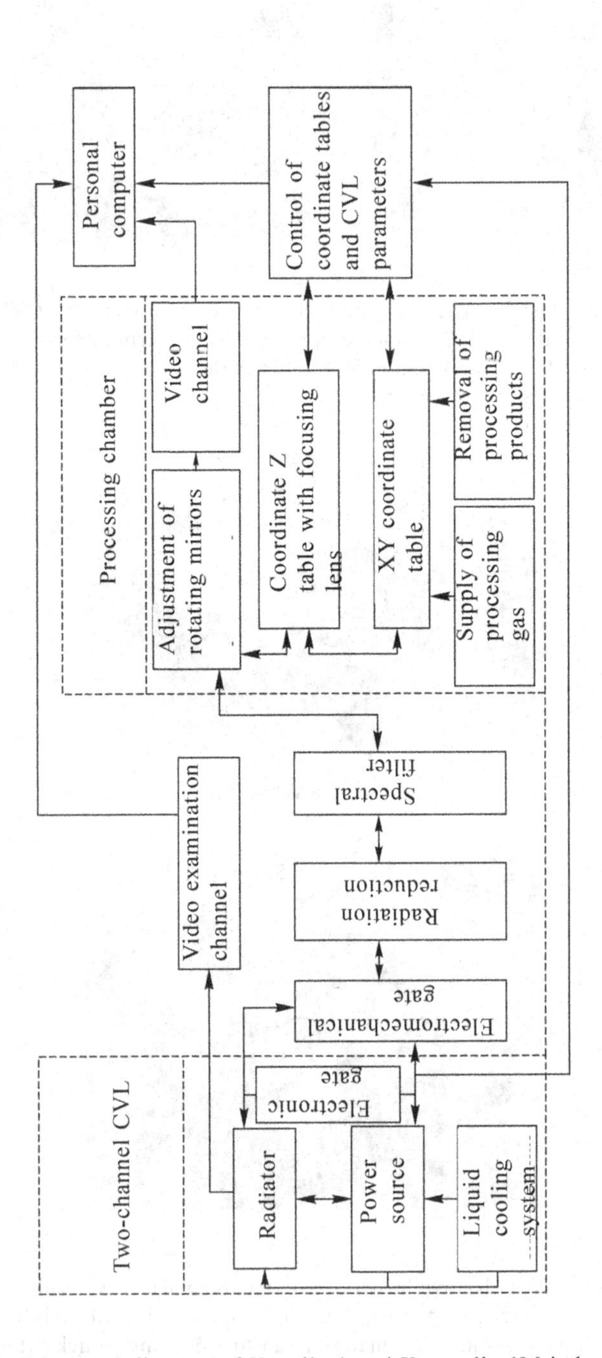

Fig. 6.3. Structural diagram of Kavella-1 and Karavella-1M industrial ALTI

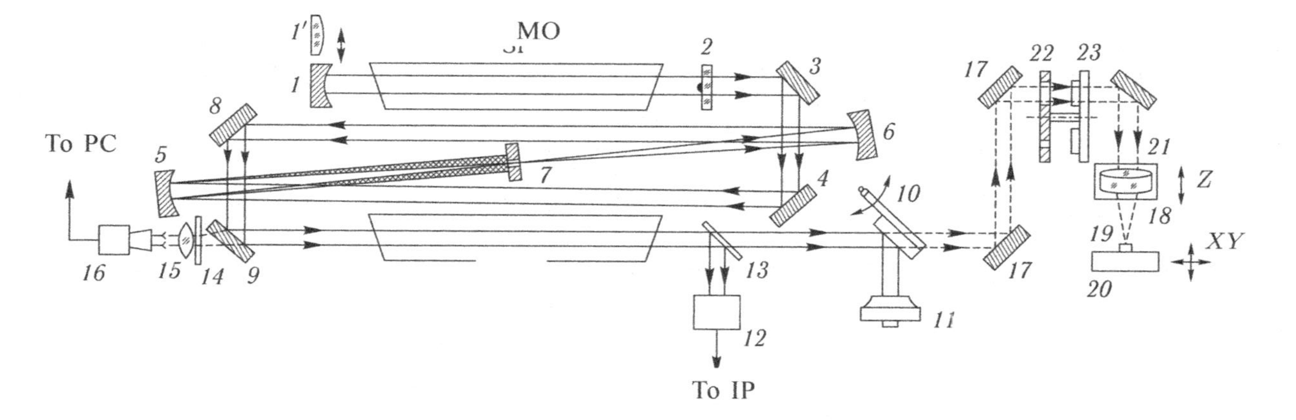

Fig. 6.4. Optical diagram of industrial ALTI Karavella-1 and Karavella-1M: MO and PA – master oscillator and power amplifier with sealed-off AE GL-206D (Kulon LT-10Cu); *1* and *2* – telescopic unstable resonator with magnification $M = 200$ with a blind concave and exit convex mirrors; 1 'and *2* – unstable resonator with two convex mirrors; *3*, *4*, *8*, *9* and *17* – flat rotary mirrors; *5* and *6* – concave spherical mirrors SFC; *7* – selective diaphragm PFC; *10* – electromechanical shutter with a flat mirror; *11* – laser radiation power sensor; *12* – optoelectronic receiver; *13* – beam splitter plate; *14* – attenuator of laser radiation intensity; *15* – matching optics; *16* – video camera; *18* – focusing anatomic lens; *19* – object to be processed; *20* – horizontal two-coordinate table XY; *21* – vertical table Z; *22* – emitter of radiation power; *23* – spectral filter

365–376, 395,397, 398]. The appearance of ALTI Karavella-1 is shown in Fig. 6.1.

The laser radiation power of ALTI Karavella-1M is 1.5–2 times higher than that of ALTI Karavella-1 and is 20–25 W. ALTI Karavella-1M is designed for efficient microprocessing of thicker materials: metallic up to 1 mm thick, non-metallic thicknesses up to 1.5–2 mm [398]. Appearance of this installation is shown in Fig. 6.2.

6.2.1. Composition, construction and principle of operation

Figure 6.3 is a block diagram and Fig. 6.4 the optical diagram of industrial ALTIs Karavella-1 and Karavella-1M.

The industrial ALTIs Karavella-1 and Karavella-1M constructively consist of 3 modules: ALTI module, power source (PS) rack, operator's desk with the PC manager (Figs. 6.1 and 6.2). The ALTI module includes the following main functional blocks and systems (Figs. 6.3 and 6.4):

1. Two-channel radiator CVLS Kulon-10 or Kulon-20 on the basis of industrial sealed-off self-heating AE models GL-206E (Kulon LT-10Cu) and GL-206I (Kulon LT-15Cu), working on the MO–SFC–PA scheme (see Chapters 2 and 3). CVLS Kulon-10 is part of ALTI Karavella-1, CVLS Kulon-20 – in the composition of the ALTI Karavella-1M.

2. A two-channel transistor synchronized power source that generates high-voltage nanosecond pulses for pumping the CVLS AE emitter with a current amplitude of up to 250–300 A for a base time of 100–120 ns with a PRF of 12–16 kHz and ensuring synchronization of the MO and PA channels within ±2 ns.

3. The motion and control system (MCS), which includes a horizontal XY table with a working field of 150 × 150 mm and a vertical table Z with a travel length of 60 mm, with an error of positioning on each axis of not more than ±3 μm and a travel speed of up to 20 mm/s and a coordinate table control unit (TCU) connected to the control computer of the ALTI.

4. Optical system for the formation of a beam of diffraction quality, its amplification and attenuation, transportation and focusing on the object being processed.

5. A system of double-circuit cooling, operating according to the water-to-water scheme.

6. Laser system of observation and adjustment of the focused radiation spot on the processed object with the display of the image on the monitor of the control PC.

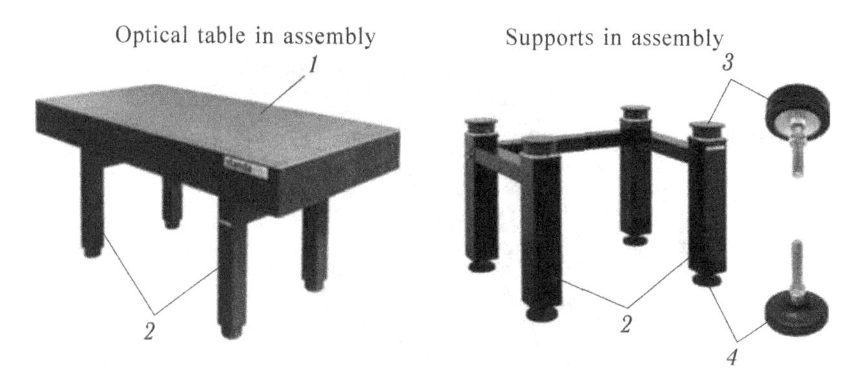

Fig. 6.5. Vibrodamping optical honeycomb table, model 1NT10-20-20 (1), metallic supports IRS05-12-05 (2) with four pneumatic insulators AP-1000 (3) and rubber insulators ITS-AS (4).

7. Adjustable systems for blowing process gas into the zone of laser treatment and extraction of the products of destruction from this zone.
8. The laser radiation power attenuator.
9. Spectral filter for selection of individual wavelengths of radiation (0.51 or 0.58 μm).
10. Support frame and hanging frame with cast iron plates, which is a carrier for the two-channel radiator CVLS Kulon-10, three-coordinate table XYZ and two high-voltage modulator power sources (in ALTI Karavella-1).
11. Optical honeycomb table with vibration-damping supports as a carrier for the two-channel emitter CVLS Kulon-10, three-coordinate XYZ table and two high-voltage modulators of the power source (in ALTI Karavella-1M).
12. Laser software.

The ALTI module is the most saturated in terms of the number of functional blocks and nodes and responsible for the accuracy of processing. When designing the module, in accordance with the requirements for the stability of the spatiotemporal and energy characteristics of the laser beam and the accuracy of the coordinate table movement, special attention was paid to providing structural rigidity, eliminating the influence of external vibrations and the effect of internal heat sources on the deformation of the responsible components and parts. The basis of the supporting structure of the module ALTI Karavella-1 is a power metal frame with two cast iron plates (of its own production), suspended through stretch springs to the lower supporting metal frame, which is mounted on the floor

Fig. 6.6. Motorized swivel platform model 8MR151-1.

on 8 adjustable legs. A two-coordinate horizontal XY table and a vertical table Z are mounted on a 600 × 600 × 80 mm board, on a 600 × 1450 × 80 mm board with stiffeners – a two-channel CVLS radiator, two high-voltage pulse modulators, under the radiator there are two high-voltage pulsed modulators for pumping the power source radiator. This design of the carrier ensures the isolation of the laser emitter, optical elements and precision three-axis table from the mechanical vibrations of the floor, which usually take place in the production premises.

In the Karavella-1M ALTI all elements of the ALTI module are mounted on the optical honeycomb table firm Standa (Lithuania) of model HT10-20-20 with dimensions 2000 × 1000 × 200 mm. The table is set to 4 supports of the type 1TS05-12–05 through passive air vibroinsulators AR-1000 with a total permissible load of 1100 kg (Fig. 6.5) [401].

On the table in the middle part there is a two-channel radiator of the CVLS, on both sides of which there are installed in parallel two high-voltage pulse modulator of the power source. Fundamental changes have been made in the latest designs of modulators. In each modulator, in order to increase their electrical strength and reliability, all elements of the high-voltage path: transformers, chokes and storage capacitors, are mounted in a sealed reservoir and are filled with silicone fluid for power transformers and other electrical equipment with an electrical strength of 35 kV grade Sofexil-TSZh. The tank has an outer jacket of water cooling. After

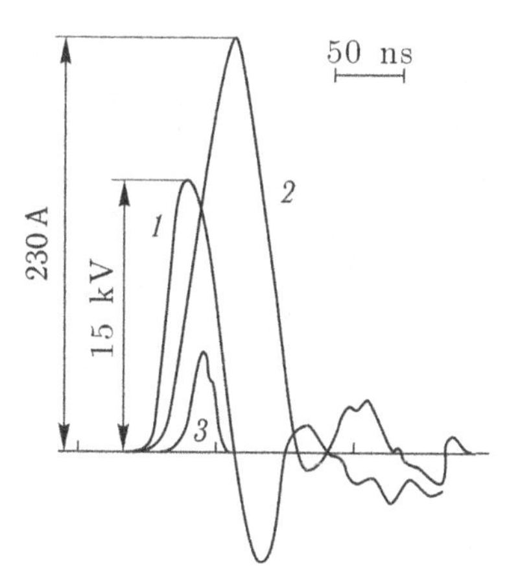

Fig. 6.7. Oscillograms of voltage pulses (*1*), pumping current (*2*), and radiation (*3*) of the AE GL-206E (LT-10Cu Kulon) with transistor power source.

the reconfiguration, the high-voltage power part of the power source was moved beyond the control rack, which reduced the level of radio interference affecting the operation of the low-voltage electrical circuits of the control system. At the same time, there was no need to use long high-voltage cables (an additional source of radio interference) connecting the control rack and modulators. The complex of structural changes made increased the reliability of the two-channel power source and expanded its functional capabilities.

Directly to the CVLS emitter from the exit side of the radiation beam there is a technological chamber with a 3-coordinate table placed in it. Inside the camera there are also systems blowing process gas into the zone of laser treatment of material and suction of the products of destruction. Under the optical table on a special frame fixed to its supports there is the CVUCT – control unit of the XYZ coordinate table and the switching unit of the ALTI.

In the ALTI Karavella-1 and Karavella-1M on the XY horizontal coordinate table there is the motorized turntable 8MR151-1 of the firm Standa with a rotation angle of 360° for processing details in the polar coordinates (Fig. 6.6).

This device, using additional equipment, is used for productive microprocessing of parts with a cylindrical surface.

Two-channel radiator and power supply of the CVLS. The two-channel radiator of the pulsed CVLS Kulon-10 and Kulon-20 in these ALTIs operates according to the effective scheme 'a master oscillator (MO) – a spatial filter – a collimator (PFC) – power amplifier (PA)', which allows to form powerful radiation beams with diffraction quality. As an AE the radiator uses industrial sealed-off self-heating AE of the Kulon series (GL 206 according to TU). The same AE models GL-206D (LT-10Cu Kulon) are used in the MO and PA radiator of ALTI Karavella-1, and the more powerful GL-206I (LT-15Cu Kulon) in the PA of the ALTI Karavella-1M (see Chapter 2). The AEs are installed in water-cooled cylindrical metal heat sinks, with reclining upper half-cylinders for convenient and prompt replacement of spent AE. Pumping of the AE is carried out from high-voltage pulse modulators of a two-channel transistor power source that generates current pulses with an amplitude of up to 300 A and a duration of 100–120 ns at a pulse repetition rate of 12–16 kHz. Such characteristics of the pumping current pulses are achieved by applying in the high-voltage modulator of each channel four matched magnetic compression cells with bias and an additional step-up transformer, which compress the input voltage pulse with an amplitude of 15–20 kV from 1.2 μs to 60 ns. Figure 6.7 shows the oscillograms of the voltage pulses (*1*) and the pumping current (*2*), as well as the radiation (*3*) for the AE GL-206E (LT-10Cu Kulon) in the steady-state operating thermal regime.

Under these pumping conditions, the AE discharge channel is heated up to the optimum operating temperature (1600°C) and the excitation efficiency of the AM is high. The duration of the radiation pulses along the base is 25–30 ns.

To obtain maximum output radiation power in a two-channel laser system, the accuracy of synchronization and retention of the MO and PA channels should be within 2–4 ns. This accuracy in ALTI is provided by the electronic device of the power source due to the introduction of feedback through the fiber-optic cables from MO and PA. Feedback light signals are fed to low-inertia and highly sensitive integral photodetectors in the power source. Channel synchronization parameters, the frequency of repetition of the pumping pulses and operating modes are set from the control unit of the power source, made in the form of a separate rack (in Fig. 6.1 on the right). The control unit in the power source rack provides interaction and control of the operation of all functional units and IP modules; there is also a rectifier unit and cooling system components: a heat exchanger, a pump and a tank with distilled water.

In the radiator mirrors of the optical system (Fig. 6.4) there are installed alignment mechanisms (ALM) with a landing diameter of 35 mm. All ALMs are fixed to the optical honeycomb table through the transitional rigid steel squares. To create radiation beams of diffractive quality in a radiator, a telescopic UR with a large magnification (M = 150–200) or UR with two convex mirrors with a small radius of curvature (R = 1–10 cm) (1, 1′ and 2 in Fig. 6.4) are used. The physical principle of the operation of these optical resonators and the advantages of their use in technological installations based on pulsed CVLS for microprocessing of materials are discussed in detail in the Chapters 3 and 4. The formation of a radiation beam in a two-channel emitter occurs in the following sequence. The output radiation of the MO is transmitted to the SFC consisting of two concave spherical mirrors (positions 5 and 6) and a selection diaphragm (position 7) using two rotary planar mirrors (positions 3 and 4). In the SFC, the diffraction beam is focused and extracted from the background of incoherent radiation and the extracted divergent beam is transformed into a cylindrical beam with a diameter equal to the aperture of the discharge channel of the PA. After passing through the PA, a power-amplified diffraction beam is guided by a rotary mirror (position 17) into the process chamber.

Optical system for the formation, transport and focusing of a laser beam. Figure 6.4 shows the diagram of the optical system of the ALTI Karavella-1 and Karavella-1M for the formation, amplification, transportation and focusing of the laser beam. The optical system includes the following interconnected functional nodes:

- two-channel radiator of the CVLS;
- attenuator with a spectral filter;
- electromechanical shutter;
- optical column with a power lens;
- axial video surveillance system – laser microscope;
- a system of axial video surveillance with a telescope.

Two-channel radiator of the CVLS. The composition, the main operating principle and the pumping conditions of the two-channel CVLS radiator are already considered in 8.2.2. In this section we consider in detail only its optical component, which is responsible for the formation of a beam of diffraction-quality radiation and the amplification of this beam in terms of power.

The master oscillator. The radiator is an AE on copper vapour GL-206E (LT-10Cu Kulon) with a discharge channel diameter of 14 mm (see Chapter 2) placed in an unstable resonator (UR) of telescopic type or with two convex mirrors (see Fig. 6.4). As detailed above, the work of these URs is described in the Chapters 4 and 5. The UR of the telescopic type has a gain M = 200. The radius of curvature of the surface of the blind concave mirror of this telescopic resonator is R = 2000 mm, of the output convex mirror R = 10 mm. But the resonator, in addition to the radiation beam with diffraction divergence, forms several beams and with greater divergence, unambiguously reducing the quality of microprocessing of materials.

In the UR with two convex mirrors (see Fig. 6.4), a blind mirror has a radius of curvature R = 10–30 mm. Directly in front of the blind mirror there is a diaphragm with a hole of 100 μm. The diaphragm faces the AE with a concave conical side and has a black light-absorbing coating. The wall thickness of the diaphragm in the hole is 300 μm. Both elements arc assembled into a single unit and secured to the alignment mechanism (ALM). The output mirror in this UR can be used in two versions. The first version is shown in Fig. 6.4 and is a convex mirror with a radius of curvature R = 30–100 mm and a diameter of 2 mm, glued to an antireflected glass thin-walled substrate at an angle of 4° to its optical axis. In this case, the radiation beam emerging from the resonator is divergent, which is not always convenient for practical applications. The second version of UR is an enlarged convex–concave meniscus (positive lens) with a focal length of 930 mm or 1070 mm, the convex surface of which with a radius of curvature of 70 mm or 40 mm faces the AE. In the centre of the convex surface, a mirror section of aluminium with aa diameter of 0.7–1 mm is dusted, with the help of which a feedback is made to the blind mirror of UR. Since the focus of the lens is aligned with the focus of the blind mirrors, the radiation beam reflected from the blind mirror with a strictly spherical wave meniscus is transformed into a cylindrical beam 15 mm in diameter with a plane wave, that is, into a beam with a diffraction divergence. (The design of UR is protected by a patent for invention No. 2 432 652 RF Pulsed laser [371].)

The output radiation produced in the UR with two convex mirrors differs from the variant with the telescopic type of UR by the presence of only one beam of diffraction quality radiation (θ_{diff} = 0.1 mrad) against the backdrop of highly divergent beams of superlight (see Chapter 2, Section 2.2). The power of this beam is insignificant and amounts to 0.1–0.2 W, but it is sufficient to introduce saturation of the PA to a level of 20 W. A beam of this quality completely meets the technical requirements for ALTI for microprocessing.

Spatial filter–collimator (SFC). Purification of the output radiation of the MO from its background components is carried out in SFC. It consists of two confocal concave spherical mirrors: an input (position 5, Fig. 6.4) and an output (position 6) with radii of curvature of the surface of 1.25 m, in the focal plane of which there is a diaphragm (position 7) with a hole diameter of 100–150 m. Such a diaphragm freely passes through itself a focused beam with a diffraction divergence.

With the telescopic UR (M = 200) in the MO, after passing the output radiation through SFC, the fraction of power in the diffraction quality beam is 23–25% (2.5 W) from the total power (10.5 W). When UR is used with two convex mirrors of the same length (935 mm), the power in the diffraction beam is 0.4 W. In the case when in UR with two convex mirrors a positive convex-concave meniscus with a focal length F = 1070 mm is used as the output mirror and in the center with a sprayed reflecting spot of 1 mm in diameter the power in the diffraction beam increases to 0.8–1 W. When the radius of curvature of a blind mirror varies within the range of 6–30 mm, the divergence of the radiation beam does not change and remains constant and equal to the diffraction limit (θ_{diff} = 0.1 mrad). Such a configuration of an unstable resonator is accepted for the basic version of the MO version when the CVLS is operated according to the MO–SFC–PA scheme. At the same time, if necessary, in the design of the radiator the traditional UR of the telescopic type can also be installed, although studies in Chapter 4 show that it is advisable to use it in single-channel radiators of the MO–SFC type without the use of the PA.

Power amplifier. The diffraction beam of the MO, separated by SFC, after reflection from the rotary planar mirrors (positions 8 and 9, Fig. 6.4) is sent to the aperture of the discharge channel (active medium) of the PA. The PA in the radiator of the ALTI Karavella-1 is the AE GL-206E, in the ALTI Karavella-1M – AE GL-206I with increased power, which allows, when using an effective pumping from the power source, the power of radiation from the active medium of the PA increases to 20–25 W (~1.5 times).

At the output of the MO and the PA, beam splitting plates (key 13) with a reflectance of ~1% are installed, directing the optical signal to the receiving sensors of the power source, which in turn uses them to ensure optimal timing of the operation of the radiator channels – MO and PA.

Attenuator with a spectral filter. The output radiation beam with a diameter of 14 mm is directed from the CVLS emitter using rotary mirrors (item 17, Fig. 6.4) to the attenuator (positions 22 and 23). The attenuator is designed to attenuate the power of the radiation beam and consists of two 100 mm diameter discs rotating on the same axis of the stepping motor that is controlled from the PC. Each disk has 8 holes 18 mm in diameter. The first, facing the incident radiation, is passive, the second is active. An active, PC-controlled disk can occupy eight discrete positions corresponding to the passage of the radiation beam through the holes. The algorithm of the attenuator operation is such that in the beginning the active disk hooks the passive and, at its rotation, takes it to the required position, and then, while rotating in the opposite direction, takes the position intended for it.

The active disk has 7 diaphragms with a diameter of 3.5 to 8.5 mm, the eighth hole of the maximum size is 18 mm. On the passive disk there are 6 diaphragms with a diameter of 6 to 18 mm, and dichroic mirrors are inserted into the remaining two holes 18 mm in diameter as light filters with a transmittance of 97–98% at wavelengths of 0.51 and 0.58 μm. By placing the discs in the

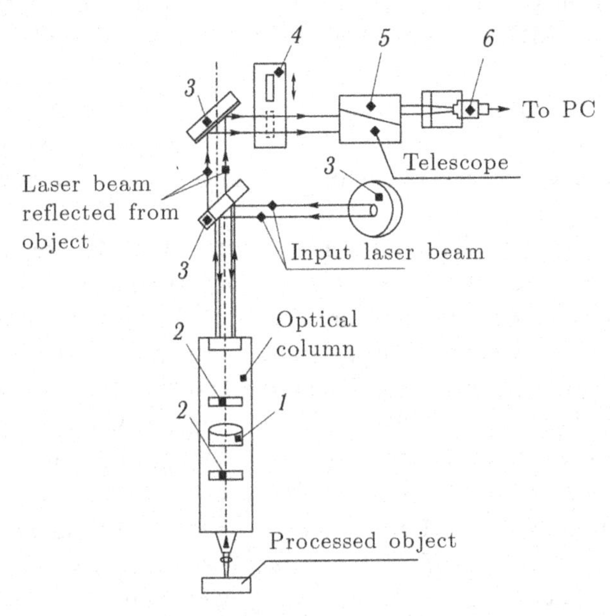

Fig. 6.8. Optical scheme of the axial channel of video surveillance: *1* – focusing achromatic lens; *2* – protective glass plates; *3* – flat rotary mirrors with dielectric coating, *3'* – flat swivel mirror with aluminium coating, *4* – electromechanical shutter; *5* – telescope; *6* – video camera.

specified positions, it is possible to obtain at the output of the attenuator the maximum or weakened radiation both at both the individual wavelengths.

Electromechanical shutter. The electromechanical shutter with a flat rotary mirror (item 10, Fig. 6.4) can be installed both in front of the attenuator and behind it. When the beam is closed, the CVLS radiation from the mirror is sent to the sensor of the TI-3 laser radiation power transducer (item 11), the signal from which is fed to the PC and the working beam power is displayed in a special window on the monitor screen. The electromechanical gate also performs the function of a safety shutter with a response time of 0.5 s.

Optical column with a power lens. The optical column is designed to focus the radiation beam on the object being processed. The column is made in the form of a tube and consists of two parts. The tube is equipped with a power achromatic lens (item 18, Fig. 6.4). Lenses are used with different focal lengths: 50, 70, 100, 150, 200 and 250 mm. On both sides it is protected from contamination by replaceable, antireflecting thin-walled plates with a diameter of 20 mm. On the tube, on the output side, the shank is additionally screwed, the length of which varies from 20 to 100 mm depending on the focal length of the lens, the height of the tooling and the workpiece (item 19). The optical column is fixed in the clamping device to the vertical coordinate table *Z*.

Optical column with a focusing lens. An axial video surveillance channel is required to adjust the position of the focus of the power lens to the surface of the object being processed. The video surveillance channel under consideration operates on the principle of a laser microscope. Its design is relatively simple and includes the minimum number of elements: a power lens (item 18, Fig. 6.4), rotary mirrors (item 17), an active medium of the PA, a radiation intensity attenuator (item 14), a matching lens (position 15) and a video camera (item 16) that is connected to the PC. As practice has shown, this version of the video channel has two noticeable drawbacks. First, these are problems that arise when adjusting the sharpness of the image on the surface of objects that are reflective in the yellow–green spectrum. When observing such mirror surfaces, the contrast of the image decreases, until it disappears. The second drawback is the loss of working time to prepare for the launch of the object processing program. First, one needs to tune into the sharpness of the image (the object is behind the focus of the lens), and only then move the lens along the vertical Z axis to the operating position (processing mode). In addition, for each individual lens with its focal length, it is necessary to determine experimentally the magnitude of this displacement.

For these reasons, another video surveillance channel was additionally introduced into the processing units, practically devoid of these shortcomings.

Axial channel of video surveillance with a telescope. The axial channel of video surveillance with the telescope (Fig. 6.8) was adopted in the ALTI for the main one and consists of the following elements: a power achromatic lens (item 1) (in Fig. 6.4 item 18) with protective clarified glass plates (item 2), rotary flat mirrors for transporting yellow–green radiation CVLS, a telescope with variable magnification – 20–50 times (item 5), a black and white video camera brand KDC-190 (item 6), and a small electromechanical shutter (item 4). This shutter performs a protective function, overlapping the radiation beam of the CVLS in the video camera at the time of opening the main electromechanical shutter (in Fig. 6.4, item 10) and starting the laser treatment.

Surveillance of the surface of the processed object occurs when the main shutter is closed (item 10) and the radiation of the CVLS does not create parasitic light in the video camera. To illuminate the object in this video surveillance device, several illumination LEDs of increased brightness are installed in a ring clip coaxial with the processing laser beam, which provides shadowless illumination of the working area. The light blue spectrum of LEDs with rather small

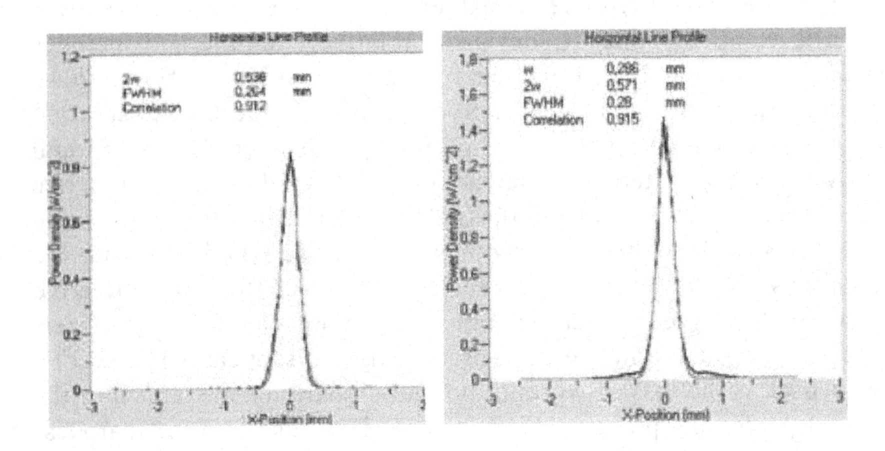

Fig. 6.9. The intensity distribution in the plane of the focused beam of radiation (in the processing spot of ALTI),

losses penetrates through the mirror with a dielectric coating (item 18) and is well reflected from aluminium mirrors.

The monitoring system functions as follows. When the focus of the power lens (position 18) is aligned to the surface of the observed object, the image of the highlighted object goes to infinity. Passing through a rotary mirror with a dielectric coating and reflecting from an aluminium-coated mirror, the light image is fed to the entrance of a telescopic telescope designed to observe remote objects. To reduce aberrations, the standard input lens of the video camera is removed, and the image is perceived without distortion with the entire working surface of the array.

To observe the image of an object on the PC monitor, the camera's position and the eyepiece of the telescope are sharpened. The total increase in the video surveillance system with the most frequently used objective with F = 100 mm is about 250. Such a video system after a single adjustment allows working with power lenses with other focal lengths (F = 50–250 mm) without additional adjustment.

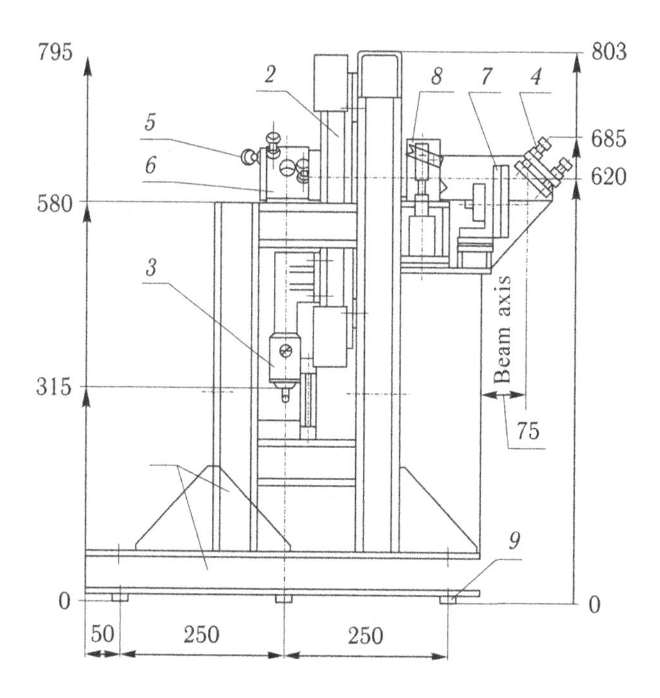

Fig. 6.10. Optical–mechanical aggregate of the processing chamber of ALTI Karavella: *1* – steel frame; *2* – coordinate table Z; *3* – optical column; *4*, *5*, *6* – adjusting mechanisms of rotary mirrors; *7* – power attenuator and spectral filter; *8* – electromechanical shutter; *9* – fastening platforms.

All other things being equal, in comparison with the laser microscope system, the contrast of the image has significantly increased. Thanks to the approximately 6 times shorter optical path from the object to the video camera, the effect of air refractions decreased and the image acquired a high spatial stability. In addition, it became possible, if necessary, to change the system's magnification.

Characteristics of the processing spot (tool for material processing). The main function of the optical system in ALTI Karavella-1 and Karavella-1M, as was already shown above, is the formation of a beam of diffraction quality, its amplification in power, transportation and focusing on the object being processed. In this case, the intensity distribution in the spot of the focused radiation beam has a very high degree of correlation with the Gaussian beam – 0.915, Fig. 6.9. The latter in turn testifies to the practical coincidence of the angular and energy divergences of the beam and, accordingly, the availability of an ideal light instrument for microprocessing materials.

Processing chamber. The basis of the processing chamber (Fig. 6.3) is an optical mechanical unit (Fig. 6.10), which is supported by a steel welded frame (item 1) from a channel of size 65 × 35 mm. A vertical coordinate table Z (position 2) with an optical column (pos.3) is fixed on two transverse beams. Transportation of the laser beam from the CVLS emitter to the optical column is performed with the help of flat rotary mirrors fixed in alignment mechanisms (position 4, 5, 6). At the rear of the frame, an attenuator (item 7) is installed on a special platform as a power attenuator and a spectral filter and an electromechanical shutter (item 8) to block the beam. Above the adjustment mechanisms (positions 5 and 6) there is a block of the video surveillance channel using a telescope with adjustable focal length.

The opto–mechanical aggregate of the processing chamber of the ALTI is installed and fixed on a honeycomb optical plate on nine milled areas in one plane (position 9). Under the coordinate table Z with a clearance of 80 mm is a horizontal XY coordinate table, the massive base of which is fixed to the optical honeycomb plate. The opto–mechanical unit is protected by steel removable casings, which allow to carry out work on the adjustment of the optical path. The front casing of the process chamber is provided with a tilting door intended for access of the operator to the working area. In the door there is an observation window with a light filter, which attenuates the scattered laser radiation from the processed object to a safe level. To prevent from accidental entry into the treatment zone, the door is equipped with a key lock.

Injection and suction systems. Elements and components of the processing gas blowing system are located in the processing chamber (Fig. 6.3). It consists of a gas valve, a rotameter and an injector. The nozzle at the end has a section in the form of a thin tube with a diameter of 2–4 mm with an oblique cut, which makes it possible to direct the jet of gas exactly to the treatment zone and at the required angle. The spatial position of the nozzle is regulated by means of a lever–screw mechanism.

Removal of the products of destruction from the processed object is carried out using a suction system. It includes a plastic duct of rectangular cross-section 120 × 60 mm with connecting bushings connected to the external exhaust ventilation system equipped with a filter. The second end of the air duct is inserted through the right wall of the protective casing into the process chamber and ends with a flexible shaped-type modular tube of the Loc-Line type (Toledo Co.) with a narrowed tip. In the processing mode, the tip is fed into the processing area of the material to remove the products of failure from the process chamber.

The water cooling system of the ULTA module. The total power dissipation in ALTI Karavella-1 is about 5 kW, in the ALTI Karavella-1M 6 kW, about 60% of it is in the AE of the CVLS emitter, the remaining power is released in the high-voltage pulse modulators of the power source. Therefore, all the fuel elements are placed in water-cooling casings. The water cooling system of ALTI is a two-circuit water-to-water type. The heat transfer from the internal circuit to the external one is effected through the Alfa Laval type heat exchanger located at the bottom of the control cabinet of the power source. The external flow rate for effective cooling is 10–15 l/min with an excess pressure of not more than 1 atm. The internal circuit is filled with distilled water, which circulates along two parallel channels: the MO and PA channel and the channel of high-voltage pulse modulators.

The use of an optical honeycomb table with vibration damping supports in the ALTI Karavella, placement of heat-loaded AE in a cooled heat sink, sealing the UR field and maintaining a constant temperature in the design have made it possible to get rid of the instabilities of the beam axis of the diffraction beam and the

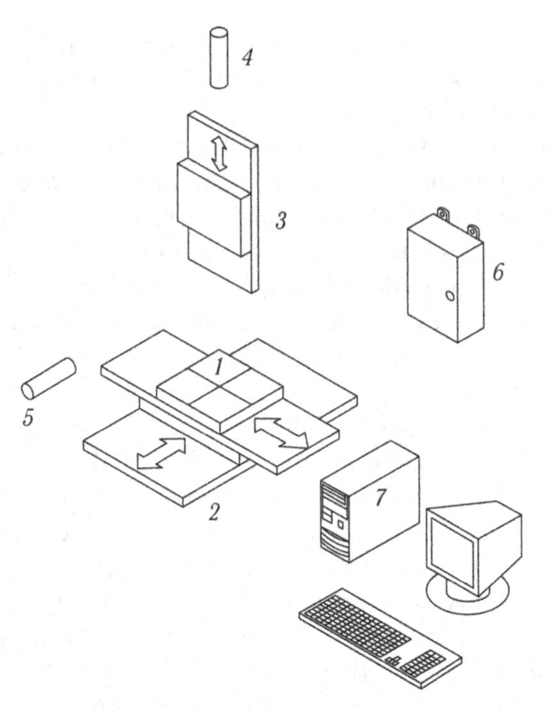

Fig. 6.11. Composition of MCS: *1* and *2* – XY horizontal table; *3* – vertical table Z; *4* – upper view camera; *5* – side view camera; *6* – CUCT; *7* – control PC.

pulsed energy. These measures maximally improved the quality of microprocessing materials.

6.2.2. Principle of construction and structure of the motion and control system

The motion and control system (MCS) is designed for laser processing according to the program realized by moving the object and the optical column with simultaneous switching of the laser beam and the necessary correction of software movements during processing. The composition of the MCS is shown in Fig. 6.11.

The MCS consists of:

– coordinate table XYZ, containing horizontal XY and vertical Z tables based on linear motors with position sensors (analog encoders);

– control unit for coordinate tables (CUCT);

– the manager of a personal computer (PC);

– a set of Laser software;

– television (TV) surveillance systems with side and top view cameras.

Table 6.1. Basic parameters of the XYZ three-axis table

Parameter	Value
Movement in the plane XY, not less than, mm	150 × 150
Movement along the vertical axis Z, not less than, mm	60
Deviation from the perpendicularity of the X and Y axes, not more than, angular. from	±10
Weight of the table's payload XY, no more than, kg	2
Maximum travel speed, not less than, mm/s	15
Resolution of the displacement reference, μm	1
Positioning error of each axis for (20 ± 1)°C, not more than, μm	±3
Mean square deviation from the given values in the axes XY of contour motion along a circle with a diameter of 1 mm at a speed of 5 mm /s at (20 ± 1)°C, not more, μm	2

The XYZ coordinate tables and the laser are connected to a CUCT controlled by the PC. The vertical coordinate table Z is equipped with a pneumatic compensator of the mass of the optical column. The pneumatic compensator together with the mass of the optical column creates a non-linear static load in the form of dry friction. Autocalibration of the electric drive provides software compensation for dry friction, and the state controller controls the deviation, which results in a trajectory error of the Z axis in the transient mode of not more than 3.5 μm and in steady state is not worse than ± 1 μm.

The dimensions and accuracy of the microprocessing of materials are determined by the parameters of the movement of the XYZ three-axis table (see Table 6.1).

The electric drive XYZ-tables in the MCS implemented on the principle of direct drive (without kinematic converters). The basis of the drive is linear synchronous motors (LSD) with optical position sensors (analog encoders). According to the principle of action – this is synchronous motors with excitation from permanent magnets (PSDM). LSD is widely used in precision machining technology, which is caused by the absence of alternating locking force, small normal forces in the supports and high linearity of the LSD force-to-current characteristic (especially in LSD of the non-band construction). In the SDE, LSD is used for a non-band construction - non-pole–pole synchronous machines without saturation. The LSD includes a mobile anchor and a stationary stator (Fig. 6.12).

The anchor consists of flat coils filled with epoxy resin, and does not have a steel magnetic core. The stator of the engine is a steel unshifted U-shaped with glued magnets of Fe–Nd–B alloy. The anchor has two phases, which allows to form in the engine calibrated (non-sinusoidal and asymmetrical) currents for compensation of non-linearities of the engine and the electric drive.

Usually, both air and mechanical linear supports are used in MCS. The latter are used in non-critical in the accuracy of evacuated systems, since the load created by them, such as dry friction, complicates control in contour regimes. Requirements for the accuracy of the MCS in ALTI Karavella-1M are provided by precision mechanical supports.

The software 'Laser', intended for precision processing of parts on a complex contour and surface processing, operates under the Windows XP operating system. Laser processing of the workpiece is carried out by the MCS and CVLS in accordance with the program

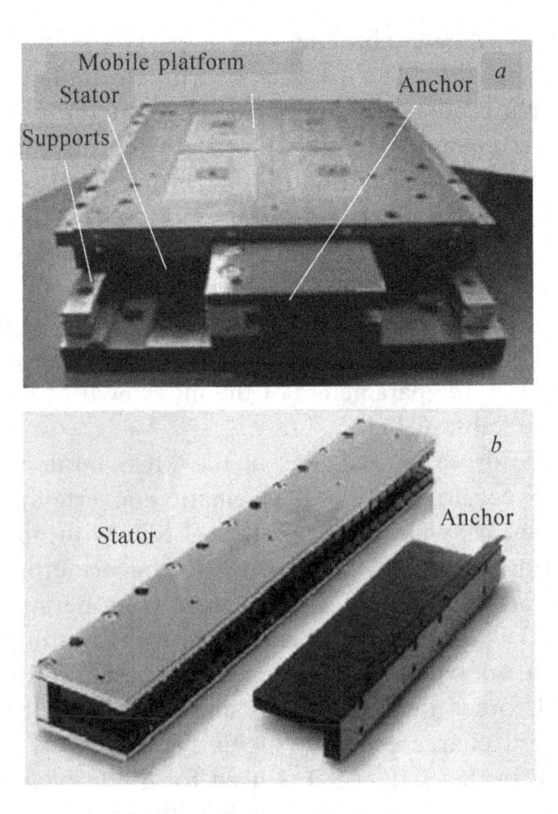

Fig. 6.12. Linear synchronous motor of the unrestricted design in assembly (*a*) and in disassembled form (*b*).

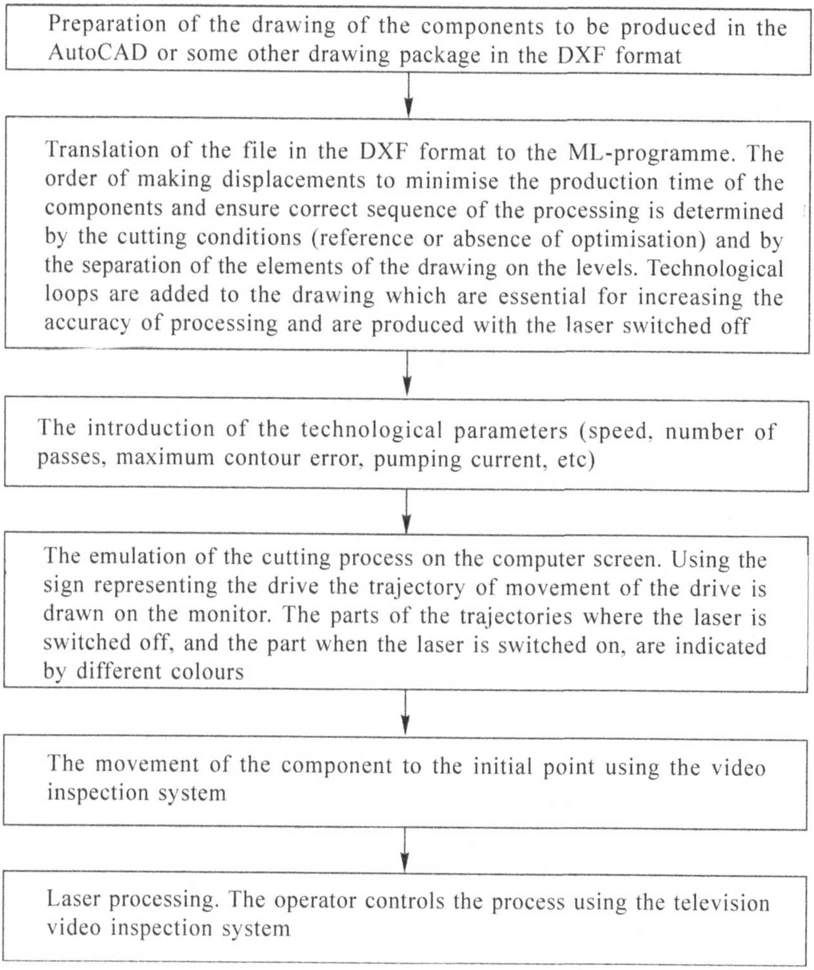

Preparation of the drawing of the components to be produced in the AutoCAD or some other drawing package in the DXF format

↓

Translation of the file in the DXF format to the ML-programme. The order of making displacements to minimise the production time of the components and ensure correct sequence of the processing is determined by the cutting conditions (reference or absence of optimisation) and by the separation of the elements of the drawing on the levels. Technological loops are added to the drawing which are essential for increasing the accuracy of processing and are produced with the laser switched off

↓

The introduction of the technological parameters (speed, number of passes, maximum contour error, pumping current, etc)

↓

The emulation of the cutting process on the computer screen. Using the sign representing the drive the trajectory of movement of the drive is drawn on the monitor. The parts of the trajectories where the laser is switched off, and the part when the laser is switched on, are indicated by different colours

↓

The movement of the component to the initial point using the video inspection system

↓

Laser processing. The operator controls the process using the television video inspection system

Fig. 6.13. Software of MCS ALTI Karavella.

(Fig. 6.13), set with a DXF file in the AUTOCAD design package. At the stage of preparation, the billet is placed on the XY table. The table moves the workpiece in the horizontal plane XY. Vertical table Z provides movement of an optical column with a power (focusing) lens. The coordinate tables X, Y, Z are connected to the CUCT controlled by a PC.

The operator specifies the workpiece movements using the mouse pointer connected to the PC. Control of movements is made on the monitor of the control computer with the help of a video camera. The camera axis coincides with the axis of the processing beam. Based

Table 6.2. The main parameters of technological installations of ALTIs Karavella-1 and Karavella-1M

Parameter name	Karavella-1	Karavella-1M
Wavelengths of radiation, nm	510.6 and 578.2	
Diameter of the radiation beam, mm	14	
Average radiation power, W	10–15	20–25
Pulse repetition frequency, kHz	13.5	14
The duration of the radiation pulse (0.5 level), ns	10 ± 1	
Instability of the average radiation power for 8 hours, %	< 5	
Pulse energy, mJ	0.1–1	0.5–1.5
Instability of Pulse energy,%	< 3	
Divergence of the radiation beam, mrad	0.1	
Instability of the beam axis position of the radiation beam, mrad	$\sim 10^{-4}$	
Focal length of the lens, mm	50–150	50–250
Diameter of the working radiation spot, μm	5–15	5–25
Density of peak power, W/cm^2	$4 \cdot 10^{10}$ –$6 \cdot 10^{11}$	$3 \cdot 10^{9}$ –10^{12}
Thickness of processed materials, mm – metal–semiconductors and dielectrics	0.05–0.5 to 1	0.5–1.0 to 2
Moving the coordinate table in the horizontal plane XY, mm	150 × 150	
Moving of the coordinate table along the vertical axis Z, mm	60	
Maximum speed of moving the coordinate table, mm/s	20	
Positioning error for each axis at (20 ± 1)°C, μm	±3	
Readiness time, min,	80	
Continuous operation time, h	Unlimited	
Power consumption from a three-phase network, kW	≤5	≤6,5
Water consumption (water–water system), l/min	≤10	
Overall dimensions, mm	2600 × 1700 × × 1350	2600 × 2100 × × 1650
Weight, kg	≤1150	≤1200
Guaranteed operating time, h	>1500	
Technical resource, years	>5	

on the technological task, the operator specifies the technological parameters: the speed of the object's movement, the number of repetitions of the program, etc. Then the choice of the reference point is made: the operator, driving the table drives, brings the required workpiece point under the crosshair in the centre of the TV monitor, combined with the laser focus, and confirms the selection by pressing the mouse button on the Reper button.

When the Emulation mode is selected, a demonstration traversal along the path specified in the program file is performed.

When the Start mode is selected, laser processing is performed. Processing can be stopped by the operator at any time. At the same time, the control PC switches off the laser beam by the electronic shutter, stops the drive and remembers the coordinates of the interrupt point. If necessary, the PC resumes processing from the same point.

6.2.3. Main technical parameters and characteristics

The main technical parameters of ALTIs Karavella-1 and Karavella-1M are presented in Table 6.2.

The main characteristic of ALTI Karavella for a given working PRF and with other things being equal is the dependence of the output power of the pulsed CVLS radiation operating according to the effective MO–PA scheme on the time mismatch of the light signal of the MO with respect to the light signal of the PA. This characteristic was studied in Chapter 4 (section 4.3) and it was established that the AM of the pulsed CVL has 4 zones: weak absorption, power amplification, full absorption and maximum transparency. Due to the presence of such zones in the active medium of th PA for the pulsed CVLS methods and electronic devices for the operational control of the radiation power and the PRF for any given algorithm, including packet and pulse impulse modulation, were developed for the ALTI Karavella-1 and Karavella-1M. The method and device are protected by the patent for invention of the Russian Federation No. 2 264 011 'Method of excitation of radiation pulses of laser systems on self-terminating transitions [365].

The operational radiation power control modes in the CVLS Kulon-10 and Kulon-20 are graphically presented in Fig. 6.14.

At the same time, it is possible to control efficiently the radiation power and pulse repetition rate from zero to the maximum operating values. The holding time of the light pulse of the MO in the absorption zone of the AM PA and, accordingly, the time it takes

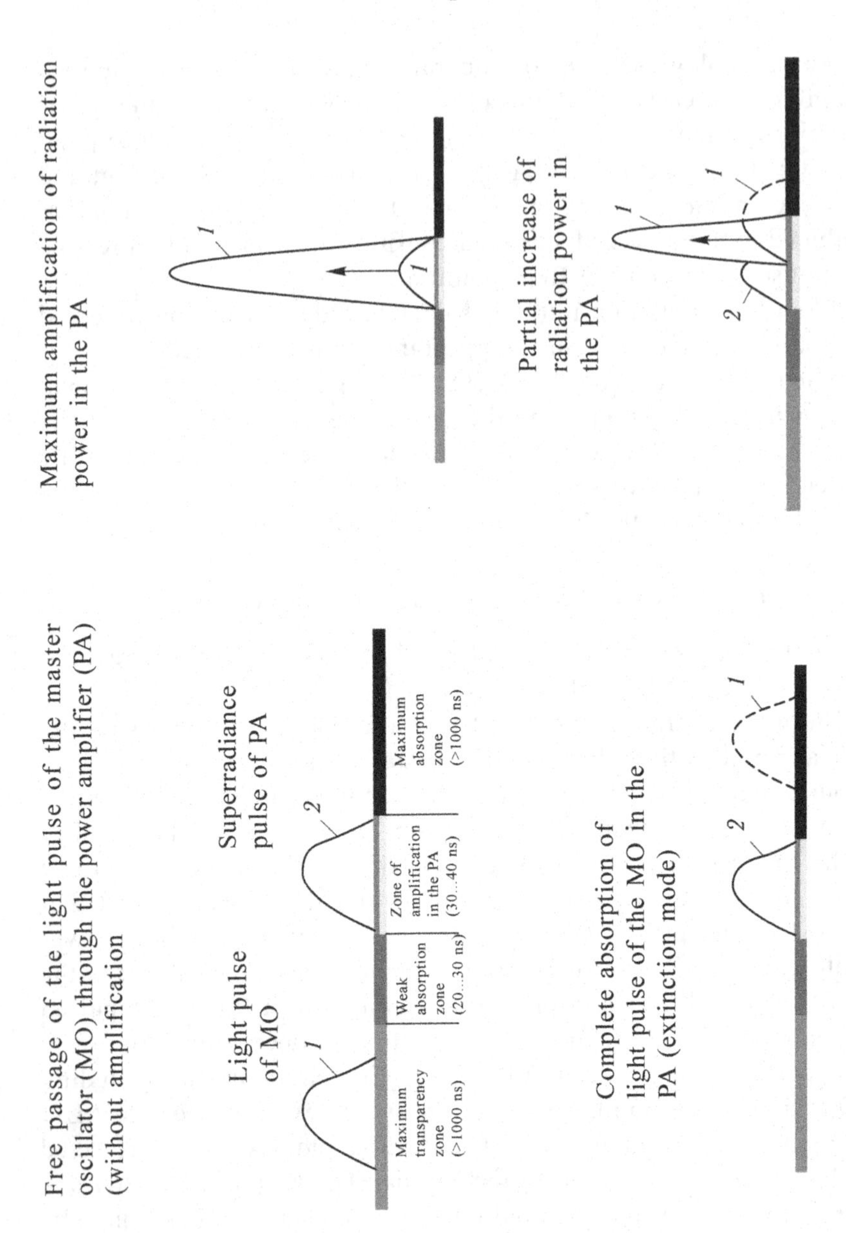

Fig. 6.14. Modes of radiation power control and PRF in CVLS, working according to the MO–PA scheme. Patent of the Russian Federation No. 2 264 011.

to find it in the amplification zone and vice versa determines the number of pulses N in the radiation packet in one second,

$$N = f(t_0 - t_1),$$

where $t_0 = 1$ s; t_1 is the time of holding the MO pulse in the absorption zone or the gain of the PA; f is the PRF. When the light signal of the MO is set to the transparency zone of the AM of the PA, we have the free-going power transmission from the MO, when the MO signal is set partially into the amplification zone of the AM – the mode of partial amplification of the radiation power.

ALTI's readiness for operation is determined by the time it takes to reach the steady-state level of laser radiation power, which in turn is determined by the warm-up time in the CVLS AE Kulon: GL-206E and GL-206I. The AEs are heat-intensive elements with an operating temperature of the discharge channel in the steady-state regime of 1500–1600°C. Figure 6.15 is an example of the dependence of the average radiation power of ALTI Karavella-1 on the warming-up time of the CVLS at the nominal power consumption and the PRF of 13.6 kHz.

As can be seen from the course of the curve, the time for the CVLS to exit to the steady-state thermal regime is about 80 min, at the level of 0.8 from the maximum power – 52 min. With such a loss of time in the case of a one-shift operation of ALTI Karavella-1, the load factor is 0.83, in the case of a two-shift operation it is 0.92, and so on.

With continuous use of process equipment in mass production, such load factors are quite acceptable.

An important practical characteristic of ALTI is the divergence of the output radiation beam (see Table 6.2), which, with the selected objective lens, determines the minimum spot diameter focusing and its intensity (power density) and, accordingly, the efficiency of material processing. The output radiation, formed in the MO by an unstable resonator, after spatial filtration in the SFC has a practically one-beam structure with a diffraction quality ($\theta = 0.1$ mrad).

Calculations show that in the ALTI Karavella-1 (see Table 6.2) with an average radiation power $P_{rad} = 10–15$ W, the duration of the pulses of radiation is $\tau_{pulse} = 10–12$ ns and the PRF $f = 13.5$ kHz with achromatic lenses with $F = 50–150$ mm, the density of the peak power in the spot of a focused beam of radiation with a diameter $d_0 = 5–15$ mm and a neck depth of 0.5 mm is $4 \cdot 10^{10}–6 \cdot 10^{11}$ W/cm² ($\rho = P_{rad}/(f \cdot \tau_{0.5} \cdot S_{spot})$) The results of practical work have shown that this power level is sufficient for productive microprocessing of the metallic materials up to 0.5 mm thick and non-metallic ones up to 1 mm thick. Tt the ALTI Karavella-1M (see Table 6.2) with a radiation power $P_{rad} = 20–25$ W and lenses with $F = 50–250$ mm

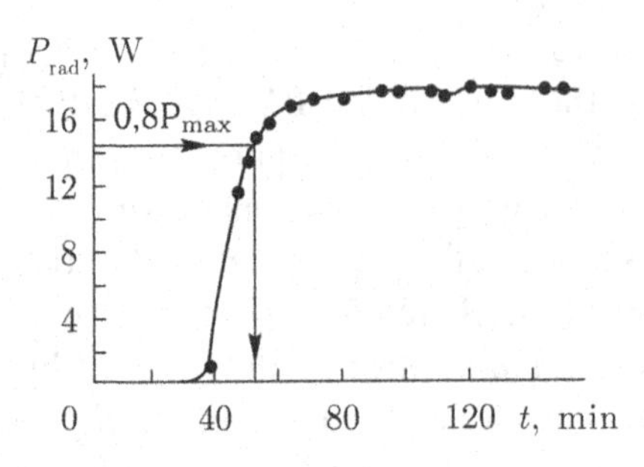

Fig. 6.15. Dependence of the average radiation power of ALTI Karavella-1 on the time of CVLS warming up at the nominal power consumption and the PRF of 13.5 kHz.

the peak power density in the spot of a focused radiation beam with a diameter d_0 = 5–25 mm and a waist depth up to 1 mm is $3 \cdot 10^9$ -10^{12} W/cm^2, is sufficient for productive microprocessing of metallic materials up to 1 mm thick, semiconductors and dielectrics up to 2 mm. The pulsed radiation of CVL with such levels of power density and the axis of the directivity diagram ($\Delta\theta < \theta_{diff}/100$) and pulsed energy ($\Delta W < 3\%$), which have an increased stability of position, are processed with high quality: the minimum HAZ (≤5–10 μm) and the roughness of the cut (≤1 –2 μm), almost any metals and their alloys.

6.3. Industrial ALTIs Karavella-2 and Karavella-2M on the basis of single-channel CVL

6.3.1. Basics of creating industrial ALTIs Karavella-2 and Karavella-2M

The main results of research and development on the development of industrial sealed-off laser AEs on copper vapour of the Kulon series (Chapter 3) [16], optical systems for the formation of single-beam radiation with diffraction divergence and with stable parameters (Chapter 4) [16, 395], methods for the operational control of the radiation power (Chapter 4, item 4.7) [16, 397] and on their basis commercial low-power pulse CVLs (Chapter 5, 5.2) [16, 158, 159, 170, 171, 179 –181, 183] became the basis for the development of modern industrial processing units with a low level of power for micromachining thin materials for electronic components and, in particular, microwave equipment [392, 399].

Many years of practical experience in the manufacture of precision parts at the first industrial technological installation of

ALTI Karavella-1, established in 2003, showed that for effective microprocessing of film coatings (≤10 μm), cutting solders (20–100 μm), thin-sheet metal (0.1–0.2 mm) and non-metallic (up to 0.5 mm) materials, marking and engraving with a high resolution, it is sufficient to have an average radiation power of a fraction and a unit of watt (with the diffraction quality of the beam) [372, 373, 375, 383, 392]. These practical results served as a stimulus for the creation of more compact, smaller mass-scale indicators and low power industrial ALTI Karavella-2 and Karavella-2M on the basis of a low-power CVL operating as a separate generator with the telescopic UR and the use of S~FC for the separation of diffraction component from the total output radiation.

The ALTI Karavella-2 was developed in 2008–2010, ALTI Karavella-2M in 2011–2012 on the basis of the CVL Kulon-06

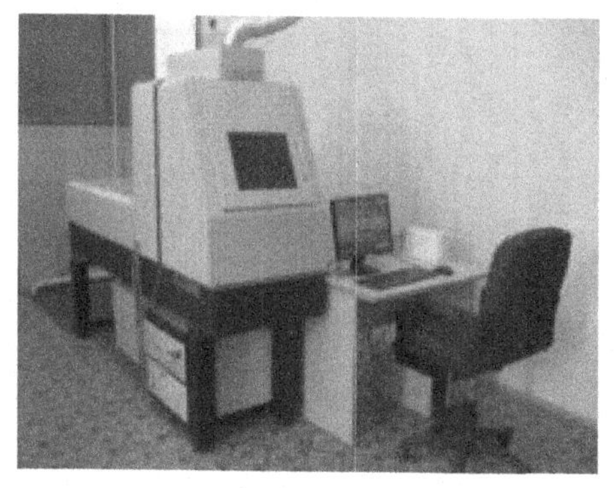

a

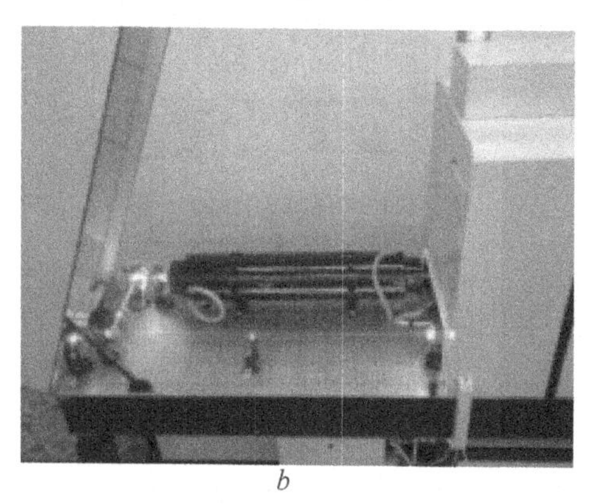

b

Fig. 6.16. The appearance of the industrial ALTI Karavella-2 and Karavella-2M on the basis of the Kulon-06 CVL, working according to the MO-SFC scheme, for precision microprocessing of metallic materials with a thickness of 0.001–0.3 mm and non-metallic materials up to 0.7 mm thick: *a*) with closed covers, b) with the cober open above the radiator.

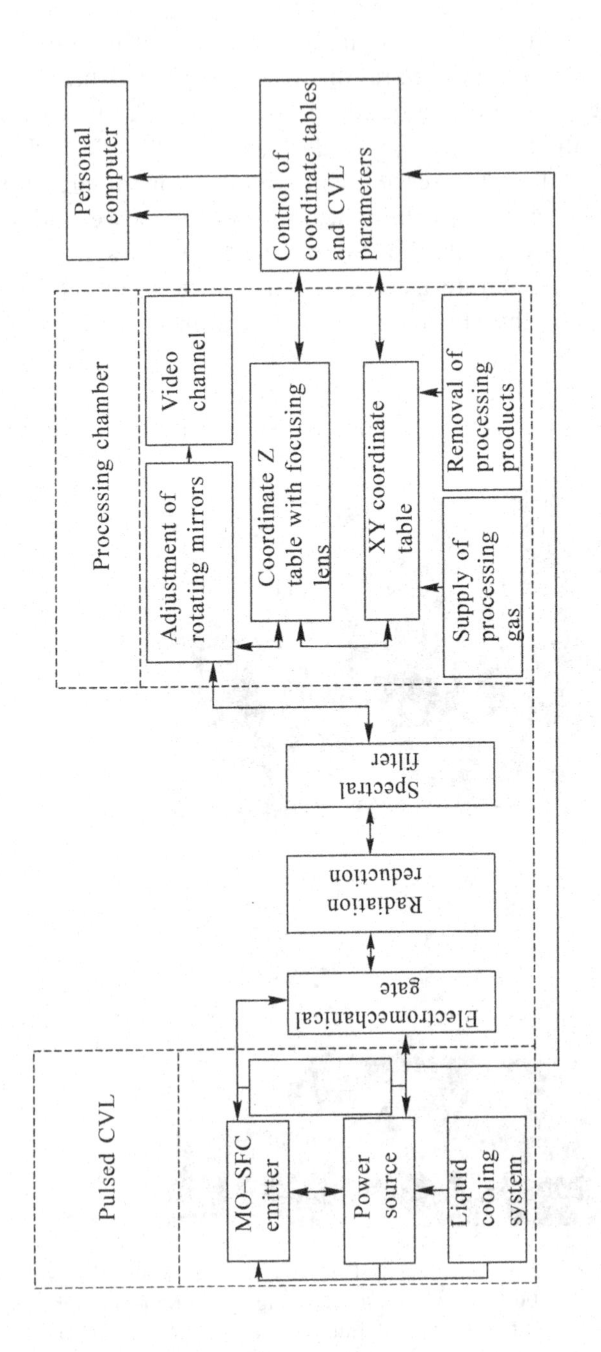

Fig. 6.17. Structural diagram of industrial ALTI Karavella-2 and Karavella-2M

with a total average radiation power of 20 W (see Chapter 5, item 5.2) and precision coordinate tables: horizontal XY and vertical Z with positioning accuracy of ±2 ns and control from the PC. These technological installations are characterized by high compactness and reliability, convenient maintenance during operation and repair. The radiation power in the diffraction beam of the CVL Kulon-06

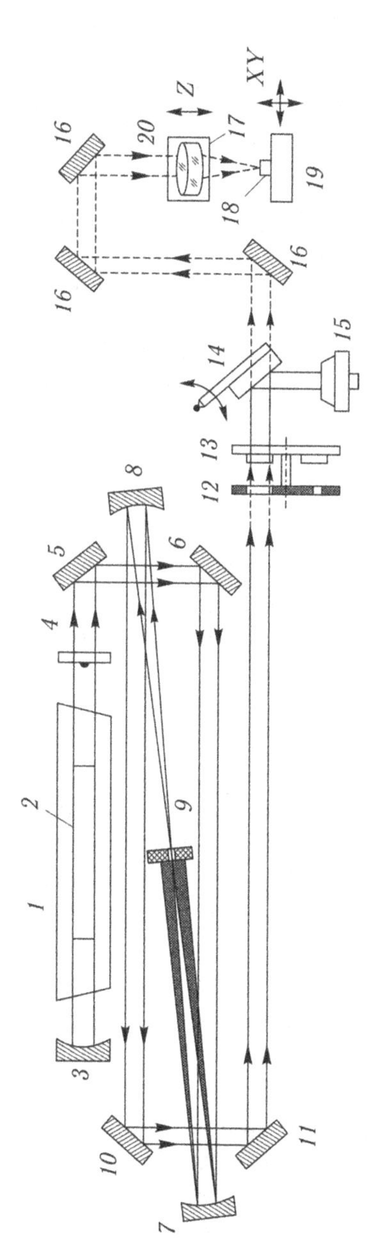

Fig. 6.18. Optical scheme of industrial ALTIs Karavella-2 and Karavella-2M: *1* – sealed-off AE Kulon, model GL-206I; *2* – discharge AE channel; *3* and *4* – hollow concave and output convex spherical mirror of telescopic UR with magnification $M = 200$; *5, 6, 10, 11, 16* – flat rotary mirrors; *7* and *8* – concave spherical mirrors SFC; *9* – selective diaphragm of the SFC; *12* – radiation power attenuator; *13* – spectral filter; *14* – electromechanical shutter; *15* – laser radiation power sensor; *17* – focusing achromatic lens; *18* – processed object; *19* – horizontal two-coordinate table XY; *20* – vertical coordinate table Z

in the ALTI Karavella-2 and Karavella-2M is the same and the maximum value is 6–8 W with a working PRF of 15 kHz, which is 2 times less than that of the ALTI Karavella-1 and 3 times that of the ALTI Karavella-1M. But as shown by experimental studies, this level of average power with a duration of light pulses of 20–30 ns is sufficient to solve many technological tasks for microprocessing. The compact ALTI Karavella-2 and Karavella-2M are the first domestic technological units intended for productive and high-quality precision microprocessing of thin-film coatings, foil and thin sheet metal materials with a thickness of 0.001–0.2 mm and non-metallic to 0.5 mm for electronic components [392, 399].

6.3.2. Composition, design and operation principle of ALTI

The exterior of the ALTI Karavella-2 and Karavella-2M is shown in Fig. 6.16.

The structural diagram of the industrial ALTI Karavella-2 and Karavella-2M (Fig. 6.17) is similar to the industrial ALTI Karavella-1 and Karavella-1M and most of the functional units and parts are unified.

The optical scheme of the ALTIs Karavella-2 and Karavella-2M is shown in Fig. 4.18.

The industrial ALTIs Karavella-2 and Karavella-2M constructively consist of 2 modules: the ALTI module and operator's desk with the control PC (Fig. 6.16 *a*). The ALTI module includes the following main functional blocks and systems.

1. A single-channel emitter of a pulsed CVL Kulon-06 (Chapter 5, item 5.2) on the basis of an industrial sealed-off self-heating AE, model GL-206I (Chapter 4, Table 16) and an unstable resonator operating according to the MO–SFC scheme (Chapter 4, Fig. 4.1 *b*).

2. The thyratron power source that generates high-voltage nanosecond pump pulses of the AE of the CVL with a current amplitude of up to 300 A for a base time of 100–120 ns and a 14–16 kHz PRF and low-power current pulses in the additional channel for high-speed power control, including monopulse and packet modulation laser radiation (Chapter 5, Section 5.2).

3. The system of movement and control (MCS), which includes a two-coordinate horizontal XY table with a working field of 150 × 100 mm (in ALTI Karavella-2) or 200 × 200 mm (in ALTI Karavella-2M) and a vertical table Z with a displacement length of 60 mm and with a positioning error for each axis of no more than ±2 μm and a

travel speed of up to 20 mm/s, a control unit for coordinate tables (CUCT) connected to the control computer ALTI.

4. The optical system for the formation of a beam of diffraction quality, its amplification and attenuation, transportation and focusing on the object being processed.

5. The system of double-circuit cooling, working on a water-to-water scheme with a temperature feedback sensor.

6. The laser system of observation and adjustment of the focused radiation spot on the processed object with the display of the image on the monitor of the control PC.

7. Adjustable systems for blowing processing gas into the zone of laser treatment and extraction of the products of destruction from this zone.

8. The disk attenuator of laser radiation power.

9. The spectral filter for laser wavelength selection (0.51 or 0.58 μm).

10. The optical honeycomb table with vibration damping supports, which performs the function of the carrier of the CVL radiator, the XY and Z coordinate tables, a video system and all individual elements included in the ALTI optical system.

11. Software 'Laser'.

In the ALTIs Karavella-2 and Karavella-2M, separate units and modules of the ALTI module responsible for the stability of the axis position of the laser radiation beam diagram are mounted on the optical honeycomb table of the Standa company (Lithuania). The table is used in models HT08-20-20 with the overall dimensions of 2000 × 800 × 200 mm [401]. These single elements include a single-channel radiator of the CVL with the telescopic UR, SFC for separation and collimation of the diffraction beam, rotary mirrors for beam transport, two-coordinate horizontal table XY and vertical Z with a focusing lens. On the exit side of the radiation there is located the optical table and on it a processing chamber with an optical-mechanical unit and a video surveillance system. The radiator and the opto–mechanical aggregate are interconnected by a common optical scheme of ALTI (Figure 6.18). The optical table with all the elements mounted on it is closed with a protective and decorative cover. The table is mounted on 4 supports of type 1TS05-12-05 through passive pneumatic insulators AR-500 with a total permissible load of 550 kg (Fig. 6.5) [401]. Supports are interconnected by a metal frame, in which are installed a pulsed high-voltage thyratron power source,

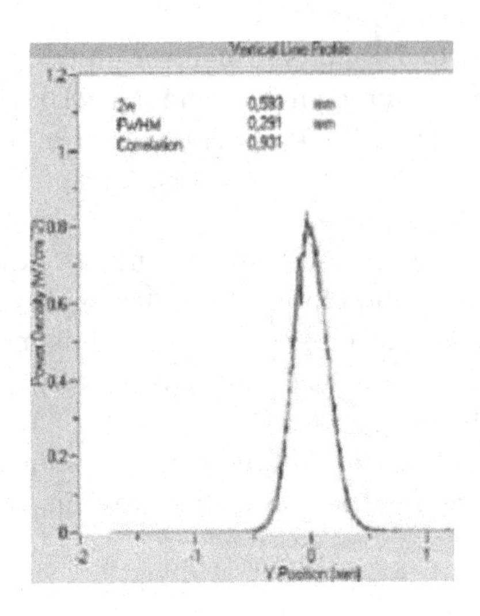

Fig. 6.19. The intensity distribution in the plane of the focused radiation beam (in the processing light spot of the ALTI).

the CUCT, a power distribution board and a water cooling system with a temperature feedback sensor.

The radiator of the pulsed CVL is an industrial sealed-off laser AE on copper vapour the Kulon model GL-206I (Chapter 4, Table 4.3), placed in a cylindrical water-cooled heat sink with a tilting top cover. The mirrors of the optical resonator are mounted along the ends of the heat detector. The telescopic UR is used with a gain of $M = 200$. The radius of curvature of the blind mirror is $R_{blind} = 2200$ mm, output $R_{out} = 10–12$ mm (3 and 4 in Fig. 4.18). To eliminate the effect of air-heat fluxes at the ends of the hot AE and cause instability of the position of the axis of the beam pattern of the laser beam, the space between the mirrors of the resonator and the heat sink is sealed by dielectric tubes.

The use of the optical honeycomb table with vibration damping supports, the placement of heat-loaded AE in a cooled heat sink, the sealing of the HP field, and the maintenance of a constant temperature in the design of the experimental setup have made it possible to eliminate the instabilities of the beam axis of the diffraction beam and the pulsed energy. These measures maximally improved the quality of microprocessing materials.

The SFC collimator is a mirror and is formed by two concave spherical mirrors with a radius of curvature $R = 1250$ mm (positions 7 and 8), in the focus of the input mirror of which is located a diaphragm with a hole of 100–150 μm (item 9). The SFC is designed to isolate a radiation beam with diffraction divergence (θ_{difr} = 0.1 mrad) with a pulse duration (at half-height) of 10...12 ns from the background of the total multibeam radiation. The average radiation power of the extracted diffraction beam is P = 6–8 W. The extracted SFC diffraction beam is transformed into a cylindrical diameter of 14 mm and is transported by means of rotary flat mirrors (Fig. 6.18, items 10 and 11) into the process chamber and in it is directed (position 16) to the focusing achromatic lens (key 17) mounted on the vertical coordinate Z table (item 20). Interchangeable lenses with F = 50, 70, 100 and 150 mm focus the radiation beam into a light spot 5–15 μm in diameter onto the object to be processed (item 18), which is placed on the horizontal XY table (item 19). Figure 6.19 shows the intensity distribution in a focused light spot, which is a contactless precision tool for processing materials with micron accuracy.

It can be seen from the radiation intensity distribution that the correlation with the Gaussian beam is high (0.93), which practically indicates the coincidence of the angular and energy divergences of the radiation and, accordingly, the effectiveness of the light spot on the material being processed.

The attenuator, consisting of two rotating disk drums (items 12 and 13), is installed at the output of the SFC. The first disk drum (item 12) is designed to attenuate the processing power of the processing beam before focusing, the second drum (key 13) is used to extract a radiation beam from one beam with a total radiation spectrum (510.6 nm + 578.2 nm) the wavelength is 510.6 or 578.2 nm). The electromechanical gate (item 14), intended as a shutter for beam overlapping, is installed directly behind the attenuator. In the closed position, the laser radiation reflected from the mirror is sent to the sensor of the power meter (item 14), by means of which the processing power of the radiation beam coming into the focusing lens (item 17) is monitored.

The motion and control system (MCS) is designed for laser processing according to the program realized by moving the object and the optical column with simultaneous switching of the laser beam and the necessary correction of software movements during processing. From the PC via the CUCT, in accordance with the specified program for the material processing process, the horizontal XY table with the processing object and the vertical table Z with the focusing lens are moved, controlled by a high-speed electronic switch

Table 6.3. The main parameters of ALTIs Karavella-2 and Karavella-2M

Parameter	Karavella-2	Karavella2M
Radiation wavelength, nm	510.6 and 578.2	
Diameter of the radiation beam, mm	14	
Pulse repetition frequency, kHz	14–16	
The duration of the radiation pulse (0.5 level), ns	11 ± 1	
Average radiation power, W	6...8	
Instability of average radiation power for 8 hours,%	<3	
Pulse energy, mJ	0.2–0.5	
Pulse energy,%	<3	
Divergence of the radiation beam, mrad	0.1	
Instability of the beam axis position of the radiation beam, mrad	$\sim 10^{-4}$	
Focal length of the lens, mm	50–150	
Diameter of the working radiation spot, μm	5–15	
Density of peak radiation power, W/cm^2	$3 \cdot (10^{10} - 10^{11})$	
Thickness of processed materials, mm - metal - semiconductors and dielectrics	0.001–0.2 to 0,5	
Moving the coordinate table in the horizontal plane XY, mm	150 × 100	200 × 200
Moving of the coordinate table along the vertical axis Z, mm	60	
Maximum speed of moving the coordinate table, mm/s	15	
Positioning error for each axis at (20 ± 1)°C, μm	±3	
Readiness time, min	60	
Continuous operation time, h	Unlimited	
Power consumption from a three-phase network, kW,	±3	
Cooling system, water flow, l/min	Water–water and air 4–6	
Overall dimensions, mm	2200 × 1700 × 1830	
Weight, kg,	≤800	
Guaranteed operating time (with replacement of active elements), h	>1500	
Technical resource, years	5	

and the electro-mechanical shutters of overlapping the radiation beam.

The MCS consists of the following elements:

– two-coordinate horizontal table XY and vertical Z table on the basis of linear motors with position sensors;

– control unit for coordinate tables (CUCT);

– the control computer (PC);

– a set of 'Laser' software;

– television (TV) surveillance systems with side and top view cameras.

The XY and Z coordinate tables and pulsed CVL are connected to the computer-controlled CUCT.

The systems of blowing the processing gas (N_2, Ne, O_2, etc.) into the processing area of the material and the suction of vapor spray products from this zone and the optical system of axial video surveillance for adjustment of the radiation spot on the object in terms of design and principle of operation are analogous to the systems installed in the ALTIs Karavella-1 and Karavella-1M.

The water cooling system consists of a water-to-water heat exchanger and is designed to effectively remove heat energy from the water-cooling casing of the high-temperature AE laser heat sink. The water flow in the external circuit of the heat exchanger is 4–6 l/min. The system allows to monitor the flow rate of the coolant and its temperature and to stabilize the temperature of the liquid with respect to the surrounding elements by means of feedback, in order to stabilize the position of the axis of the processing beam of laser radiation. In ALTI, the temperature of the cooling water is tied to the temperature of the optical table, which is the carrier of the entire structure.

6.3.3. Main technical parameters and characteristics

The main technical parameters of the ALTIs Karavella-2 and Karavella-2M are presented in Table. 6.3.

As can be seen from this table, with a diffraction divergence of the laser beam of radiation (0.1 mrad) in a focused light spot 5–20 μm in diameter and pulse width at a half-height of 11 ± 1 ns, the peak power density is 10^{10} –10^{11} W/cm^2, so that microprocessing of metallic materials in the evaporative mode is possible.

Experimental studies have shown that at such levels of power density and high stability of the output radiation parameters, the

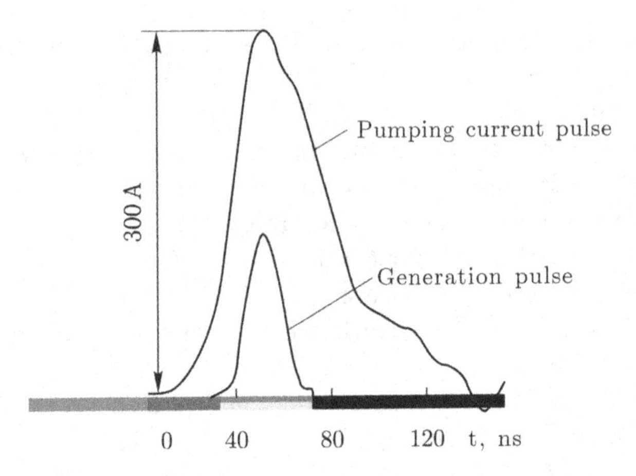

Fig. 6.20. Current and radiation pulses with different colours of the time zones of the AM of the pulsed CVL Kulon-06.

quality of the microprocessing of materials is high: the surface roughness of the cut does not usually exceed 1–2 μm, and the heat-affected zone is 5–10 μm.

It should be noted that at present the development of the prototype ALTI Karavella-2S with a high-speed laser beam scanner and a focusing F-Theta lens model Superscan-14 is continuing. The ALTI Karavella-2S also uses the pulsed CVL Kulon-06 with the AE model GL-206I, telescopic UR with $M = 200$ and SFC. The radiation power in the diffraction beam is 6–8 W. The minimum processing spot is 30 μm, the scanning field is 90 × 90 mm. The main purpose of ALTI Karavella-2S is high-speed cutting of solders with a thickness of 20–50 μm, perforation of stencils on 100 μm material and milling of hemp structures by the method of pulse processing at speeds up to 1–2 m/s.

Methods for controlling the radiation power. Heating and excitation (pumping) of AE GL-206 and pulsed CVL Kulon-06 in the ALTIs Karavella-2 and Karavella-2M is carried out from a pulsed high-voltage thyratron power source with an amplitude of 18 kV and 300 A (Chapter 5). Figure 6.20 shows the current and generation pulse, indicating the time zones of the AM of the pulsed CVL Kulon-06 in different colours. Each zone has specific properties with respect to interaction with its own radiation (Chapter 4).

In the figure, four distinct consecutive time zones of the AS pulsed CVL: a zone of weak absorption (brown colour), a zone of amplification (yellow–green), a zone of total absorption (black) and a zone of maximum transparency (blue) are clearly distinguished

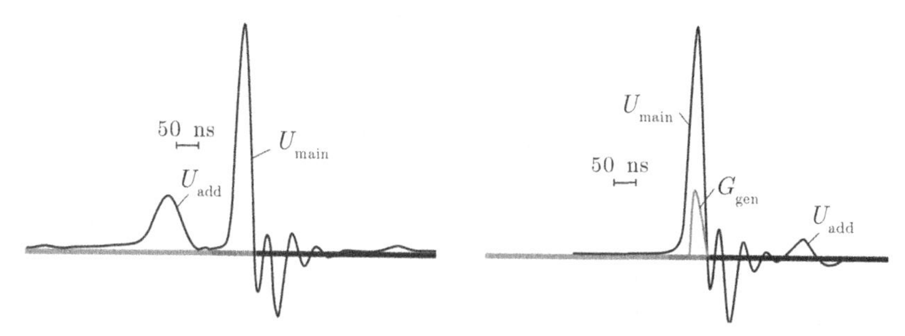

Fig. 6.21. The oscillograms of the main (U_{main}) and additional (U_{add}) pulses of the pumping voltage of the pulsed CVL Kulon-06 in the modes of quenching and generation, G_{gen} is the generation pulse.

one after another. Due to the presence of such zones in the CVL, methods and electronic devices were developed to control the power and repetition frequency of the radiation pulses in the process units, including packet and pulse impulse modulation. For this purpose, an additional channel with a low-power generator of nanosecond pulses was introduced into the pulsed LP Kulon-06. This generator generates pump current pulses with an amplitude 4–6 times smaller than the amplitude of the main pump current pulses. To induce the generation mode in the laser an additional low-power current pulse is formed after the main excitation pulse – in the absorption zone, and to provide a mode of total or partial extinction of the generation, an additional pulse is formed in front of the main pulse– in the transparency zone. This method of controlling the radiation parameters of the CVL is protected by patent No. 2 251 179 'The method of exciting pulsed lasers on self-limited transitions of metal atoms operating in self-heating mode and the device for its implementation' [161]. Figure 6.21 shows the oscillograms of the additional and fundamental pulses of the pump voltage (excitation) at the electrodes of the AE, explaining the operation of the CVL in the modes of quenching and generation. The energy of the additional current pulses should be sufficient only for populating the metastable (lower) laser levels of the active substance (copper atoms) and should not affect the processes of their relaxation in the plasma during the interpulse period. Therefore, an additional pulse should be located near the main pump current pulse, which excites the resonant (upper) levels of copper atoms. Naturally, an effective control of the output energy characteristics of the laser is provided to the greatest extent when the time difference between the additional current pulse and the main excitation pulse is less than the lifetime

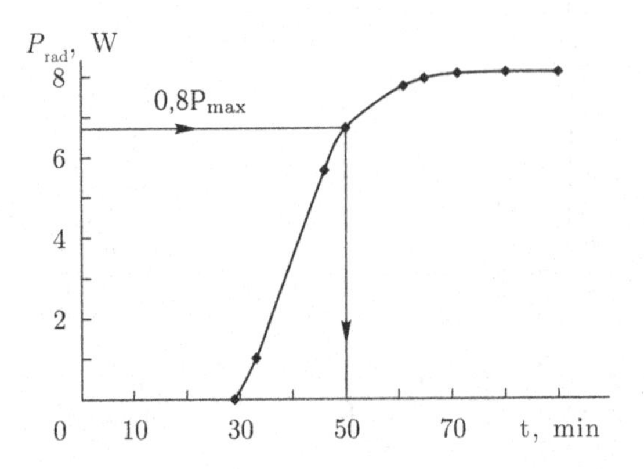

Fig. 6.22. Dependence of the average radiation power of ALTIs Karavella-2 and Karavella-2M on the warm-up time of the CVL Kulon-06 at the nominal power consumption and 15 kHz PRF.

of metastable levels (≥1–2 μs). With the experimental optimization of the CVL, the indicated time detuning was ≤1 μs. From the point of view of stabilizing the parameters of the plasma of the active medium, the optimum operating mode of CVL is optimal, when the power consumed from the electric network during generation (with a lagging additional pulse) is equal to the power consumed by the laser in the case of extinction of the generation (with an advanced additional pulse). This mode is achieved by adjusting the phase and the amplitude of the additional current pulse.

This technique became the basis for the development of industrial CVL Kulon-06 with the AE GL-206I with controlled power characteristics. In this CVL, the delay lines provide a range of time control of the pulses of the main and additional pumping channels up to 1300 ns. Adjustment of the amplitude of the main current of the excitation pulse is ensured up to 400 A with the duration of its front not exceeding 50 ns, the voltage amplitude up to 20 kV. The amplitude of the additional current pulse is regulated to 60 A, the amplitude of the voltage is up to 5 kV. The duration of the main excitation current pulse along the base is about 120 ns, the additional pulse duration is up to 250 ns. The change in the response time of adjustable delay lines in the power source, as well as the formation of single pulses according to a predetermined law, allows high-speed pulse modulation of laser radiation with an accuracy of up to one pulse, changing the generation PRF, and performing any of their sequences, specifying certain values of the pulse energy, And more importantly, in the design of the CVL, it is possible to control the operating modes from an external personal computer.

Readiness of the ALTIs Karavella-2 and Karavella-2M for work is determined by the time it takes to reach the steady-state level of the laser radiation power, which in turn is determined by the warm-up time for the steady-state thermal conditions of the AE GL-206 and the pulsed CVL Kulon-06. The AEs are a heat-consuming element with an operating temperature of the discharge channel in the steady state of 1600°C. The readiness time of the ALTIs Karavella-2 and Karavella-2M is the same, because they use the CVL of one model. Figure 6.22 shows the dependence of the average radiation power on the warm-up time for these ALTIs with the LP Kulon-06 at the nominal power consumption and the 15 kHz PRI.

As can be seen from the course of the curve, the time for reaching the steady-state regime by the radiation power is about 70 minutes, at a level of 0.8 from the maximum power 50 minutes. At 70 minutes the loss of time in the case of a one-shift operation of the ALTIs Karavella-2 and Karavella-2M the load factor is 0.85, in the case of two-shift 0.93, etc., which is quite acceptable with continuous use of process equipment in series production.

6.4. Conclusions and results for Chapter 6

1. To achieve high-quality (5–30 kHz), short-pulse (10–40 ns) and short-wave (510.6 and 578.2 nm) high-quality CVL radiation, it is necessary to have radiation with a single-beam (single-mode) structure, diffraction divergence ($\theta_{diff} = 2.44\lambda/D_{chan}$), high position stability of the axis of the radiation pattern ($\Delta\theta = \theta_{diff}/10^3$) and pulsed energy ($\Delta W \leq 3\%$), a clear boundary of the focused spot and a Gaussian intensity distribution in the far zone. At the same time, the peak power density in the focused radiation spot should be 10^9–10^{12} W/cm^2, and to smooth and polish the surface of the cut, the treatment should be performed in several passes.

2. The modern ALTIs Karavella-1 and Karavella-1M with a diameter of the processing light spot of 10–20 μm and a density of peak power of 10^9–10^{12} W/cm^2 at 14 kHz for fine and high-quality precision microprocessing of thin sheet materials of electroniocic componets respectively, with a thickness of 0.3–0.5 mm and 0.6–1 mm. ALTIs are developed on the basis of industrial CVL Kulon-10 with an average radiation power of 10–15 W and Kulon-20 with a power of 20–25 W, new highly selective optical systems and methods

for controlling parameters and the use of precision XYZ three-axis tables with positioning accuracy along the axes ±2 μm and the working field of the horizontal XY-table 150×150 mm.

3. The modern ALTIs Karavella-2 and Karavella-2M with a light spot diameter of 10–20 μm and a density of peak power of 10^9–10^{11} W/cm^2 with a 15 kHz PRF are created and investigated for productive and high-quality precision microprocessing of foil materials The IET is 0.01–0.2 mm thick. The ALTIs are developed on the basis of industrial CVL Kulon-06 with an average radiation power of 6–8 W, new highly selective optical systems and methods for controlling parameters and the use of the precision XYZ three-axis tables with positioning accuracy of the axes of ±2 μm axes, the working field of the horizontal XY-table of 150 × 100 mm and 200 × 200 mm.

4. In the industrial ALTIs Karavella-1 and Karavella-2M, a motorized, programmable, rotary platform (model 8MR151-1 of the firm Standa) is additionally built in the design of the XY horizontal coordinate table with a 360° rotation angle for machining of parts in the polar coordinates. With this device, high-quality microprocessing of parts with a cylindrical surface is ensured.

5. The operative control of the output power and PRF radiation, and by any predetermined algorithm, including packet and monopulse modulation, in the ALTIs Karavella-1 and Karavella-1M is produced by the method of dissynchronization in the CVLS of the light signal of the MO relative to the signal of the PA (using an electronic key switch). When the light signal of the MO is established in the zone of amplification of the PA, the generation mode is realized, and the absorption mode is the radiation damping mode. The time delay of the MO signal within each zone and the sequence of its transitions are determined by the specified power control law and the PRF of radiation, and the time for the MO signal transition from one zone to the other is the pulse–pulse interval of the laser working FPI time. (For example, with a working PRF of 15 kHz, this transition time, which corresponds to the response time of the electronic switch, is 70 μs.)

6. Operational control of the output power and PRF of radiation, and for any predetermined algorithm, including packet and monopulse modulation, is provided in the ALTIs Karavella-2 and Karavella-2M by creating in the active environment of the AE CVL Kulon-06 conditions for full or partial extinction of generation and maximum generation due to the formation of an additional low-power pump current pulse. When a low-power current pulse is set in advance of

the main pulse of the pumping current, the mode of full or partial extinction of the generation is ensured (metastable levels of copper atoms are populated), and in the case of a lag, the mode of full generation. For high-speed displacement of a low-power pump current pulse, a high-speed electronic switch with a minimum response time of up to 30 μs is created in the laser IP.

7. The possibility of operational control in the ALTI Karavella with the power of radiation according to a given algorithm, including packet and monopulse modulation, allows to significantly increase the efficiency of selecting optimal processing parameters for each particular part with its material and determine the optimal sequence of the manufacturing process, to make cuts and holes with minimum dimensions, minimum roughness and the heat-affected zone.

7

Laser technologies of precision microprocessing of foil and thin sheet materials for components for electronic devices

Laser technologies of the precision microprocessing of foil and thin sheet materials of electronic components were developed on technological installations ALTI Karavella. The units are based on industrial CVL and CVLS with high radiation quality and precision XYZ three-axis tables with accuracy of positioning along the axes of $\pm$ 2 μm (see Chapter 6): ALTI Karavella-1 with a radiation power of 10–15 W, Karavella-1M with a power of 20–25 W, Karavella-2 and Karavella-2M with a power of 6–8 W, in which the processing tool is a focused light spot 5–20 μm in diameter and a peak power density of 10^9–10^{12} W/cm^2 [382 , 383, 389, 390, 392, 394–400]. The ALTI Karavella-1 with the density of peak power in the processing spot $4 \cdot 10^{10}$–$6 \cdot 10^{11}$ W/cm^2 is intended for productive and high-quality precision microprocessing of thin sheet metal materials, as well as a number of semiconductors and dielectrics 0.1–0.5 mm thick [382, 383, 389, 390, 392, 398), the ALTI Karavella-1M with power density $3 \cdot 10^9$–10^{12} W/cm^2 – microprocessing of materials with a thickness of 0.5–1 mm [392, 398], the ALTIs Karavella-2 and Karavella-2M with a power density of $3 \cdot (10^{10}$ –10^{11} W/cm$^2)$ – foil materials with a thickness of 0.01–0.2 mm [392, 399] for electronic components.

In parallel with the creation of thye modern ALTI Karavella, laser technologies were developed on their basis for the production of precision parts with high quality: the minimum surface roughness of the cut (≤1–2 μm) and the heat-affect zone (≤5–10 μm) for the electronic components and, in particular, microwave technology.

7.1. The threshold densities of the peak and average radiation power of CVL for evaporation of heat-conducting and refractory materials, silicon and polycrystalline diamond

Knowing the threshold (minimum) peak power density of evaporation of materials is necessary to determine the laser radiation power level, both when creating modern laser processing equipment, and when developing laser technologies for processing materials, in our case, for microprocessing, on its basis.

Calculation of the threshold peak power density of evaporation was carried out for a large group of metallic and non-metallic materials widely used in modern production of the components for electronic devices. These include heat-conducting metals: copper, aluminium, gold and silver and refractory: molybdenum, tungsten and tantalum, nickel, stainless steel, silicon, polycrystalline diamond, sapphire, ceramics and others. The theoretical calculation was carried out according to the formula given in [402, 405]. It includes the main thermophysical and optical parameters of the material:

$$q_u = \frac{kT_u}{2A}\sqrt{\frac{\pi}{\delta\tau}}, \tag{7.1}$$

where k is the thermal conductivity coefficient, T_u is the evaporation temperature (boiling point), $A = 1 - R$ is the absorption coefficient (R is the reflection coefficient) [403], δ is the thermal diffusivity, and τ is the laser pulse duration. For pulsed CVLs used in the ALTI Karavella, the value of pulse duration for calculation according to formula (7.1) is taken at half-height, since the pulse has a shape close to triangular and is 10^{-8} s. All the data on the thermal and optical parameters for calculating the evaporation power density and the corresponding average radiation power of the CVL for the above materials are given in Table 7.1.

Substituting the data of the Table and the real duration of the CVL pulses (10^{-8} s) into the above formula (5.1), we obtain the value of the calculated threshold peak (impulse) power density for evaporation of the substance (Table 7.1). For an example, we give

Table 7.1. The main thermophysical and optical parameters, the threshold densities of the peak evaporation power and the corresponding average radiation power of the CVL for heat-conducting and refractory metals, nickel, stainless steel, silicon and artificial polycrystalline diamond

No.	Material	Evaporation (boiling) temperature (T_i), K	Absorption coefficient (A)	The coefficient of thermal conductivity (k), W/(m · K)	Coefficient of thermal diffusivity (δ), cm²/s	Threshold power density (q_i), 10^9 W/cm²	The threshold average thickness of the CVL (P_{cr}), mW at d = 10 and 20 μm	
				Heat conducting materials				
1	Cu	2840	0.368	401	1.17	0.25	30	120
2	Au	3080	0.15	317	1.28	0.5	60	240
3	Al	2792	0.086	236	0.938	0.7	80	320
4	Ag	2485	0.045	429.5	1.74	1.6	190	750
				Refractory materials				
1	Ta	5731	0.59	55.2	0.237	0.1	12	36
2	Mo	4885	0.41	140	0.54	0.21	24	96
3	W	5828	0.484	162.8	0.64	0.22	26	104
				Stainless steel				
1	Steel	3000	0.5	32	0.035	0.09	11	43
				Nickel				
1	Ni	3005	0.384	90.4	0.229	0.13	15	60
				Silicon and artificial polycrystalline diamonds				
1	Si	3543	0.5	150	1.48	0.08	9	36
2	diamond	4273	0,15	1600	7	1,5	180	720

here calculations of the power density for silver and gold, which have relatively high reflection and thermal conductivity [400].

$$q_u(\mathrm{Cu}) = \frac{4.01\ \mathrm{W/(cm\cdot K)\cdot 2840\ K}}{2\cdot 0.368}\sqrt{\frac{3.14}{1.17\mathrm{cm}^2/\mathrm{s}\cdot 10^{-8}\mathrm{s}}} = 0.25\cdot 10^9 W/\mathrm{cm}^2,$$

$$q_u(\mathrm{Mo}) = \frac{1.4\ \mathrm{W/(cm}\cdot K)\cdot 3885\ \mathrm{K}}{2\cdot 0.41}\sqrt{\frac{3.14}{0.54\ \mathrm{cm}^2/s\cdot 10^{-8}\mathrm{s}}} = 0.21\cdot 10^9\ \mathrm{W/cm}^2.$$

On the other hand, the experimental values of the peak radiation power density are easily determined by the well-known formula:

$$q_u = \frac{P_{av}}{f\cdot \tau \cdot S}, \tag{7.2}$$

where P_{av} is the average power of laser radiation; f is the repetition rate of pulses; τ is the duration of radiation pulses at half-height; $S = \pi d^2/4$ is the area of the light spot of the focused laser beam (d is the diameter of the light spot). Using formula (7.2), knowing the threshold peak radiation power density, we determined the minimum average radiation power of the CVL necessary for the microprocessing of the material in the evaporation regime.

In the ALTI Karavella the pulse repetition frequency (PRF) of CVL is 14–15 kHz. The most common focusing lenses in the ALTI are achromatic objectives with a focal length F = 100, 150 and 200 mm, when the diameter of the processing light spot is d = 10, 15 and 20 μm. Table 7.1 shows the threshold average radiation power for the diameters of the light spot of 10 and 20 μm (the last column). For example, for silver and gold with a peak evaporation power density of $1.5\cdot 10^9$ and $0.5\cdot 10^9$ W/cm² for a 15 kHz PRF and an objective with F = 100 mm, the threshold mean radiation power for silver and gold is 0.2 and 0.06 W:

$$P_{\mathrm{av\ Cu}} = q_{\mathrm{pulse}}\cdot f\cdot\tau\cdot\pi\cdot d^2/4 = 0.25\cdot 10^9\ \mathrm{W/cm}^2\times$$
$$\times 3.14\cdot(10\cdot 10^{-4})^2\mathrm{cm}^2/4\cdot 15\cdot 10^3\mathrm{s}^{-1}\cdot 10^{-8}\mathrm{s} = 0.03\ \mathrm{W},$$

$$P_{av\ \mathrm{Mo}} = q_{\mathrm{pulse}}\cdot f\cdot\tau\cdot\pi\cdot d^2/4 = 0.21\cdot 10^8\ \mathrm{W/cm}^2\times$$
$$\times 3.14\cdot(10\cdot 10^{-4})^2\mathrm{cm}^2/4\cdot 15\cdot 10^3\mathrm{s}^{-1}\cdot 10^{-8}\ \mathrm{s} = 0.024\ \mathrm{W},$$

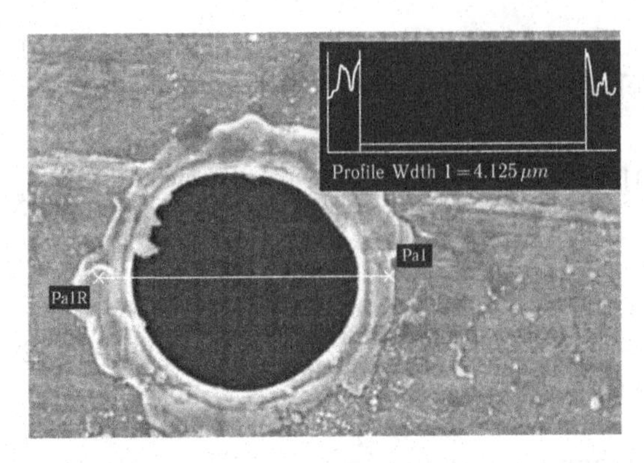

Fig. 7.1. The image of a hole in a sample of stainless steel 50 μm thick formed by the action of the radiation from the CVL with a threshold density of peak evaporation power (~10^8 W/cm^2). Magnification $\times 10^4$, F = 50 mm.

where $f = 15 \cdot 10^3$ Hz; $\tau = 10 \cdot 10^{-9}$ s; d = 10 μm.

Experimental rescarches on the microprocessing of materials with pulsed CVLs in the ALTI Karavella showed that a bump is formed from the cooled molten metal at threshold radiation power levels on the side of the laser beam exit along the edge of the cut. This testifies to the formation in the process of laser action of a large amount of liquid phase flowing down to the bottom of the material, which is not permissible in the manufacture of high-quality precision parts for the components for electronic devices. This is clearly shown in the example of drilling holes in stainless steel by direct flashing with a threshold power density of evaporation ~10^8 W/cm^2 (see Fig. 7.1). When the electron microscope magnification of $\times 10^4$ the presence of the melt around the perimeter of the hole is clearly visible. But when the density of the peak radiation power is increased by an order of magnitude and higher, $\geq 10^9$ W/cm^2, which corresponds to the average radiation power of CVL ≥ 1 W, the holes are almost perfectly clean with a submicron roughness.

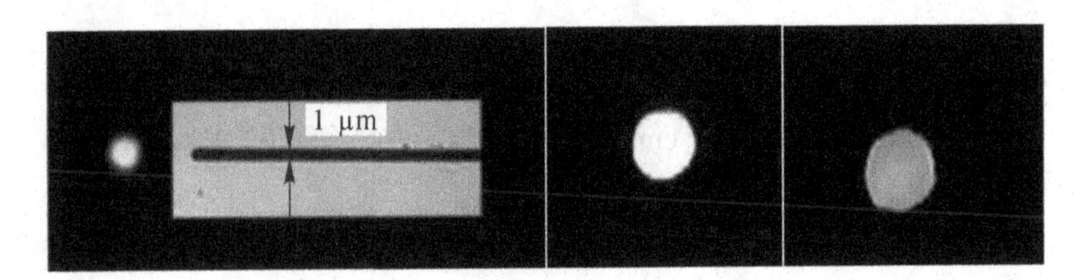

Fig. 7.2. Images of holes in samples of stainless steel 50 μm thick in transmission moze. The peak power density ρ = (1–5) $\cdot$ 10^9 W/cm^2.

Figure 7.2 shows the images of holes on samples of stainless steel with a thickness of 50 μm without a noticeable melt and a roughness of ≤1 μm at a peak power density in the processing light spot ρ = (1–5) × 10^9 W/cm^2, which is greater by an order of the threshold power density of evaporation of steel.

Our research in the field of pulsed microprocessing with the use of laser radiation was well coordinated with the works of a number of foreign authors presented in [20]. It follows from these studies that a mean radiation power of about 1 W at a peak power density of ~10^9 W/cm^2 is sufficient to cut such thin materials. When using foil materials with a thickness comparable to the width of the cut (10–50 μm), the vapours and drops of metal that scatter as a result of the microexplosion from the radiation zone are practically unobstructed and completely removed from this zone.

Thus, to ensure the processing of materials in the evaporation mode with a minimum of the liquid phase and accordingly a high quality of microprocessing, a peak power density of not less than 10^9 W/cm^2 is necessary in the processing light spot with a diameter of 5–20 μm.

7.2. Effect of the thickness of the material on the speed and quality of the laser treatment

When cutting materials with pulsed radiation of CVL with a thickness of ≥50 μm, the expansion of metal vapors occurs mainly between the walls of the already formed section and the volume of the liquid mass increases. The resulting liquid metal under the pressure of the vapor spreads out of the zone of action. This effect manifests itself the more strongly, the more the thickness of the material exceeds the diameter of the focused spot. The main reasons for the increase in the amount of liquid phase are as follows:

– decrease in the density of the light flux due to gradual defocusing of the beam with increasing depth of the hole;

– a slow decline in power at the end of the pulse, which contributes to an increase in the volume of the remainder of the liquid phase in the hole after the end of the pulse;

– increase in the duration of the pulse.

The efficiency of removal of a substance depends significantly on the laser power, which determines the initial temperature and energy of the scattered particles [328]. It was established in [328] that for a material thickness of more than 200 μm and an average radiation power of up to 20 W, no complete removal of matter occurs in one pass. It was also noted that with increasing power the width of the cut increases due to the fact that the side 'wings' of the focal spot are already involved in the evaporation [337]. With higher power, it is possible to cut the material in one pass, but the roughness can be too large and does not differ from the roughness in the treatment with a solid-state or CO_2 laser [326, 333]. Therefore, in order to ensure a high quality of cutting, it is necessary to repeatedly pass the beam. Multipass cutting reduces the processing speed, but it is necessary to minimize the zone of thermal influence and surface roughness of the cut.

The longer the exposure time, the greater the volume of the liquid phase, the depth of the heat-affected zone increases, where oxidation and structural changes occur and defects appear on the surface of the hole. It is also essential that with increasing thickness of the material, the processing time increases and the speed decreases.

Experimental studies on the effect of the thickness of the material on the time, speed, and quality of the treatment at the initial stage were carried out on copper, molybdenum, aluminium, tungsten and a number of nonmetallic materials with achromatic lenses with F = 110 and 230 mm, when the diameter of the working light spot was equal to 15 and 25 μm. The average radiation power was 20 W, the pulse repetition frequency was 8 kHz and the pulse duration was 15 ns at a half-height. The time and the processing speed for copper, molybdenum and tungsten proved to be the same, which is explained by practically identical values of the threshold density of peak evaporation power for these materials (see Table 7.1). The processing speed for aluminum and silver, with a higher threshold density due to the higher reflection coefficient, was noticeably higher.

Figure 7.3 shows the dependences of the time and drilling speed on the thickness of the material for copper and molybdenum, widely used in the production of components for electronic devices by the method of direct piercing using an achromatic objective with F = 110 mm (d = 15 μm). The peak power density with such a light spot was $\rho = 10^{11}$ W/cm^2, which is two orders of magnitude greater than the threshold power density. In this case, the curves for molybdenum and tungsten completely coincided.

It follows unambiguously from the course of the curves that when the metal thickness is greater than 1 mm, the drilling time (t_{cv}) increases sharply, and the drilling speed is correspondingly reduced (v_{cv}) and it turns out that the pulsed radiation of CVL is

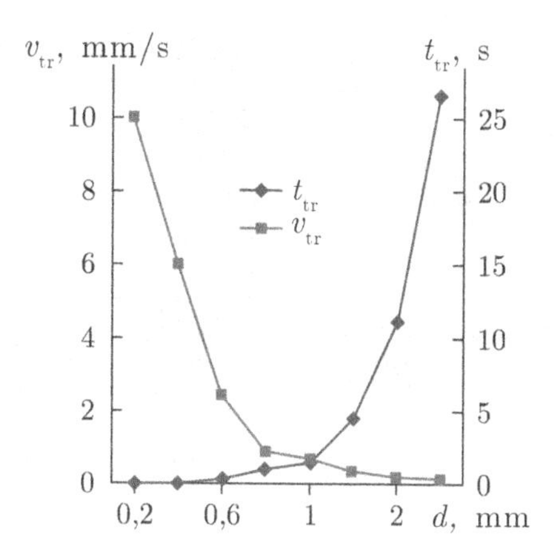

Fig. 7.3. Dependence of the time (t_{tr}) and the speed (v_{tr}) of treatment for copper (molybdenum) on the thickness of the material when drilling holes with pulsed radiation of CVL with a spot diameter of 15 μm. P_{av} = 20 W, f = 8 kHz, τ = 15 ns and $\rho = 10^{11}$ W/cm^2.

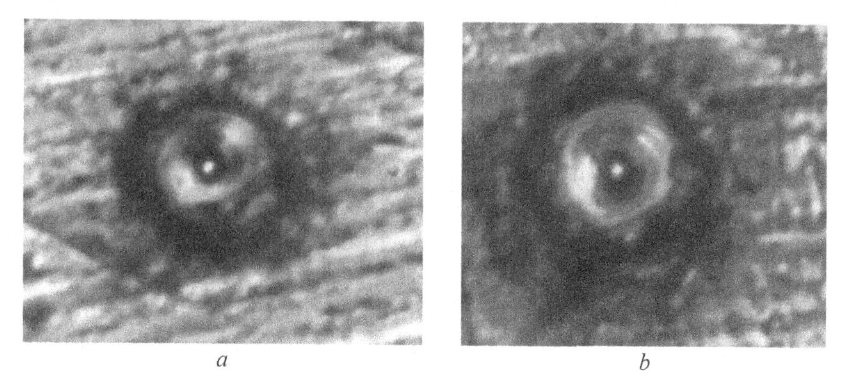

Fig. 7.4. Formation of holes in copper (*a*) and molybdenum (*b*) under the influence of focused radiation of CVL with a light spot diameter of 15 μm (F_{tr} = 150 mm). P_{av} = 2 W, f = 15 kHz, t_{pulse} = 11 ns and $\rho = 2 \cdot 10^{10}$ W/cm^2, magnification ×1000.

advantageous for processing thicknesses less than 1 mm. And it is even more advantageous to use CVL in microprocessing of metallic materials with a thickness of less than 0.6 mm, when the processing time is a fraction of seconds, and the speed is one and tens of mm/s, which is an order of magnitude greater than traditional processing methods, including electroerosion machining. The process itself is described at the beginning of this section, and is also discussed in detail in Chapter 2.

At the initial moment of drilling, when the depth of the hole is unity and tens of micrometers, traces of the melt and condensed metal

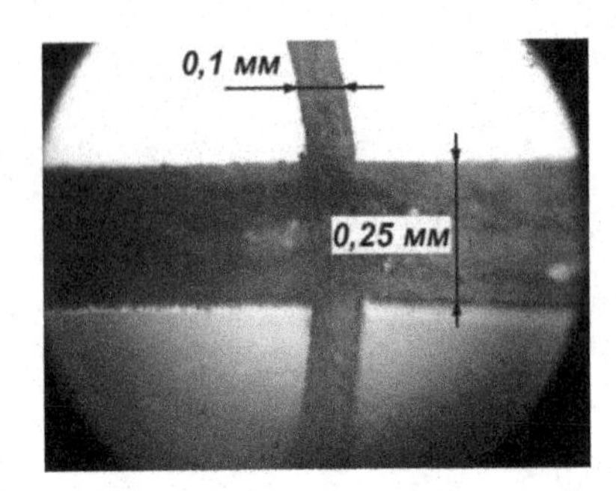

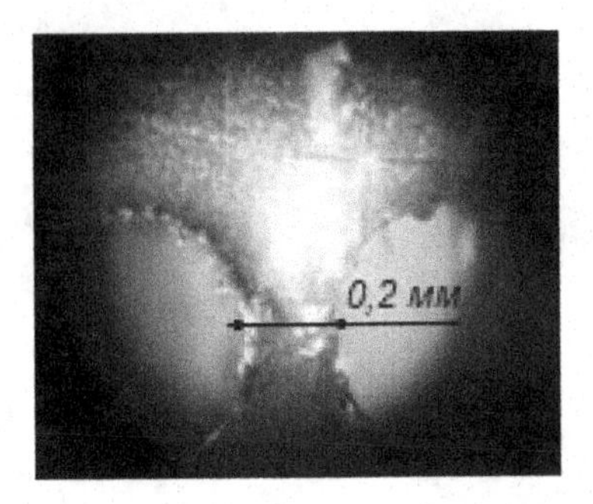

Fig. 7.5. Fragment of a spherical molybdenum grid 0.07 mm thick (left) and a copper diaphragm with holes 0.3 mm thick.

droplets, as well as slags in the treatment zone, are not observed (Fig. 7.4). For the removal of the vaporous substances dispersed from the treatment zone, the treatment was performed with a suction system.

When the thickness of the processed material is increased by more than 50 μm, the droplet and slag are already beginning to appear in the cutting zone, which is unacceptable in the manufacture of precision parts for the components for electronic devices (see Figure 7.4).

Figure 7.5 on the left presents a fragment of a spherical molybdenum net 0.07 mm thick, on the right a diaphragm with copper apertures 0.3 mm thick, on which the presence of a grate and slags in the material processing areas is clearly visible.

Therefore, the problem of developing methods for cleaning the contaminated zone formed by microprocessing with pulsed radiation of CVLs was sharply raised.

7.3. Development of the technology of chemical cleaning of metal parts from slag after laser micromachining

The microprocessing of metallic materials by radiation from CVL is usually carried out in an air atmosphere, since in the processing light spot with a diameter of 10–20 μm the peak power density reaches very high values, 10^9–10^{11} W/cm^2. At the same time, the processing is carried out in the evaporation mode, accompanied by the formation of a small amount of liquid mass and heating the cut wall. As a result, the slag originates in the cutting zone, mainly from their oxides and nitrides and grate, in the form of condensed fine droplets on the adjacent side surface. Precision parts for real components for electronic devices were manufactured for working out the cleaning modes from slag and grate on the ALTI Karavella.

First of all, the technology was developed on the widely used components for electronic devices materials: copper and molybdenum. After the laser treatment, experiments were conducted to anneal the parts in the reducing hydrogen and vacuum furnaces and in the air atmosphere under different temperature conditions, but no positive results were obtained for the purification. The next stage was the use of the ultrasonic bath type UZV-1.3 l with an operating frequency of 35 kHz and deionized water with heating. The parts were ultrasonically processed immediately after the end of laser micromachining. The effectiveness of the ultrasonic action was high enough: most of the grate and slag were removed, but there remained a small adherent bead and a thin layer of cinders. To remove the latter, new chemical methods of cleaning were developed by the specialists of the chemical department of the Istok Company.

The chemical purification regimes for such electrovacuum metals as copper, silver, molybdenum, tungsten, tantalum, titanium, nickel, kovar, pseudoalloys of MD type and other metals and alloys are fully developed. Figure 7.6 shows enlarged fragments of molybdenum and copper parts at each stage of processing.

Technology of chemical cleaning of copper parts. Requirements for cleaned parts include, in addition to removing carbon deposits from copper at the laser exposure sites, also maintaining the structure

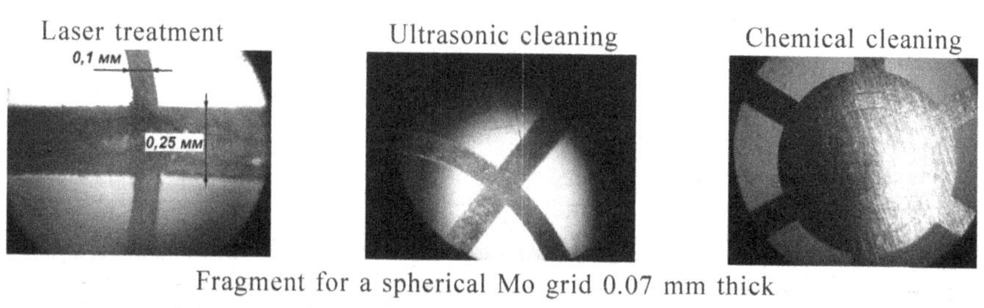

Fragment for a spherical Mo grid 0.07 mm thick

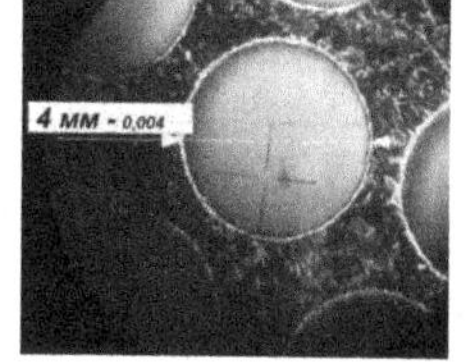

Fragment of a flat Cu diaphragm 0.3 mm thick

Fig. 7.6. Stages of technology of fabrication of precision components in laser processing.

of the copper surface and the dimensions of the cut holes in copper. Cleaning of copper diaphragms after laser cutting in the modes used to remove carbon from such diaphragms after electrospark processing did not lead to success – the deposit was not completely removed from the cutting places. Carbon from the cutting site is well removed if the etching route involves the operation of etching the diaphragms in a solution of hydrochloric acid with the imposition of an ultrasonic field. However, in this case, the copper surface is irritated. The optimal solution was obtained using the purification route given in Table 7.2. After such processing, the cuts are clean, the surface condition of the diaphragms does not differ from the original one, and the dimensions of the holes in the diaphragms exactly correspond to the drawing.

The diaphragms were made not only from pure copper, but also from a copper alloy of the MAG brand. Purification of the diaphragms from the MAG alloy according to the regime given in the table showed that the surface of copper is considerably distended. This is probably due to the interaction of aluminium present in the MAG alloy with the alkali melt. Modification of the regimes for cleaning such diaphragms will be carried out.

Technology of chemical cleaning of molybdenum components. Shadow and control grids made of molybdenum can be divided into two types: the first – the grids after cleaning are coated with an anti-emitter and the grade of their surface cleanliness need not be high (roughness improves the adhesion of the anti-emitter); the second – grids with a polished surface.

Table 7.2. Modes of chemical cleaning of diaphragms from copper after electroerosion machining (EEM) and laser processing

Treatment mode after EEM	Cleaning mode after laser treatment
1. Degreasing in trichlorethylene	1. Degreasing in trichlorethylene
2. Decarbonization in a solution of hydrochloric acid	2. Processing in a melt of alkali
3. Annealing in hydrogen	3. Processing in an aqueous solution for polishing copper
	4. Processing in water in an ultrasonic field at a frequency of 35 kHz

For the grids to which no requirements are made to maintain a high class of surface cleanliness, a cleaning regime was developed using an etching solution with hydrogen peroxide (Table 7.3). This technology provides a complete removal of carbon from the laser cutting sites while maintaining the grid dimensions.

The above technology can not be used to clean the grids with a polished surface, since the quality of the surface is not preserved in such a treatment. For this reason, another processing technology using an electropolishing operation was used for grids with a polished surface (Table 7.3, mode II-1). Processing of such grids using polished Austrian molybdenum showed that the deposit is removed from the cutting site, but matte areas appear on the surface of the grid, as well as the points of excavation.

The initial polishing of Austrian molybdenum was preserved when using a large current density during electropolishing – 200–300 A/dm^2 (Table 7.3, mode II-2). However, in this case, it is not possible to completely get rid of corroded points. It should be noted that the quality of the surface of domestic molybdenum during

Table 7.3. Chemical cleaning of molybdenum grids after laser processing

I. Requirements for maintaining the initial state of the surface are absent	
Processing technology	Result
Degreasing in trichlorethylene. Processing in the melt of alkali. Etching in a solution based on hydrogen peroxide. Processing in water, in an ultrasonic field at a frequency of 35 kHz	Degreasing in trichlorethyl-ene. Processing in the melt of alkali. Etching in a solution based on hydrogen peroxide. Processing in water, in an ultrasonic field at a frequency of 35 kHz
II. It is necessary to maintain the original surface purity	
Processing technology	Result
1. Degreasing in trichlorethylene. Processing in the melt of alkali. Electropolishing. D_k = 50 A/dm^2. Treatment in water, in an ultrasonic field with a frequency of 35 kHz	Carbon residue removed. The surface is rasterized.
2. Same as electropolishing with D_k = 200–300 A/dm^2	Carbon residue removed. The surface is polished. There are separate corroded points
3. Degreasing in trichlorethylene. Application of varnish. Laser cutting Electropolishing. D_k = 200–300 A/dm^2. Treatment in water in ultrasound field frequency of 35 kHz. Removing the varnish	Carbon residue. Initial polishing is not disrupted

electropolishing in such modes is significantly increased: the risks of rolling disappear, the surface becomes shiny, but corroded spots also appear.

Due to the fact that the initial state of the polished Austrian molybdenum can not be completely maintained, the cleaning technology was tested using a chemically resistant varnish film to protect molybdenum from aggressive media (Table 7.3, mode II-3). The varnish was applied to the preforms of polished molybdenum before the cutting operation. The operation of loosening of carbon products in the alkali melt was eliminated, since the lacquer in the alkali is not stable. However, as shown by experiments, high current density during polishing and subsequent ultrasonic treatment in water at a frequency of 35 kHz allow to remove carbon from the laser cutting sites. After removing the varnish, the surface of molybdenum is clean, there are no matte areas or melted points.

Since the current density during electropolishing is quite high, it is necessary to ensure a good contact between the polished part and the polishing electrodes in order to avoid current oscillations. To do this, special mandrels have been made, which ensure a hard contact between the part and the electrodes.

Technology of chemical cleaning of products made of titanium and other materials. Optimization of finishing treatment for titanium products after laser micro-processing, when carbon deposits and slags are formed, was carried out in the following directions:

– choice of conditions for loosening and processing of carbon and slag generated during laser cutting;

– choice of the composition of the solution for bleeding cut products after their loosening.

It should be noted that in addition to removing the products of destruction during laser cutting, the developed final cleaning technology should ensure the preservation of the surface structure of the material and, in some cases, the preservation of dimensions in accordance with the requirements of the drawing.

In the beginning, loosening of the deposit and slag was carried out by the method of keeping the product in a boiling aqueous solution of alkali (NaOH). Two solutions were used for bleeding products:

No. 1 – aqueous solution of nitric acid with potassium fluoride, the thickness of bleeding in such a solution can be from 3 to 10 μm;

No. 2 – a solution of hydrogen peroxide with ammonia, a bleeding thickness <1 μm.

However, even after repeating the opening and etching operations, it was not possible to completely remove the cut products.

Therefore, a more rigid method was tested – processing (soaking) in the melt of alkali. This method of loosening gave a positive result – after pickling in solution No. 1 and No. 2 and additional washing in deionized water with ultrasound (35 kHz) – a clean surface was obtained, without traces of deposit in the holes.

The choice of a solution for cleaning the laser treatment products should be determined by testing the final cleaning technology for specific parts. Based on the results of the experiments, the following route of titanium purification has been worked out:

1) control of appearance (microscope);
2) soak in the melt of alkali;
3) washing in hot running water;
4) etching in solution No. 1 and No. 2;
5) washing in running cold water;
6) washing in deionized water with ultrasound;
7) drying;
8) control of appearance (microscope).

In addition to the technologies presented here, chemical purification technologies have also been developed for other metals widely used as heat conducting (Al, Au, Ag) and refractory (W, Ta, Re) metals, their alloys and a number of other materials.

7.4. Investigation of the surface quality of laser cutting and the structure of the heat-affected zone

In this section, the influence of speed, number of passes and radiation power on laser processing on the structure, roughness and hardness of the material in the region adjacent to the cutting zone is investigated. All the investigated objects were previously subjected to complete chemical treatment. It is established that an increase in the cutting speed and in the number of passes at constant power leads to a decrease in the surface roughness less than 1 μm. It is shown that in samples of refractory materials: molybdenum and tantalum, the hardness in the zone adjacent to the cut increases. In high-conductivity materials (copper, bronze), hardening or softening in the zone adjacent to the laser cut was not detected. There were no structural changes in all samples [394].

In the production of components for electronic devices, there is a wide range of precision parts the manufacture of which requires the production of not only simple but also complex configurations with high processing quality. Because of the problems that arise in the manufacture of a special tool necessary for this, the implementation of such operations by traditional methods of processing is associated with certain difficulties and great labour costs. Therefore, of considerable interest is the use of fundamentally new technological processes, in particular, laser microprocessing. In addition, when using laser radiation, the processing capacity of a material can exceed by an order of magnitude or more in comparison with traditional methods [16, 25].

The task of this part of the experimental work is to investigate the effect of parameters of focused laser radiation with a diameter of 10–20 μm and a peak power density of 10^{10} –10^{11} W/cm^2 on the structure and properties of near-surface cutting regions [394].

Cutting of materials was carried out at the ALTI Karavella-1 and Karavella-2 (see Chapter 8) with the following modes: focal length of achromatic lens – F = 100 and 150 mm; the average radiation power P_{rad} = 2.4–9.6 W; the wavelength of the radiation λ = 0.51 and 0.58 μm (with 50% power content at each wave); processing speed V = 1–8 mm/s; number of passes during cutting N = 1–120.

The study of the surfaces of the cutting of thin sheets of copper, molybdenum, tantalum and beryllium bronze by means of optical microscopy (Leica microscope) has shown that for copper, an increase in the cutting speed (from 2 to 8 mm/s), and for molybdenum an increase in the thickness of the plate (with 0.1 up to 0.2 mm) lead to a certain decrease in the roughness from 2–3 μm to about 1 μm. For a more detailed analysis of the cutting zone, samples of copper

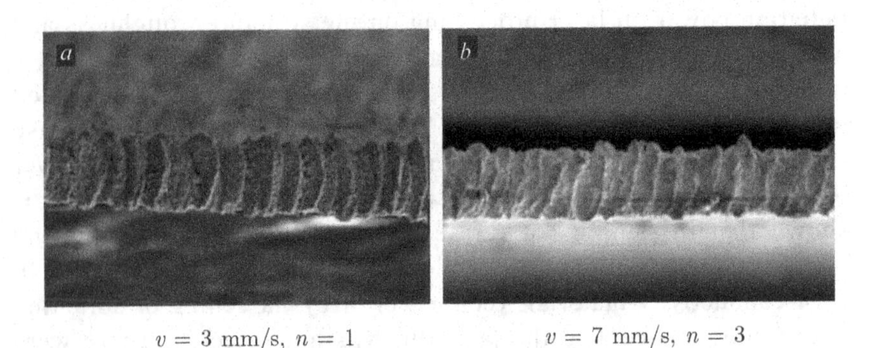

Fig. 7.7. Image of the cut surface of tantalum with a thickness of 0.03 mm in secondary electrons, magnification × 600. P_{rad} = 1.5 W; V – processing speed, n – number of passes.

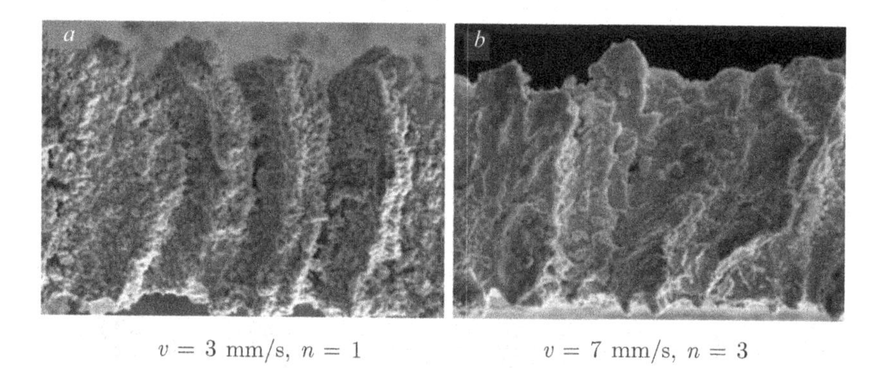

$v = 3$ mm/s, $n = 1$ $v = 7$ mm/s, $n = 3$

Fig. 7.8. The image of the cut surface of tantalum with a thickness of 0.03 mm in secondary electrons, magnification × 2000. $P_{rad} = 1.5$ W; V – processing speed, n – number of passes.

and tantalum were studied using scanning electron microscopy (SEM). Figure 7.7 are images of the surface of a tantalum cut with a thickness of 0.03 mm in secondary electrons, showing the presence of protrusions and valleys, and an increase in the cutting speed leads to an increase in the cutting length per unit length of cut (at a constant power $P_{rad} = 1.5$ W), i.e., grinding and reducing the surface roughness of the cut to values less than 1 μm. The average distance between the protrusions on the surface of the cut at a velocity $V =$ 3 mm/s is 14.5 μm, an increase in the cutting speed to 7 mm/s leads to a decrease in this distance to 8–10 μm, i.e., almost 1.5 times.

Figure 7.8 shows the microstructures of the same samples at a magnification of ×2000, where there is a large amount of crystallized droplets, probably resulting from the reflow of the surface of the cutting zone and subsequent rapid crystallization. It should also be noted that when the cutting speed is increased to 7 mm/s, a noticeable decrease in the fraction of such drops on the surface of the cut zone (both in the region of the depressions and in the region of the protrusions) is observed.

The topography of the cut surface of the copper samples was studied in detail at $P_{rad} = 3.3$–9.6 W and a constant velocity $V = 5$ mm/s. From the images in Fig. 7.9 show that with an increase in power from 3.3 to 9.6 W, the surface of the cut acquires a more precise form of the depressions and protrusions. It should also be noted that, at all powers of laser radiation, a noticeable difference in the structure of the cut surface in the upper, central and lower parts can be observed (the upper part is the radiation entrance region, the lower part is the emission exit region).

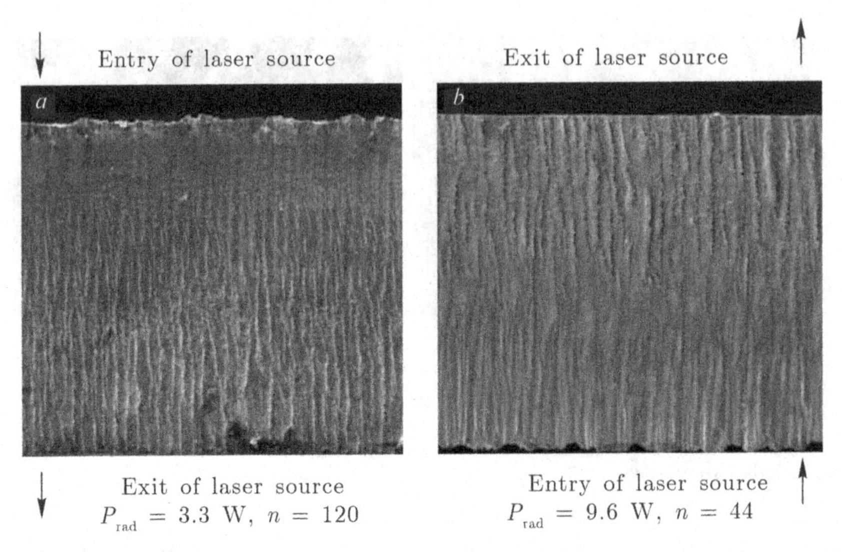

Fig. 7.9. The image of a copper cut surface 0.5 mm thick at a velocity V = 5 mm/s in secondary electrons. n is the number of passes.

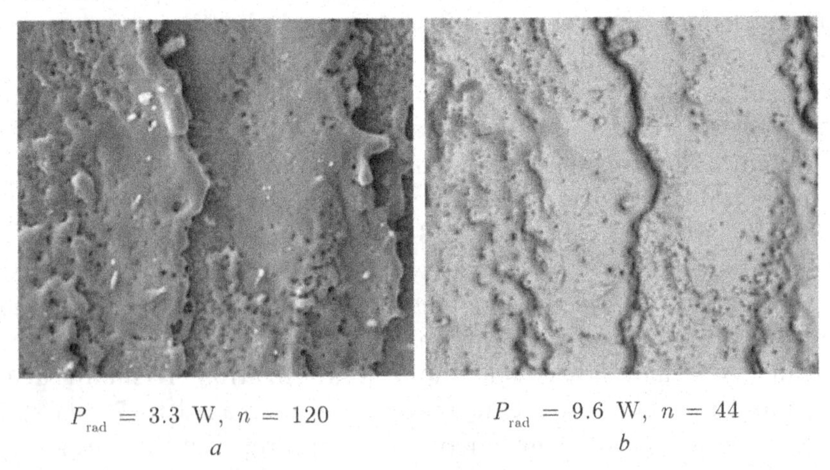

Fig. 7.10. The image of a copper cut surface 0.5 mm thick in its central region at a processing speed of V = 5 mm/s in secondary electrons. n is the number of passes.

A more thorough study of individual sections of the upper, lower and central regions shows that in all regions of the cut one general pattern is observed: the crystallized droplets formed have a definite orientation. In addition, inclusions can be observed on the surface of the cut (Fig. 7.10).

An assessment of the nature of the heterogeneity of the surface of copper cutting was carried out along the surface structure in the upper, central, and lower regions in both reflected and secondary

a

b

Fig. 7.11. Microstructure of BrB2 alloy (0.1 mm), V = 2 mm/s, P_{rad} = 2.4 W, n = 6; *a*) zone adjacent fo the cut area; *b*) central zone of the sample.

electrons. In all three regions, a difference in the contrast of individual areas is detected, which indicates a certain chemical heterogeneity. It is also possible to observe individual fine-dispersed regions, which are presumably copper oxides. It should be noted that images in secondary electrons are more distinct.

An analysis of the microstructure of copper samples, the BrB2 alloy, tantalum and molybdenum showed that there is no significant difference in the zones adjacent to the surface of the cut and the central zones. For example, the structure of the BrB2 alloy in the cutting zone, shown in Fig. 7.11 *a* (V = 2 mm/s, material thickness 0.1 mm), does not differ from the structure in the centre of the sample (Fig. 7.11 *b*). It can be seen that the structure of the alloy in both zones is crystallized and has no noticeable changes.

a

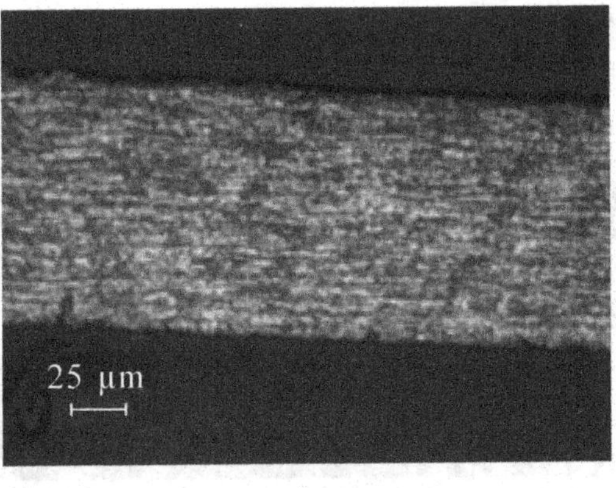

b

Fig. 7.12. Microstructure of molybdenum (0.1 mm), V = 3 mm/s, P_{rad} = 2.4 W, n = 8; *a*) zone adjacent to the cut area; *b*) central zone of the sample.

The microstructure of one of the samples of molybdenum subjected to laser cutting is shown in Fig. 7.12. The structure of the sample itself is deformed – elongated grains are visible, but there are no structural changes in the zone adjacent to the cutting area (Fig. 7.12 *a*) (a narrow region of recrystallized sections could be expected). The same pattern is observed on samples of a different thickness and other processing regimes. Thus, no noticeable structural changes were observed in the zone adjacent to the laser exposure site at various processing parameters and peak power density levels in a focused spot 10–20 μm in diameter 10^{10}–10^{11} W/cm^2.

In addition to structural studies on samples subjected to laser cutting, microhardness measurements were made in the zone adjacent

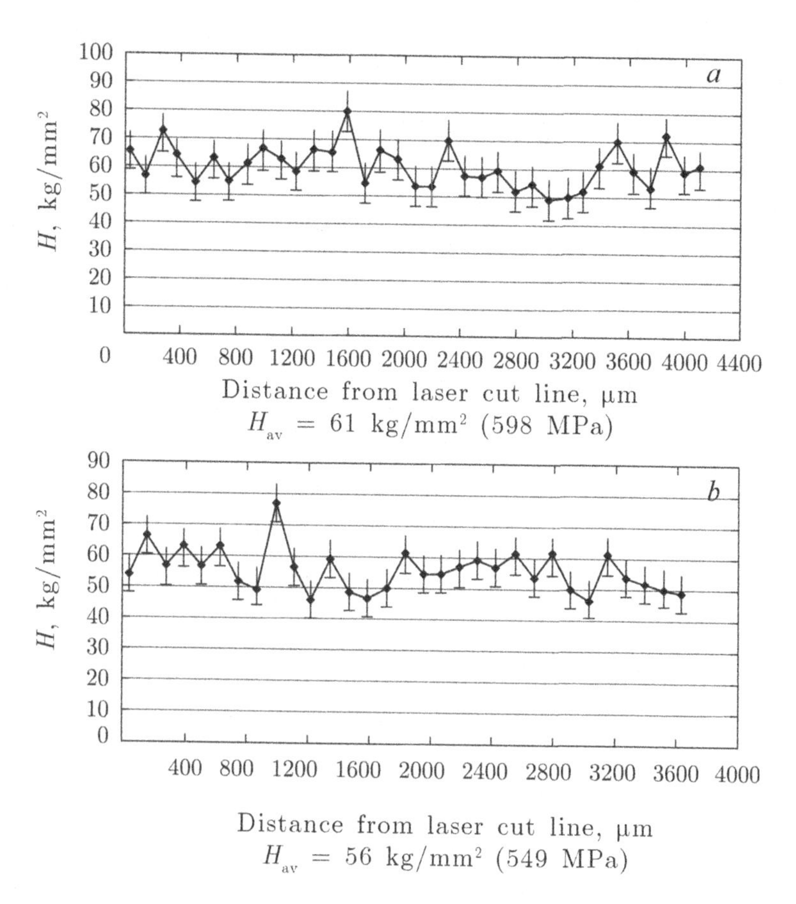

Fig. 7.13. A change in the microhardness of a copper MB sample 0.1 mm thick along (*a*) and in a perpendicular direction (*b*) from the laser cutting line. P_{rad} = 2.4 W, C_{cut} = 6 mm/s, n = 3.

to the surface of the cut and, for comparison, in the central zone. Figures 7.13 and 7.14 show the measurements of the microhardness of copper MB of 0.1 mm thickness along the cutting line and towards the center of the sample from the cutting line at an average radiation power of 2.4 W and processing speed of 6 and 2 mm/s. The error in measuring the hardness is not more than ± 50 MPa.

The microhardness of bronze samples of BrB2 slightly varies both along the direction of the cut and in the direction of the center of the sample. The change in the cutting speed and the radiation power do not influence the changes in the hardness values at the sections adjacent to the cutting zone and the inner regions of the samples. For molybdenum with a thickness of 0.1, 0.15 and 0.2 mm, a certain increase in hardness is observed in the zone adjacent to the surface

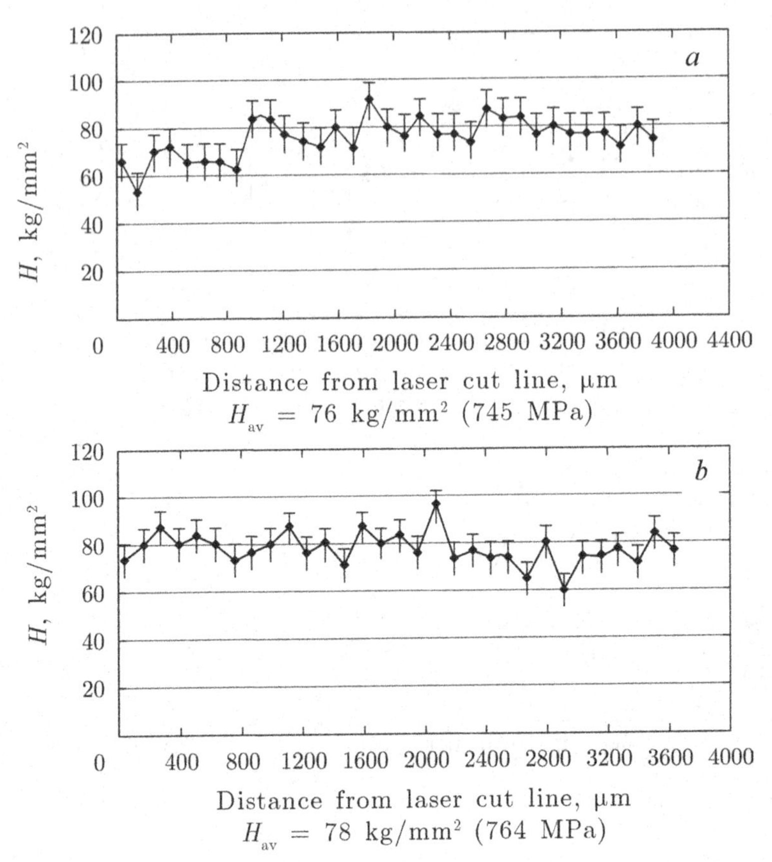

Fig. 7.14. A change in the microhardness of a copper MB sample 0.1 mm thick along (*a*) and in a perpendicular direction (*b*) from the laser cutting line. P_{rad} = 2.4 W, C_{cut} = 2 mm/s, n = 3.

of the cut, which also indicates an insignificant strengthening of the cutting zone (Figure 7.15, Figure 7.16).

Measurements of the microhardness of tantalum samples showed that the hardness of the zone adjacent to the cut, both at a power source of 2.4 W and 1.5 W, is significantly higher than the central zone of the sample (1650 and 1410 MPa for 2.4 W power). It should also be noted that increasing the cutting speed from 3 to 7 mm/s leads to an increase in the hardness value in the zone of action of the laser cut, which is about 15–20 μm. Thus, hardening of the material is observed for tantalum with a significantly lower thermal conductivity (55–60 W/(m · K)) in the zone adjacent to the laser cut.

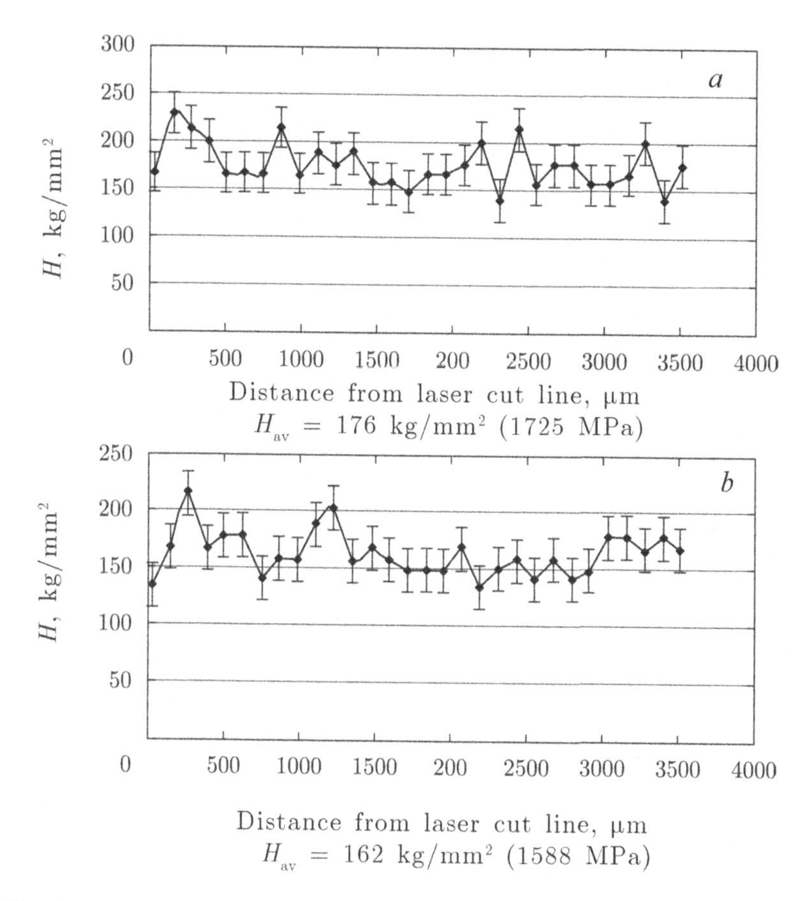

Fig. 7.15. A change in the microhardness of a Mo sample 0.1 mm thick along (*a*) and in the perpendicular direction (*b*) from the laser cutting line. P_{rad} = 2.4 W, C_{ucut}= 3 mm/s, n = 8.

7.5. Development of microprocessing technology in the production of LTCC multi-layer ceramic boards for microwave electronics products

The possibility of using the laser micro-processing method in the technology of formation of thick-film silver conductors and gaps between them in the width of 20 to 100 μm on the domestic glass-ceramic material (SCM) is shown. The regimes of laser microprocessing of conductor coatings and the cleaning of the laser processing zone on boards from slags that are formed at the ceramic-silver interface are studied. The use of laser microprocessing for cutting foil from beryllium bronze made it possible to create a

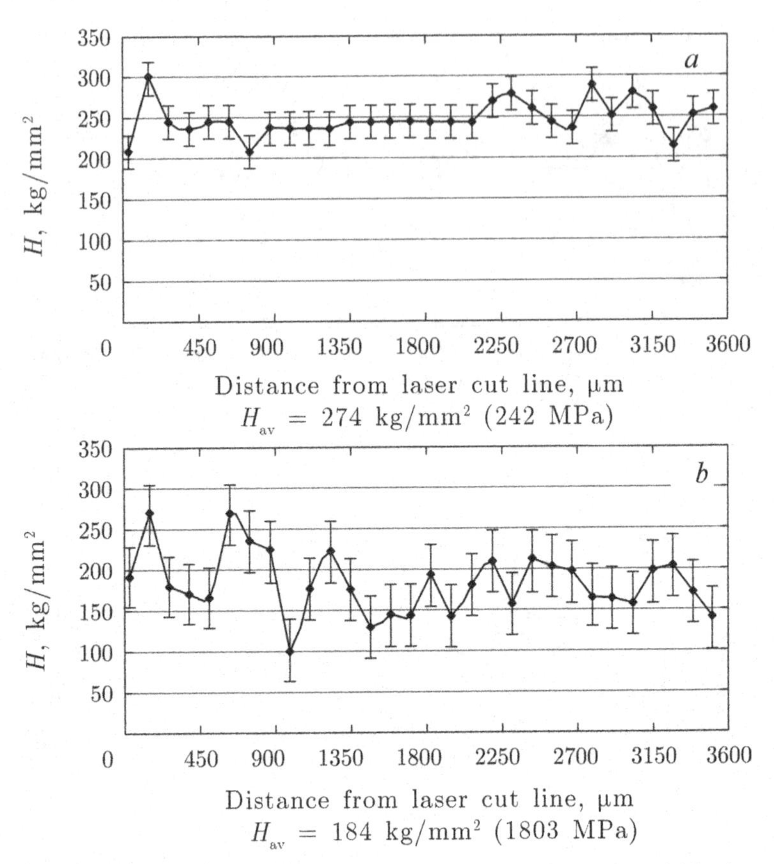

Fig. 7.16. A change in the microhardness of a Mo sample 0.1 mm thick along (*a*) and in the perpendicular direction (*b*) from the laser cutting line. P_{rad} = 4.7 W, C_{cut} = 5 mm/s, n = 100.

technology for manufacturing foil stencils used in the production of multi-layer ceramic boards [400].

For miniature components of electronic products based on multi-layer LTCC ceramic cards, new highly efficient materials processing technologies with micron accuracy are required. There is even a new term HDI (High Density Interconnect Board). The production of such boards is no longer possible without the formation of printed conductors of a width less than 50 μm, as well as of transient apertures with a diameter of less than 200–300 μm. An express and cheap method of forming the topology of the conductors is needed to select the best option for the circuit in the stage of prototyping thick-film ceramic boards.

In this paper, using the laser microprocessing method, a task was set to form the topology of the conductors on the outer surface

Table 7.4. Requirements for parameters of microprocessing of LTCC multilayer boards based on SCM glass-ceramic material

Parameter	Value
– processing accuracy, μm	±3
– roughness, μm, not more than	1–2
– heat-affected zone, μm, not more than	3–5
– minimum size of the topology element of the outer layer of the conductor/gap board, μm	30/20

of LTCC multilayer boards based on the SKM domestic glass-ceramic material. Requirements for the microprocessing parameters re presented in Table 7.4.

At the first stage, work was carried out to determine the optimum parameters of laser precision microprocessing regimes of external thick-film conducting coatings of silver and gold up to 20 μm thick, at which the sections of the conducting layer were removed with a width (gap) in the range 20–100 μm with minimal damage to the glass–ceramic substrate material. For this purpose, an automated laser technological installation (ALTI) Karavella-2 based on pulsed CVL with emission wavelengths in the visible yellow–green region of the spectrum (λ = 0.51 and 0.58 μm) was useed with an average power up to 7 W and high repetition rates of radiation pulses (14–16 kHz) [392, 400].

Optimization was carried out according to the following parameters: laser radiation power, focal length of achromatic lens, processing speed, wavelength of laser radiation, number of passes for complete removal of conductive coating. But at the beginning, the threshold (minimum) density of the peak evaporation power for the materials (silver and gold) to be processed and the corresponding average threshold radiation powers of the CVL were determined. These parameters were calculated using the formulas (6.1) and (6.2) and are given in Table 7.1. The threshold density of peak evaporation power for gold is $q_{Au} = 0.5 \cdot 10^9$ W/cm^2, for silver – $q_{Ag} = 1.6 \cdot 10^9$ W/cm^2 and the corresponding values of the average radiation power of CVL are $P_{av. Au} = 0.06$ W and $P_{av. Ag} = 0.19$ W.

But the technology was tested at the emissivity of the CVL somewhat higher than these values, since at a minimum power level the complete removal of the diffused metal from the ceramic was not ensured. For experimental studies, a range of capacities for silver from 0.2 to 1 W and for gold from 0.1 to 0.6 W was chosen. To obtain a processing spot of radiation with a diameter of 10 μm, sufficient to form gaps with a width of not more than 50 μm, an achromatic objective with F = 100 mm was used as the focusing element. In the experiments, the processing speed varied from a minimum value of 0.2 mm/s to 5 mm/s. The minimal heat-affected zone during microprocessing was provided in the mode of operation at a shorter wavelength (λ = 0.51 μm). Two programs were compiled in the .dxf (AutoCAD) format. The first program made it possible to form a conductor/gap topology on the glass–ceramic substrate surface with the dimensions of 100/100 μm. According to the second program, a conductor/gap topology with the dimensions of 40/40 μm was formed.

Experimental results. Experiments to develop the technology of precision microprocessing of coatings of silver and gold up to 20 μm thick on a glass–ceramic SCM substrate were carried out:

a) with a conductor/gap topology of 100/100 μm at an average radiation power in the range of 0.2 to 1 W, processing speed 0.2 to 5 mm/s and the number of passes from 1 to 5. Figure 7.17 shows a fragment of the image of the conductor/gap topology with the dimensions of 100/100 μm on a glass–ceramic SCM substrate with a silver coating 10 μm thick, manufactured according to the following mode: P_{rad} = 0.3 W, the processing speed varied from 5 mm/s (right) to 1 mm/s (on the left).

b) with a 40/40 μm conductor/gap topology with an average radiation power in the range of 0.2 to 0.6 W, processing speed 0.2 to 8 mm/s and the number of its passes from 1 to 5. Figure 7.18 shows a fragment of the image of the 40/40 μm conductor/gap topology on a 5 μm glass–ceramic SCM substrate with a silver coating 10 μm thick, manufactured according to the following mode: radiation

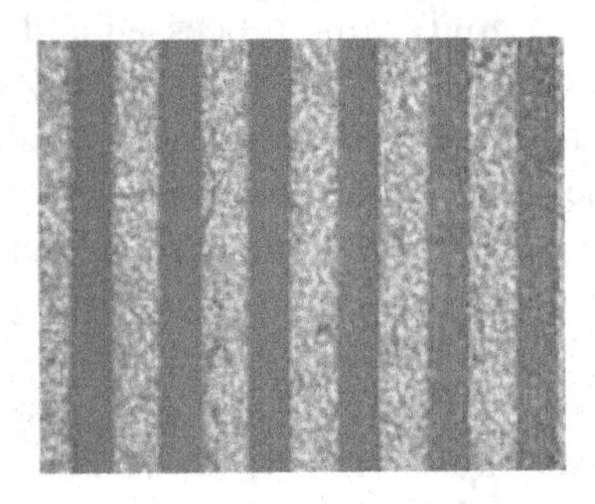

Fig. 7.17. Fragment of the image of the conductor/gap topology with dimensions of 100/100 μm.

Fig. 7.18. Fragment of the image of the conductor/clearance topology with the dimensions of 40/40 μm.

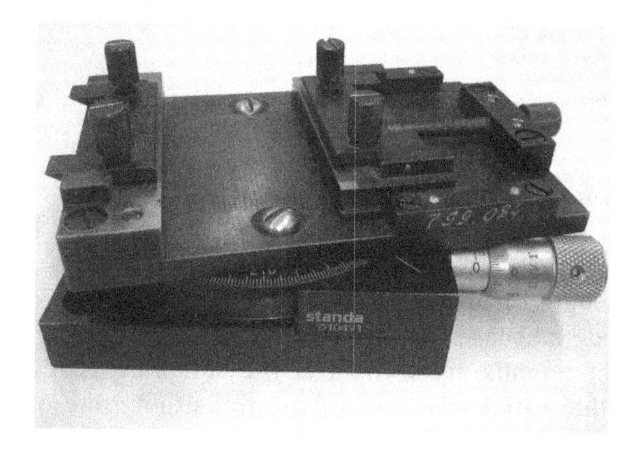

Fig. 7.19. Precision turntable supplied by the Standa company with a special device for securing processed samples.

power $P = 0.3$ W, the processing speed varied from 2 mm/s (right) to 0.2 mm/s (left).

Samples were attached to a specially made mandrel, which was installed on a precision turntable of the firm Standa, Fig. 7.19. The turntable made it possible to position the sample with micron accuracy in relation to the coordinate table in the ALTI Karavella.

As a result of the experimental work, the LTCC ceramics with silver coating microprocessing regimes were determined, allowing the formation of a conductor/gap topology with dimensions of 40/40 μm and 100/100 μm:

– the average radiation power $P_{av} = 0.3$ W;

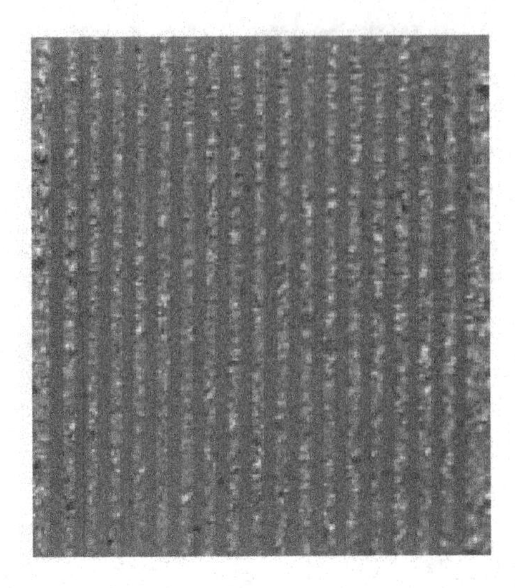

Fig. 7.20. Image of the fragment of the sample after chemical etching.

– radiation wavelength $\lambda = 0.51$ μm;
– processing speed $V = 1$ mm/s;
– number of passes $n = 3$;
– focal length of the lens $F = 100$ mm.

In the course of laser processing, oxides, silver borates and other chemical compounds are formed at the ceramics–silver boundary. To restore the initial structure of the transition zone of the metal ceramics, various ways of additional processing of the samples were used. The use of chemical purification methods, as seen in Fig. 7.20, did not yield positive results, since the products of laser processing (slags) were partially removed. Annealing at 200°C in a reducing hydrogen medium also failed to produce a positive result.

The complete removal of slags formed during the interaction of laser radiation with the sample occurred after annealing in air at 800°C.

The results obtained in the formation of the conductor/gap topology with dimensions of 40/40 μm and 100/100 μm, followed by a cleaning annealing in air at 800°C are shown in Fig. 7.21.

After determining the optimal modes of laser precision microprocessing of external thick-film conducting coatings on LTCC ceramics using ALTI Karavella-2 and the method of their final cleaning, prototypes with a conductor/gap topology with dimensions ranging from 20 to 100 μm were made, Fig. 7.22.

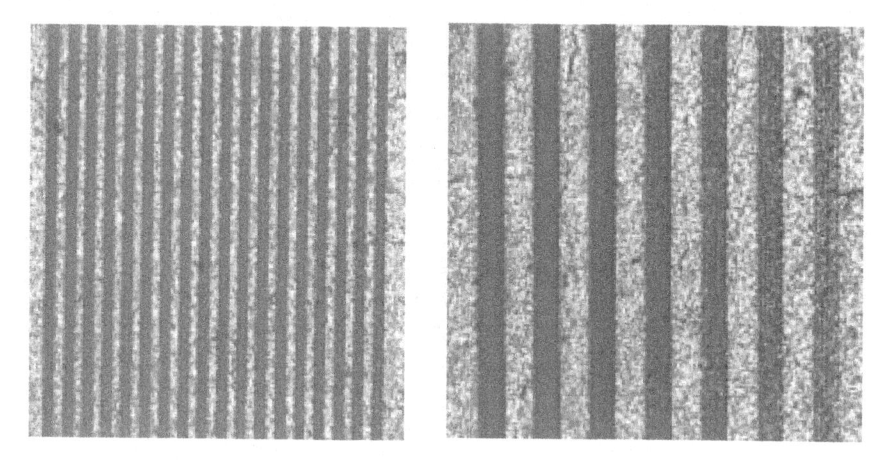

Fig. 7.21. Image of a fragment of the sample with the conductor/gap topology with the dimensions 40/40 and 100/100 μm after laser processing and annealing in air at 800°C.

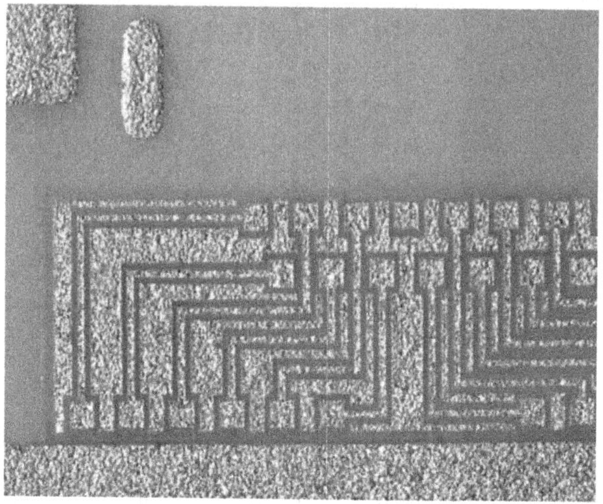

Fig. 7.22. Image of a fragment of the topology of the scheme of conductors formed on LTCC ceramics with a silver coating. The dimensions of the conductor/gap from 20 to 100 μm

Using the low-power ALTIs Karavella-2 and Karavella-2M, the technology of making stencils made of beryllium bronze 0.05 mm thick was successfully used, which is now employed in the production of multilayer boards for filling the holes of interlayer joints with conductive paste in moist ceramic layers, Fig. 7.23.

The mode of processing of stencils differs from the modes of microprocessing of conducting coatings on ceramics of SCM and it is the following:

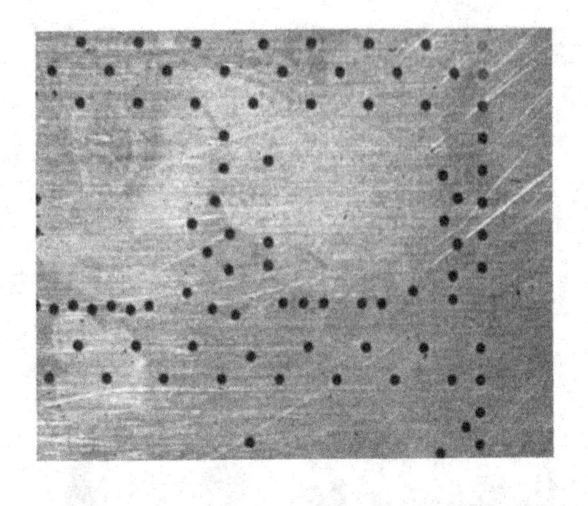

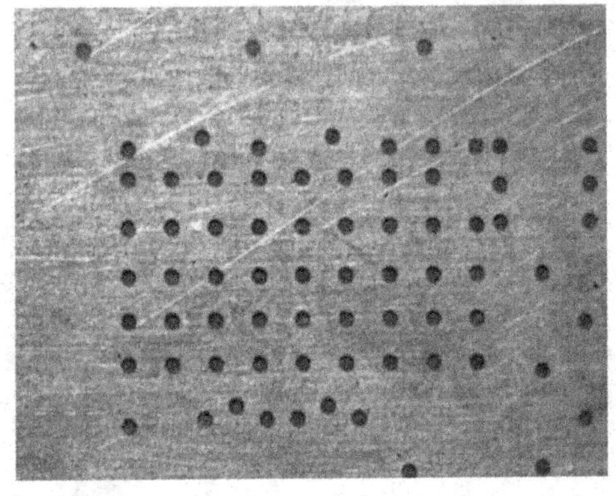

Fig. 7.23. Fragments fo stencils of Be bronze with a thickness of 0.05 mm with the holes with a diameter of 0.18 and 0.25 mm.

– average radiation power P_{av} = 2 W;
– radiation wavelength λ = 0.51 μm + 0.58 μm;
– processing speed V = 1.5 mm/s;
– number of passes n = 4;
– focal length of the lens F = 150 mm.

The accuracy of the positioning of the holes relative to the given coordinates is not worse than ±3 μm. The diameters of the openings in the size range from 150 to 300 μm were performed with an accuracy of ±3 μm.

Within the framework of this work, with the use of a more powerful ALTI Karavella-1M (20–25 W), technologies were developed for processing three-layer copper–aluminium nitride–copper structures 0.3–1–0.3 mm thick and forming contacts on this structure.

7.6. Conclusions and results for Chapter 7

1. Calculations of the threshold density of the peak evaporation power and the corresponding values of the average radiation power of the CVL for heat-conducting and refractory metals and other materials widely used in the production of components for electronic devices were carried out. The calculations showed that the threshold power density for metals is in the range 10^8–10^9 W/cm^2, and the corresponding average CVL power when focusing radiation in a spot with a diameter of 10–20 μm is equal to a fraction of watts (0.012–0.75 W).

2. Experimental studies show that in order to ensure a high quality of microprocessing, it is necessary to have a density of peak power in the processing light spot one or two orders of magnitude above the threshold density of 10^{10}–10^{11} W/cm^2, when the processing is carried out in the evaporation mode (with a minimum of the liquid phase).

3. With an increase in the thickness of the processed metal, the drilling time increases with pulsed radiation of the CVL, and the speed decreases. With metal thicknesses greater than 1 mm, the drilling time increases sharply, and the rate decreases sharply and it turns out that it is advantageous for CVL radiation to process materials with a thickness of less than 1 mm. And it is even more advantageous to use the CVLs at thicknesses less than 0.6 mm, when the processing time is fractions of seconds, and the speed is one and tens of mm/s, which is an order of magnitude greater than in electroerosion machining.

4. With an increase in the thickness of the metal parts produced in the laser cutting zone, the amount of slag and grate increases, which prevents their direct use in real products.

5. Technologies of chemical cleaning with superposition of the ultrasonic field for the removal of slag and grate for both heat conducting (Cu, Al, Au, Ag) and refractory (Mo, W, Ta, Re) metals, their alloys and a number of other materials are widely developed used in the components for electronic devices after laser micromachining.

6. The quality of the surface of the cut and the structure of the HAZ for high heat-conducting and refractory materials were studied in the course of processing by radiation with a laser spot diameter of 10–20 μm and a peak power density of 10^{10}–10^{11} W/cm^2 and the results show that:

– the surface of the laser cut has a multichannel structure with a period of repetition of the channels of 10–15 μm and depth (roughness) to 3–4 μm;

– increasing the processing speed and number of passes leads to a decrease in the roughness of the cutting surface to submicron values – <1 μm;

– no structural changes were found in the zone adjacent to the laser cut (10–20 μm) and it is similar to the central zone of the material;

– formation of cast sections (in the form of crystallized droplets) is observed on the surface of the cut, indicating the existence of a liquid phase and its crystallization at the final stage of processing;

– in high heat-conducting metals, hardening or softening of the material does not occur in the zone adjacent to the laser cut (10–20 μm);

– in refractory metals with significantly lower thermal conductivity a slight hardening occurs in the zone adjacent to the laser cut (10–20 μm).

7. The technology for the formation of thick-film (10 μm) silver conductors on a glass–ceramic material with a conductor/gap of 20 to 100 μm was developed with a low level of the average radiation power of the CVL (0.2–0.3 W).

8. The technology of manufacturing foil stencils made of beryllium bronze 0.05 mm thick was used to fill the holes of interlayer connections with conductive paste in the moist ceramic layers of multilayer boards.

8

Using industrial automatic laser technological installations Karavella-1, Karavella-1M, Karavella-2 and Karavella-2M for the fabrication of precision parts for electronic devices

The positive results of research on the development of microprocessing technology for pulsed radiation by CVL of foils (0.01–0.2 mm) and thin sheets (0.3–1 mm) of materials (Chapter 7) became the basis for the application of industrial technological equipment of the ALTI Karavella (Chap. 6) for manufacturing a wide range of precision parts of simple and complex configurations and high resolution for components of electronic devices.

8.1. The possibilities of application of ALTI Karavella for the manufacture of precision parts

The possibilities of using technological equipment for the production of precision parts at the first stage were determined at the industrial ALTI Karavella-1. At the beginning, the parts were made of molybdenum (Mo), copper (Cu) and stainless steel (12Cr18Ni10Ti) 0.05–0.5 mm thick, widely used in vacuum and electronic engineering. To assess the accuracy and quality of the microprocessing, the following measuring equipment was used: a measuring microscope of the VMM200 model from UHL Co, a projection microscope PV5100 (Mitutoyo Co.) and a Leica microscope (Carl Zeiss). In the process of optimizing the technological cutting modes for ALTI Karavella-1, all the main processing parameters changed in a wide range: the focal length of the power achromatic objective was F = 50, 70, 100, 150 and 200 mm, the average radiation power was P_{rad} = 0.4–10 W with the pulse repetition frequency (PRF) f = 14 kHz, cutting speed V = 0.5–5 mm/s and number of passes N = 1–50. Under these conditions, the width of the cut is in the range H = 8–70 μm, and its roughness is not more than 1...2 μm. On these materials, the same processing parameters were used in the experiments for the direct piercing of microholes. The exposure time of the focused spot of radiation at the 'point' with the working PRF was t = 0.07–1000 ms, which corresponds to the number of pulses of radiation per second within the range of 1–14000. The diameter of the holes to be pierced was in the range d = 5–45 μm, roughness <1 μm, deviation from the circumference <10%. Studies were also carried out at separate wavelengths of radiation, 510.6 and 578.2 nm, but the processing efficiency decreased by a factor of 2. The best processing quality is achieved in the mode of operation at a shorter wavelength – a green line with λ = 510.6 nm.

A number of cutting and drilling modes were developed for other materials: aluminium (Al), nickel (Ni), pseudoalloys (MD-50 and MD-80) and titanium (Ti) up to 0.5 mm thick, for artificial polycrystalline diamond and silicon (Si) with thickness up to 1.2 mm and other materials. Based on the results of the experimental studies, a computer database in the form of spreadsheets has been created for these materials, which makes it possible to quickly select the optimal processing parameters for manufacturing specific parts.

Figure 8.1 shows images of laser cut fragments of samples of copper grade MB 0.1 mm thick in the mode of operation at both wavelengths – 510.6 and 578.2 nm and on one yellow line – 578.2 nm (F = 100 mm, magnification ×500).

Figure 8.2 shows images of microsections of laser cutting molybdenum; Fig. 8.3 – a fragment of a jumper plate made of nickel; Fig. 8.4 – a fragment of the cutting contour of a disk made of silicon; Fig. 8.5 – the image of a fragment of the cut contour of an artificial polycrystalline diamond plate.

Visual analysis of microsections of laser cutting of copper and molybdenum plates with a thickness of 0.05 to 0.5 mm at a magnification of 500 showed that the zone of thermal action is very

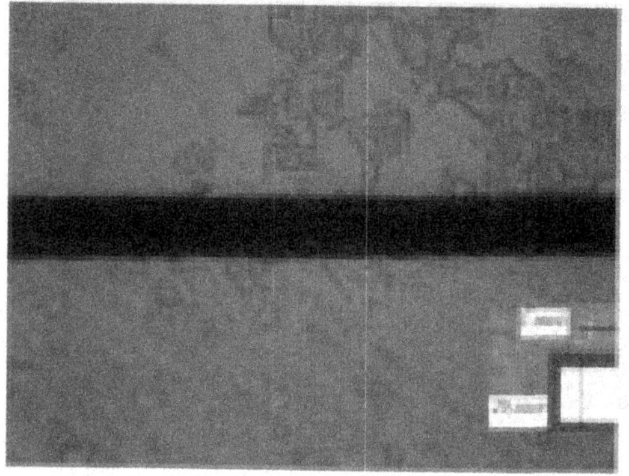

λ = 510.6 and 578.2 nm, P_{rad} = 5.4 W,
n = 6, V = 2 mm/s

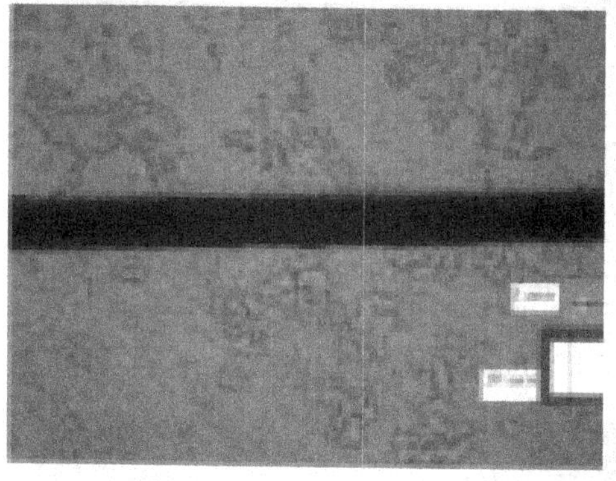

λ = 578.2 nm, P_{rad} = 1.6 W,
n = 12, V = 2 mm/s

Fig. 8.1. Images of laser cut fragments of samples from copper at both wavelengths (left) and on one yellow line (right); thickness t = 0.1 mm, F = 100 mm, magnification ×500.

small and the structure of materials in the zone of the lag behind the boundary is more than 10–20 μm shows no significant changes. The quality of laser microprocessing is much higher in comparison with electroerosion, the roughness and the zone of thermal impact are smaller and there is no stratification of the material, Fig. 8.6. This is due to a higher 'sensitivity' of electroerosion machining to heterogeneities and defects in the material.

The production of precision parts on the technological equipment of ALTI Karavella is performed with optimized processing parameters and in accordance with the program for working drawings, specified with the help of a file in DXF format in the AutoCAD design package. At present, tens of hundreds of types of flat and volumetric

Molybdenum with a thickness of 0.3 mm, ×250

Copper with a thickness of 0.3 mm, ×250

Fig. 8.2. Images of microsections of the laser cut.

Fig. 8.3. The image of a fragment of a jumper plate made of nickel 0.5 mm thick, magnification ×150.

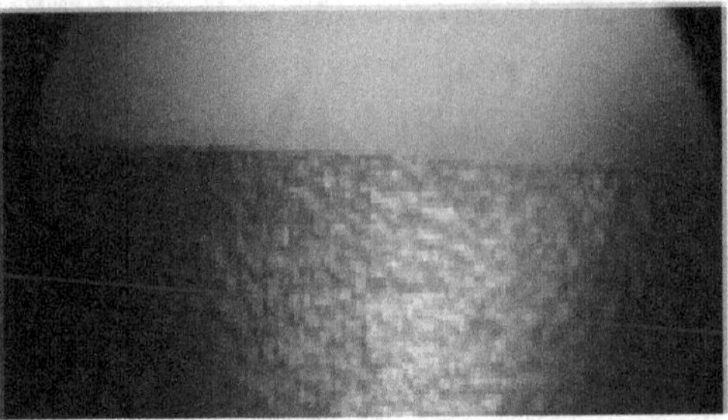

Fig. 8.4. The image of a fragment of the contour of a cutting disk made of silicon (Si) 1 mm thick. Magnification ×150.

Fig. 8.5. The image of a fragment of a contour of a cut of a plate made of an artificial polycrystalline diamond 0.65 mm thick. Magnification ×150.

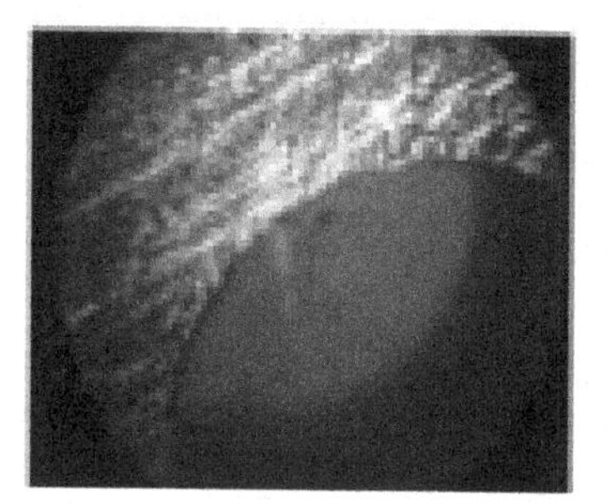

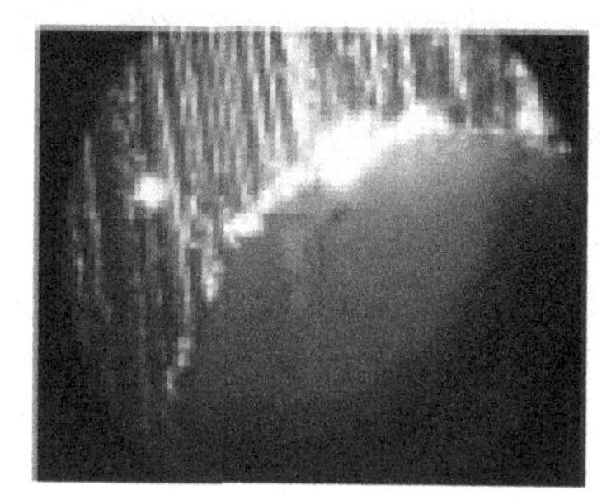

Fig. 8.6. The image of a fragment of aperture of a diaphragm from Mo (t = 0.16 mm) at laser and electroerosive processing. Magnification ×500.

precision metal parts of simple and complex configurations are manufactured at the Istok at ALTI Karavella, in providing single, small-lot and serial deliveries of microwave equipment and other IET devices. processing speed 0.5–5 mm/s: diaphragms, grids, anodes, electrodes, slats, cups, screens, contacts, petals, measuring probes, etc. thickness 0.1–1 mm, and also cutting various solders thickness of 20–50 μm with a speed of up to 15 mm/s. With a maximum capacity (5–15 mm/s), disks, windows and chips made of artificial polycrystalline diamond and silicon with a thickness of up to 1.5–2 mm are cut out and volume images are formed in transparent media such as glass and quartz. Productivity of ALTI with one-shift operation is, depending on the complexity of the parts being manufactured, 500–10000 parts/month. Compared with the traditional methods of processing, including electroerosion treatment, the manufacture of precision parts at ALTI Karavella is reduced by an order of magnitude or more.

8.2. Examples of the manufacture of precision parts for electronic components at ALTI Karavella

Optimal processing parameters and the manufacturing time of precision parts are determined by the requirements of the working drawing. But in terms of time and quality of processing of both simple and complex parts, laser technology surpasses all known traditional manufacturing methods. For example, in Fig. 8.7. The parameters for three volumetric details are presented: a rectangular configuration with an axial slit, a spherical mesh with annular grooves, a cylindrical glass with 600 grooves and a flat diaphragm with 12 holes.

As a comparative analysis Fig. 8.8 shows an example of the technological process and the duration of fabrication of flat nets from molybdenum MCHVP for powerful klystrons based on the basic technology and laser processing technology. It should be noted that the processing of the focused radiation beam is not only of working holes and grooves, but also of the outer contour of the grids, i.e. the entire part. The main advantages that were obtained: high productivity, elimination of many operations inherent in machining, absence of stratification of sheet molybdenum, high accuracy, small roughness (≤1–2 μm) and thermal exposure zone (≤10–20 μm).

Table 8.1 shows the geometric dimensions of a multi-beam flat molybdenum grid with 25 holes 0.45 mm thick (Grid No. 1) and a miniature grid with 19 holes 0.4 mm thick (Grid No. 2) made at ALTI Karavella. When processing grids on a boring machine or in

Processing parameters				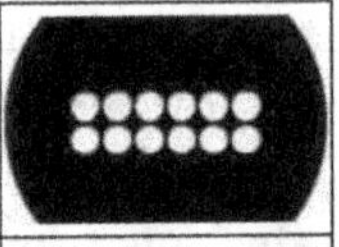
Thickness, mm	0,2 мм	0,1 мм	0,16 мм	0,45 мм
Average radiation power, W	4,8	2,4	6	8,2
Focusing distance of lens, mm	150	100	100	150
Treatment speed, mm/s	1,5	1,3	5,0	3,5
Number of passes	11	7	8	22
Manufacturing time, min	7,5	25	150	45

Fig. 8.7. Modes of laser microprocessing and the manufacturing time of volumetric precision molybdenum parts of complex configuration,

No.	Basic technology		Laser cutting method
1.	Turning, elspark. tr.	Deflection 0.15–0.3	Laser processing
2.	Boring holes	D = 3÷5(+0.025) 1.25√–2.5√	Entire grid $^{2}\surd$accuracy 3 μm
3.	Industrial chemistry (burr), 180	$R_{chamf} \approx 0.03$ $^{0.32}\surd$	Ultrasound treatment (preliminary)
			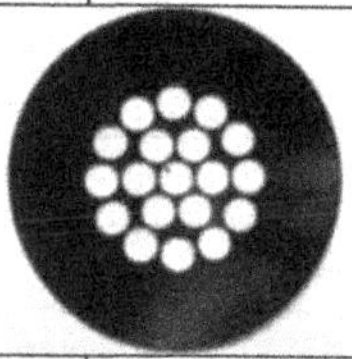
4.	Straightening grids	$T = 1100\,°C$ ▱ 0,01	Production of polished surfaces $^{0.1}\surd$
5.	ECHT of flat surfaces	$^{0,15}\surd$ ▱ $^{0,1}\surd$	–
6.	Finishing dressing	Industrial chemistry	Finishing dressing Industrial chemistry
	Total production time (batch 5 pieces)	~1.5 months	2–3 days

Fig. 8.8. Basic and laser technologies for fabricating flat grids for high-power molybdenum klystrons 0.45 mm thick (ECHT – electrochemical treatment).

conductors, there is a break in the bridges (0.15–0.3 mm) between the holes and yields are less than 50%. The laser method of manufacture excludes completely this defect.

The analysis of the measurement results shown in this table shows that the accuracy of making grids on ALTI Karavella is very high and fully corresponds to the dimensions given in the drawing. The metallographic analysis of one of these grids showed that the molybdenum structure of the cut zone in the narrow part of the bridges does not differ from the structure of the base metal, which indicates a minimal defective cutting zone (Fig. 8.9).

Figures 8.10–8.12 present enlarged fragments of precision parts from various materials: molybdenum, tungsten, copper, pseudoalloy, polycrystalline diamond and silicon with a precision of 4–10 μm and a roughness of ≤2 μm.

The laser microprocessing method proved to be very effective for the production of spherical shadow and control grids of various configurations for a multi-path klystrod with low-voltage control. Figure 8.13 shows spherical meshes with radial-circular bridges. The fabrication of such grids by the method of electrospark machining involves a very laborious process of obtaining a complex electrode.

Table 8.1 The results of measuring the dimensions of nets made of molybdenum made by laser micro-processing

Dimensions according to drawing	Actual size	Taper of the hole	Surface finish
Grid No. 1			
32 ± 0.028	31.997	2–3 μm	2
16 ± 0.028	16 ± 0.01		
72° ± 25′	72° ± 6′		
36° ± 12′	36° ± 3′		
18° ± 6′	18° ± 3′		
diam. 8.5 ± 0.028	diam. 8.5 ± 0.001		
diam. 18 ± 0.028	diam. 18 ± 0.002		
Grid No.2			
60° ± 14′	60° ± 7′	2–4 μm	2
30° ± 14′	30° ± 8′		
diam. 3.4 ± 0.016	diam. 3.4 ± 0.002		
diam. 1.7 ± 0.016	diam. 1.7 ± 0.028		
19 holes diam.0.6H9$^{+0.025}$	diam. 0.6H9$^{+0.015}$		

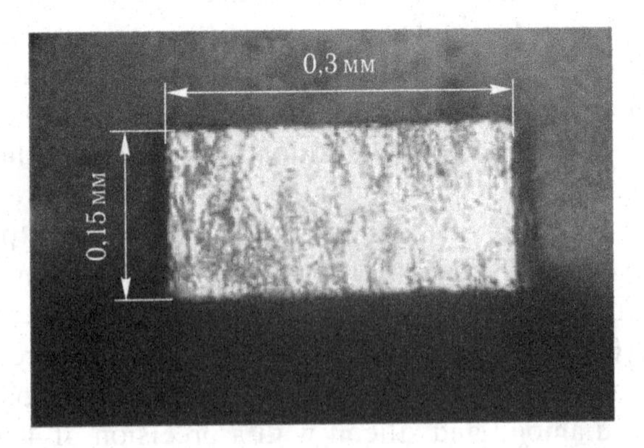

Fig. 8.9. The image of the laser cut of the web of the molybdenum grid at a magnification of ×300.

At the same time, after 5–7 operations, the elements of the electrode are thinner, deformed and, accordingly, the accuracy and quality of the part decreases. The laser method is devoid of these drawbacks.

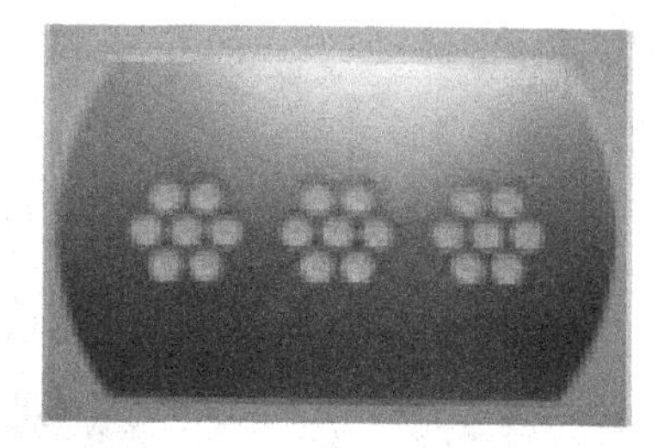

Fig. 8.10. Molybdenum (Mo) grids 0.3 and 0.6 mm thick for powerful multipath klystrons.

Fig. 8.11. Cylindrical glass of molybdenum (Mo) with a wall thickness of 0.3 mm with 360 grooves.

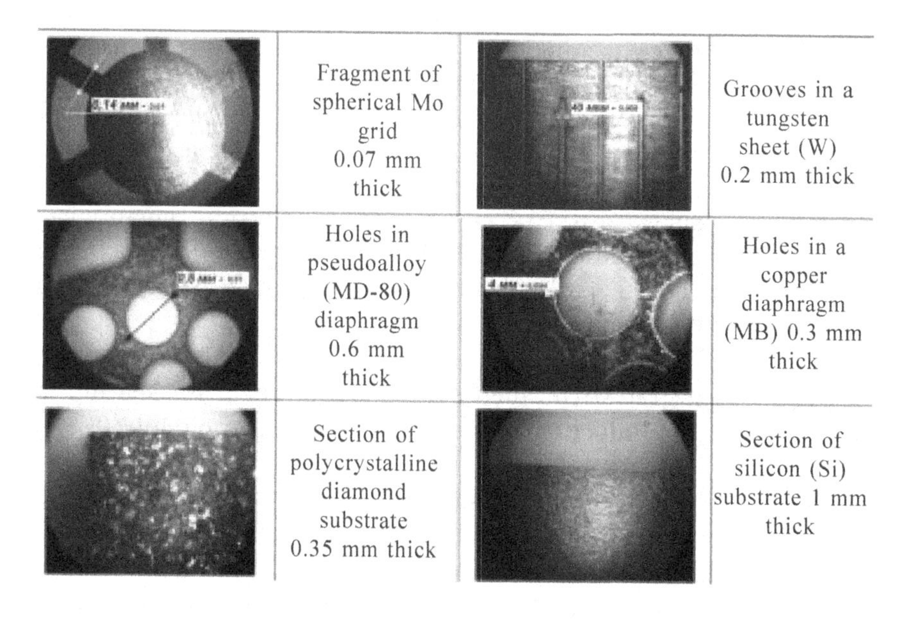

	Fragment of spherical Mo grid 0.07 mm thick		Grooves in a tungsten sheet (W) 0.2 mm thick
	Holes in pseudoalloy (MD-80) diaphragm 0.6 mm thick		Holes in a copper diaphragm (MB) 0.3 mm thick
	Section of polycrystalline diamond substrate 0.35 mm thick		Section of silicon (Si) substrate 1 mm thick

Fig. 8.12. Increased fragments of precision 12™ parts from different materials.

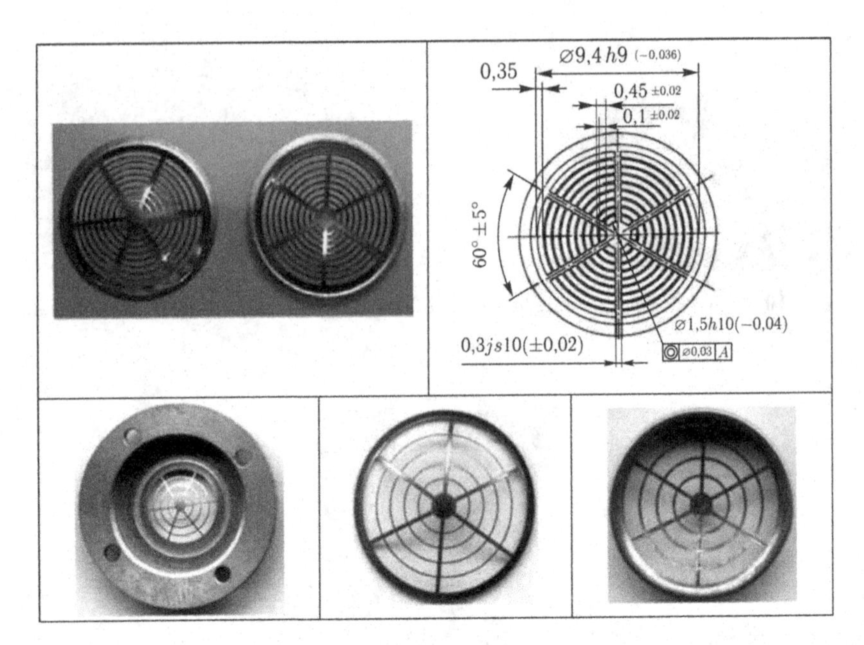

Fig. 8.13. Spherical grids with radial-ring bridges of molybdenum 0.1 mm thick.

Also, parts for electron guns of cyclotron protective devices and electrostatic guns from thin-sheet molybdenum (0.1–0.2 mm) are made at ALTI Karavella: slats, screens, anodes, electrodes, etc. A common requirement for these parts is in particular a high precision of manufacturing with tolerances of not more than 10 microns. The traditional technology of manufacturing parts in addition to preparatory operations for chemical cleaning and thermal annealing blanks involves a number of fairly time-consuming electrospark, marking and locksmith operations. This requires a high-precision special equipment for most operations, a number of works must be performed under a microscope. Figure 8.14 shows a molybdenum anode 0.2 mm thick, the coaxiality of the slit aperture (6 × 0.3 mm) of which to external base dimensions is not more than 12 μm.

A high-performance and high-quality process is the laser cutting of sapphire substrates with their subsequent mechanical shearing to the chip for the production of monolithic integrated circuits (MIC) (Fig. 8.15).

Cutting out filaments of tantalum foil 30 μm in thickness (Fig 8.16) for the receiver–ionizer of the atomic beam tubes (ABT) on cesium vapour, which are used in the time standards, is carried out with high accuracy, a minimum roughness (≤1 μm) and a thermal effect zone (10–15 μm). This laser cutting method (also in the array

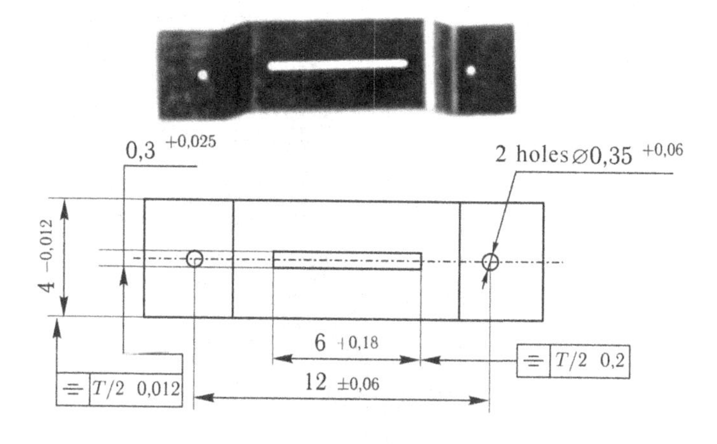

Fig. 8.14. Molybdenum anode 0.2 mm thick.

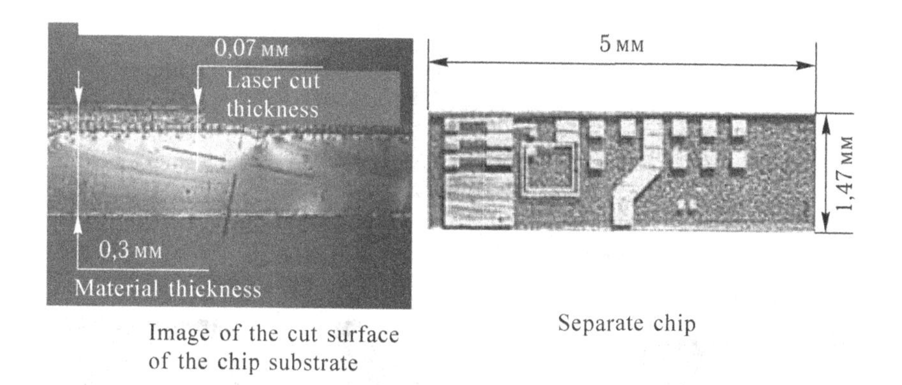

Image of the cut surface of the chip substrate

Separate chip

Fig. 8.15. Cutting sapphire substrates into chips.

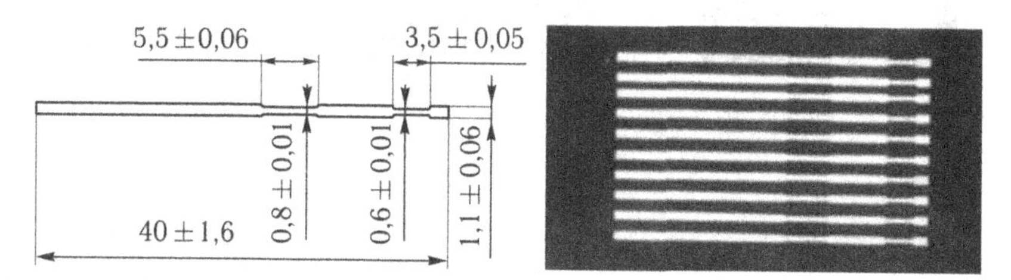

Fig. 8.16. Tape of incandescent tantalum foil 0.03 mm thick for ALT

mode) ensures the stable output parameters of the ionizer and, in general, for the ABT.

The production of nickel probes for measuring current-voltage and other characteristics of microwave transistors turned out to be promising (Fig. 8.17).

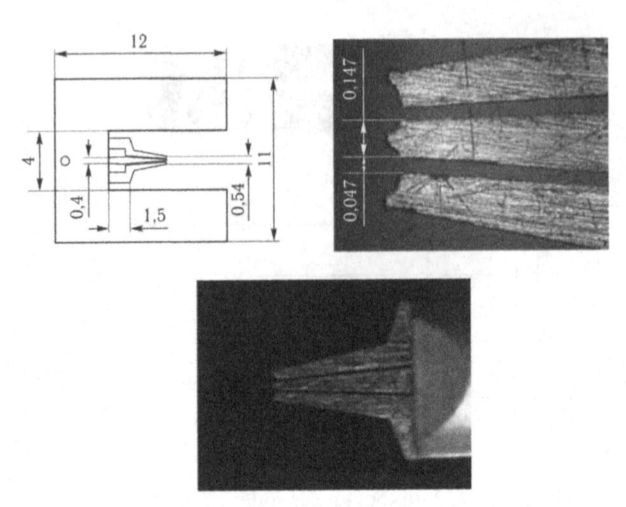

Fig. 8.17. Measuring probe made of Ni 0.1 mm thick for microwave

Fig. 8.18. Cleaning of the surface of ceramic plates (22XC) with metallized contact

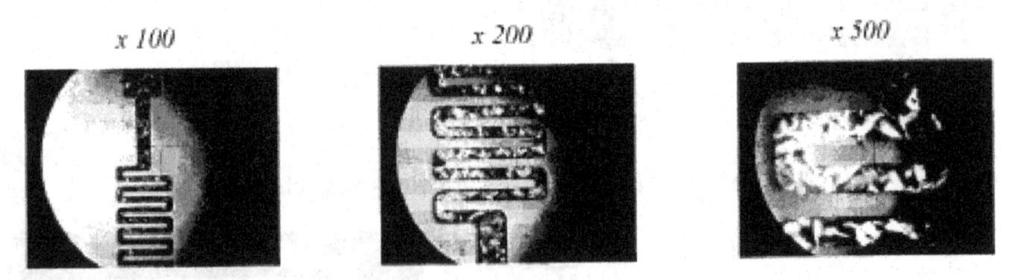

Fig. 8.19. Images of fragments of a pattern of synthetic polycrystalline diamond 0.2 mm thick.

Figure 8.18 shows an example of high-speed cleaning of the surface of a ceramic plate of 22XC from impurities at a peak power density of 10^5 W/cm^2 in the radiation spot, which is four to five orders of magnitude lower than when cutting metal materials.

Figure 8.19 shows images of meander fragments made of 0.25 mm thick polycrystalline diamond. Precision of meander manufacturing corresponds to the given drawing.

Since the thermal conductivity of polycrystalline diamond is high – 1600–2000 W/cm^2, which is 4–5 times larger than that of copper,

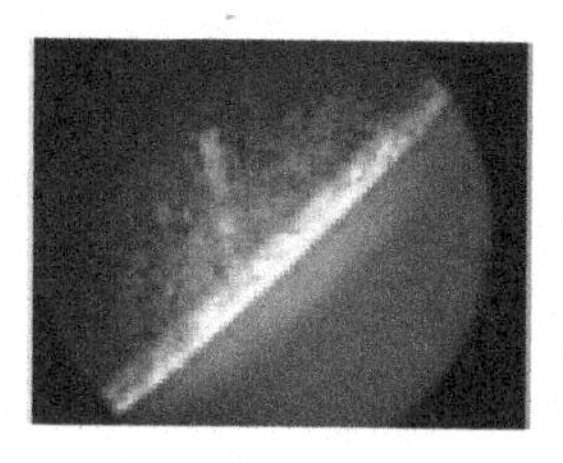
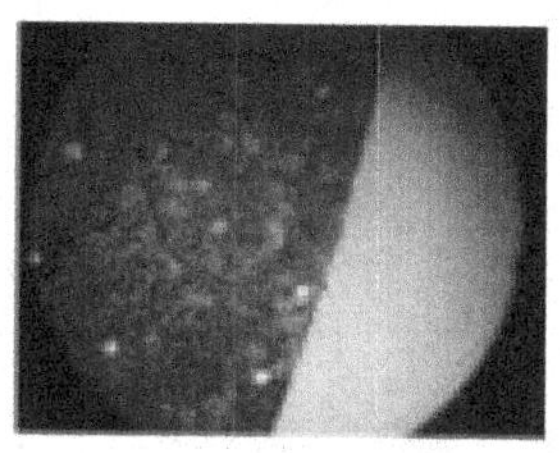
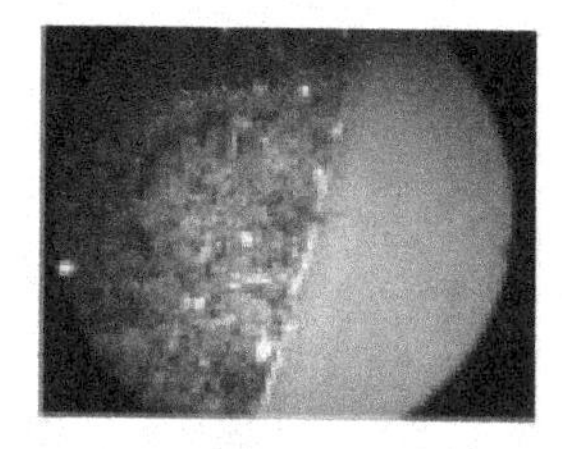

Fig. 8.20. Images of cut pieces of a silicon wafer with a thickness of 0.3 mm (×200)

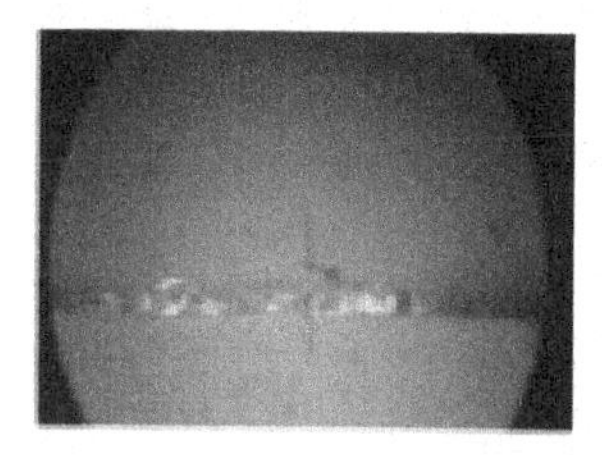
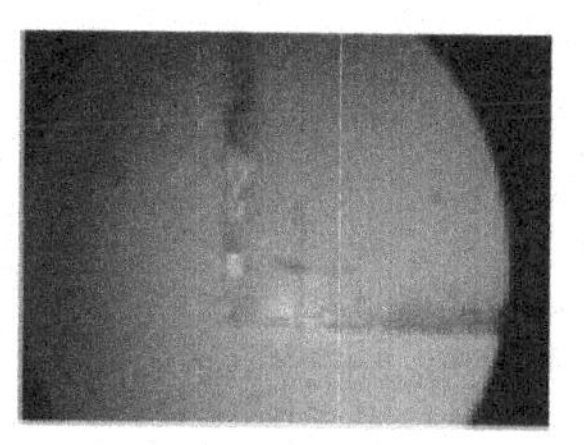

Fig. 8.21. Images of fragments of a plate of quartz with a thickness of 0.4 mm (x200).

Fig. 8.22. Heat sinks made of artificial polycrystalline diamond 0.1 mm thick

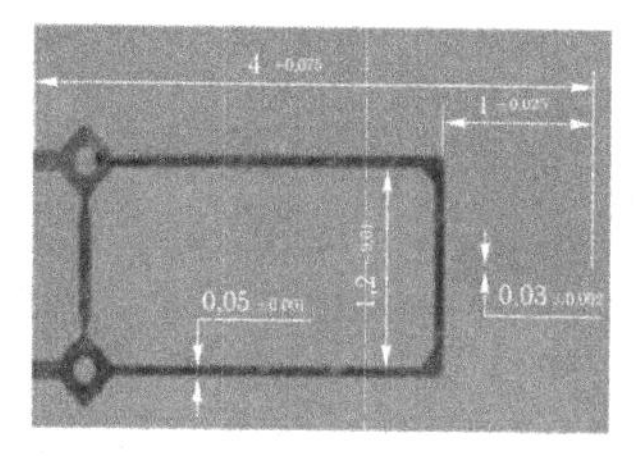

Fig. 8.23. Cathode from the W–Re alloy 30 and 50 μm thick.

its processing is efficiently performed at high levels of average radiation power (10–20 W) and high speeds (5–10 mm/sec).

The Karavella technological installations used for high-quality processing substrates of silicon and transparent materials with characteristic chipping (up to 50 μm). Figure 8.20 shows images of fragments of silicon cut 0.3 mm thick, Fig. 8.21 (color insert) – images of fragments of quartz 0.4 mm thick cut to chips.

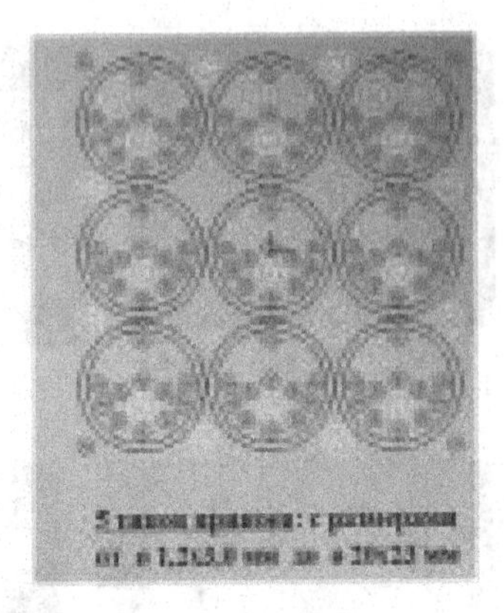

Fig. 8.24. Examples of the manufacture of a stencil for marking a packaging container and optimal cutting of solder from a material 50–150 μm thick.

Figure 8.22 shows a heat sink made of an artificial polycrystalline diamond 0.1 mm thick, Fig. 8.23 is a precision cathode assembly of W–Re alloy with 1 mm protruding to the right side with a 30 μm thick bead as an emitter, made in a gentle mode at a low laser power level.

The developed ALTI on the basis of pulsed CVLs with high productivity, quality, resolution and material saving make it possible to produce various stencils for marking packaging containers and cutting out solders from materials of thickness 20–150 μm, Fig. 8.24.

The relatively low-power Karavella-2 process unit with a single emission wavelength (λ = 510.6 μm), was used to develop the technology of applying a topology with a conductor/gap dimension of 20–100/20–100 μm on metallized LTCC ceramics. The thickness of the metal coating on the ceramic is 10 μm, the coating material is silver or gold. For topology application with such materials without significant damage to the ceramic, as it turned out, the average radiation power of 0.2–0.3 W is sufficient.

Technological installations based on pulsed CVLs have ample opportunities for adjusting and fitting the ratings of various elements for radio electronic equipment and microwave devices. Figure 8.25 shows an example of adjusting the topology of band-pass filters (BPF) on coupled resonators of microwave devices due to expansion of the gap from 0.15 to 0.22 mm.

The technology of manufacturing precision plates of rectangular and circular configurations with dimensions of 122.5 × 98.1 mm and 163.5 mm and 0.2 mm thickness from constantan (MN-45) and

Fig. 8.25. Correction of topology of the band pass filter.

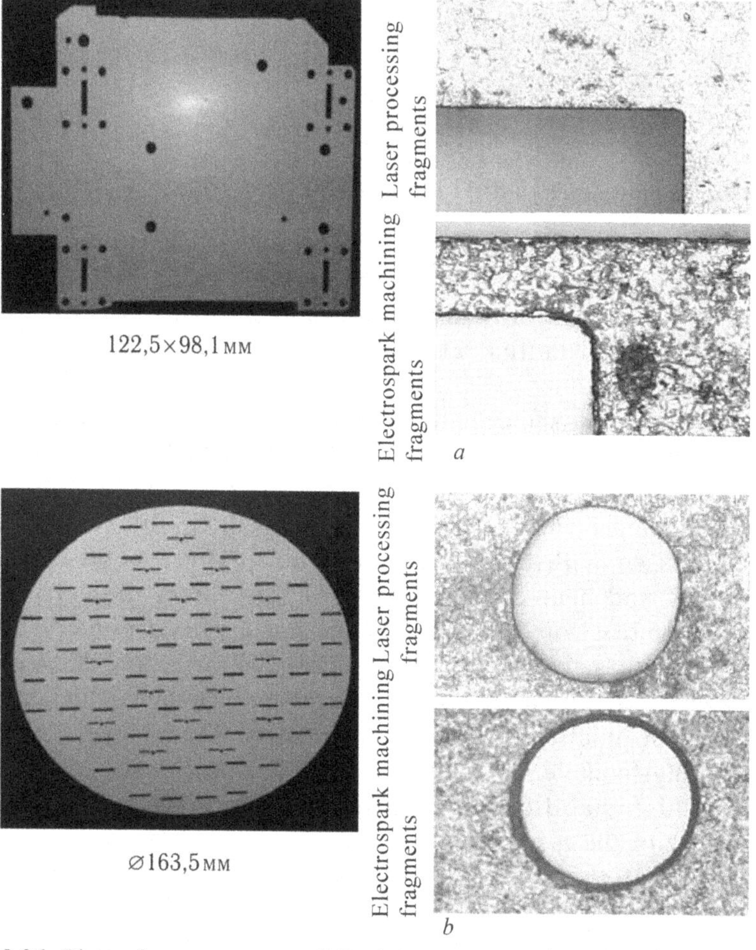

Fig. 8.26. Plates from constantan (MN-45) and steel (8UPVB-B-1) for antenna systems: *a* – holes: diam. 0.5; 2.0; 2.1; 2.8; 4.0 mm and slits 2.0 × 11.2 mm. t = 1 hour; *b* – holes: diam. 1.52 and slits 4.6 × 0.7; (8.58–9.16) × 1. t_{prod} = 2 hours

Fig. 8.27. Examples of images of fragments of flat details 0.6–1 mm in thickness made on the most powerful ALTI Karavella-1M (P_{prod} = 20–25 W).

steel (8YuPV-V-1) was developed for antenna systems used in active radar homing heads (Fig. 8.26). In addition to the high quality of processing and gain in time, this technology provides for the holes and slots a high mutual alignment, which does not exceed 3–5 microns.

Separately, one should note additional opportunities for precision microtreatment in the ALTI Karavella-1M, which is currently the most powerful (20–25 W) of all ALTIs. At the ALTI Karavella-1M, in comparison with the ALTI Karavella-1, thicker materials are processed with high efficiency and quality: metal ones like refractory and heat conductive 0.6–1 mm thick, dielectrics and semiconductors, for example, silicon and artificial polycrystalline diamond with a thickness of up to 1.5–2 mm. Figure 8.27 shows enlarged fragments of flat parts made of molybdenum, tungsten and copper with practically ideal cylindrical walls, achieved by using longer long-focus lenses (F = 200 and 250 mm) in ALTI Karavella-1M and high stability of the position of the axis of the diagram directivity of the laser beam.

An essential element is the use of laser technology in the manufacture and adjustment of single parts and assemblies for new devices being developed, since the creation and adjustment of programs for working drawings does not require significant expenses and does not require sophisticated technological equipment necessary for traditional processing methods. The gain in time in this case is the main factor in favor of laser microprocessing.

8.3. Advantages of the laser microprocessing of materials on ALTI Karavella in comparison with traditional processing methods

Today, the industrial ALTIs of the type Karavella based on the sealed-off pulsed CVL and CVLS in the automatic mode, in accordance with the programs for working drawings, are used to produce a large volume of precision parts both of simple and complex configurations in R & D, single, small-lot and serial deliveries of electronic components, in particular devices of microwave equipment produced by the Istok Co. Both high-temperature materials (Mo, W, Ta) and high thermal conductivity (Cu, Al, Ag, Au) materials and their alloys, and other metals (Ni, Ti, Zr, Fe, Pb) are involved in the sphere of precision microprocessing. a large circle of semiconductors and dielectrics (Si, Ge, GaAs, GaN, Al_2O_3, 22XC, SiC, sapphire, artificial polycrystalline diamonds, graphite glass, quartz, etc.) [392, 394–400]. Depending on the complexity, thickness and composition of the material, the requirements for precision and quality of machining of parts, the time of their manufacture can range from fractions and units of minutes to several hours. This time is in turn determined by the optimized processing parameters of the processing: the focal length of the power lens (F = 50–250 mm), the radiation power (P_{rad} = 0.2–25 W), the processing speed (V = 0.2–15 mm/s) and the number of passes (N = 1–50). Based on the results of the research, a computer database was created for technological parameters for precision microprocessing of the main materials of the electronic components. The availability of an electronic database significantly reduces the time for optimization and selection of microprocessing regimes for new parts. The laser method of microprocessing, in comparison with traditional methods, makes it possible to shorten the manufacturing time of precision parts by an order of magnitude or more and simultaneously improve their quality.

A large amount of experimental and practical work carried out at ALTI Karavella showed a number of significant advantages of the laser microprocessing method over the known traditional methods for manufacturing precision parts (including electrospark treatment):

1. A wide range of processed structural metallic and non-metallic materials;

2. Reduce the number of operations and transitions by several times;

3. Saving materials in the manufacture and cutting of blanks from the sheet;

4. High productivity of manufacturing parts, both simple and complex configurations (by an order or more);

5. High efficiency of manufacturing precision parts in R & D, in single and batch production;

6. Non-contact processing (without mechanical pressure);

7. Effective processing in air atmosphere (without blowing process gas);

8. High accuracy of manufacturing parts with tolerances in the range of 4-10 microns;

9. Small roughness of the surface of the cut ($\leq$1–2 μm);

10. Small heat-affected zone (10–20 μm);

11. Absence of stratification and microcracks of metal (molybdenum, tungsten);

12. High percentage of yield of useful parts (up to 100%);

13. Low labour intensity of manufacturing and versatility of technological equipment;

14. The presence of a computer database on the modes of laser microprocessing of materials;

15. Control of the technological process from the computer;

16. Significant reduction in the influence of the human factor on the reproducibility of the technological process.

8.4. Perspective directions of application of ALTI Karavella

More than ten years of experience with the industrial ALTI Karavella-1, Karavella-1M, Karavella-2 and Karavella-2M clearly outlined the following promising areas of the use of pulsed radiation CVL in the technology of microprocessing of materials of the electronic components:

- precision cutting,
- drilling of microholes using the direct piercing method,
- scribing,
- cutting of solders,
- processing of film coatings,
- surface treatment of parts,
- the formation of images in the volume of transparent media,
- marking and engraving with high resolution.

Precision cutting is the most capacious direction of technological application of ALTI Karavella. For the production of precision cutting of metallic materials with a thickness of up to 0.3 mm, it is necessary to use pulsed CVL in installations with a level of an average radiation power of 3–8 W (ALTI Karavella-2 and Karavella-2M), 0.4-0.0 mm thick, 5 mm – power 10–12 W (ALTI Karavella-1), thickness 0.6–1 mm – power 20–25 W (ALTI Karavella-1M) and thickness 1.5–2 mm – as shown by experiments and calculations with a power of 40–60 watts (Karavella-3). The cutting speed, depending on the processing quality requirements, can range from 1–15 mm/s at a peak power density of 10^{10} –10^{11} W/cm^2. The laser cutting process is equally effective both for refractory metals (Mo, Ta, W, Nb, Zr), and for metals with good thermal conductivity (Al, Cu, Ag, Au) and their alloys and other metals. The roughness of the surface of the cut can be less than 1 μm, and the zone of thermal impact – 10...20 μm. In addition to metals, practically any opaque materials for visible light, semiconductors and dielectrics, various composites, cermets, graphites, etc., can be processed by pulsed radiation. Polycrystalline diamond and silicon plates can be machined at maximum velocities (10–15 mm/s) and powers (10–25 W). The width and quality of the material cut, as well as the diameter of the microholes depend on the optical system used to form the radiation beam and its focusing on the object being processed and the properties of the material. For an achromatic lens with a focal length in the range F = 100–200 mm, it is 10–30 μm, and for a lens with $F \leq 100$ mm and at a material thickness ≤ 0.1 mm is 3–10 μm.

The second significant technological direction of the application of pulsed radiation of CVL in the ALTI is the drilling of an array of microholes. In the manufacture of nets and stencils of metallic materials 50–150 microns thick, the number of microholes can be hundreds and thousands. At the same time, the productivity of laser holes is in excess of traditional drilling methods by orders of magnitude. Holes with a diameter of up to 50 microns are effectively drilled using the direct flashing method. The drilling speed of thin samples exceeds 3–5 mm/s, the shape factor is 20–40, the minimum hole size is 3–5 μm. At relatively high levels of pulsed energy (3 mJ or more), holes are drilled in a few seconds in thick materials (2–3 mm). The form factor can be 50–100. Holes with a diameter of more than 50 microns are usually drilled by contour cutting. Precision and quality of manufacturing of such holes is determined by the parameters of the motion system – in practice the accuracy of the horizontal XY table.

The third promising technological direction for ALTI is the processing of materials by the scribing method. Scribing is usually done on flat parts in parallel and mutually perpendicular directions. It is especially effective for cutting chips of semiconductor and dielectric substrates, for example, from artificial polycrystalline diamond, sapphire, silicon, gallium arsenide, various ceramics. Scribing is usually performed at maximum speeds and maximum permissible power levels, determined by the thermophysical properties of the materials.

The fourth promising technological direction for laser processing of materials is the high-performance cutting of solders of simple and complex configurations. The cutting of solders, usually made on the basis of alloys of copper, silver, tin, gold, platinum or other materials, is carried out at the maximum processing speeds (10–15 mm/sec) provided at levels of an average radiation power of 1–7 W. The thickness of the solders used in the electronic components is usually 20–100 μm.

A notable (fifth) technological application for radiation of CVL is the microprocessing of film coatings. For the treatment of film coatings having a thickness of a fraction and a micron (up to 10 μm), it is sufficient to have an average CVLS radiation power of 0.1 to 1 W. The minimum cutting width is achieved by processing at one wavelength, namely, at a green wavelength $\lambda = 0.51$ μm and is determined by the radiation quality and focusing optics and can reach 1.5–2 μm, which is only 3–4 times the length wave. The speed of such processing can be tens to hundreds of millimeters per second. Regardless of the substrate material, both metallic and non-metallic as well as high-temperature superconducting films are treated equally effectively. The use of laser technology in single and small-scale production of film materials can have great advantages, as the processing time is significantly reduced, 'wet' operations are excluded, scarce materials are saved, and the ecology of production is generally improved. The formation of complex topologies with dielectric gaps (conductor/gap) in a continuous metallic film deposited on a substrate is one of the promising directions of using CVL radiation in the production of electronic components.

The sixth perspective direction of the technological application of pulsed radiation CVL is the surface treatment of materials and finished parts. With the help of the surface treatment method, qualitative cleaning of both surfaces of materials from contamination and the creation of structural changes in a thin near-surface layer of various alloys are performed with the aim of changing their physical properties [404]. In the first case, high-quality processing is achieved at a peak power density of 10^4–10^5 W/cm^2, in the second – at a power density of 10^6–10^7 W/cm^2. The surface treatment method is effective for both metallic and non-metallic materials. For example, at a laser radiation power density of 10^5 W/cm^2, the surface treatment of ceramic substrates from 22XC and A995 before applying the thin-film conductive coating results in improved quality of both local and continuous metallization. The use of CVL in nanotechnology and in other fields of microelectronics is promising [406].

The formation of images in the volume of transparent media is carried out at a peak power density in the spot of radiation focused in it 10^{10}–10^{11} W/cm^2 (the seventh technological direction). Moving the focused radiation spot along the vertical Z axis, and a transparent object in the perpendicular plane along the two XY axes, one can obtain any three-dimensional image without disturbing its surface. The divergence of the laser radiation is several times smaller than that of solid-state Nd:YAG lasers, which determines its more sharp focusing. The latter determines the smaller sizes of microdispositions and, correspondingly, the higher resolution of the 3D image. This method of processing can be used to protect against falsification of glass and quartz products in mass production and manufacture of various voluminous souvenir products.

In the production of electronic components, there is a mass production of small parts, assemblies and boards that require marking and engraving with high resolution. For this technological operation, the pulsed radiation of the CVL with a 5–20 μm processing spot is an almost ideal tool. When labeling, the level of radiation power density in the treatment spot of 10^9 W/cm^2 is sufficient. Deep laser engraving is usually performed on assembly metal housings that do not allow deformations, but even at a radiation power density of 10^{10} W/cm^2.

Promising is the use of laser radiation for the formation by laser milling of 'hemp' structures on cathodes for multipath klystrons of small and medium power, high-speed cutting of solders and foil materials. At the completion stage is the development of a specialized ALTI "Karavella-2S with a fast scanning device of the type 'SuperSkan' (up to 1–3 m/s), designed to perform this class of technological tasks.

In the nearest future plans it is planned to create an ALTI of the Karavella-3 type with an average output power of 50–60 W. The use of long-focus power lenses (200–300 mm) in this device will allow solving the problem of efficient and high-quality processing of refractory and heat-conducting materials with a thickness of 1–2 mm, for example, the production of precision parts from molybdenum and copper for microwave technology.

In addition to the electronic industry, the exact areas of application of ALTI Karavella include the precise instrument making, automobile (for example, the production of injectors for injection of fuel), chemical (production of dies for drawing thin fibers), medical (stents for expanding arteries), jewelry (cutting and processing precious materials, making souvenirs) industry, etc.

In Russia, the main developer and consumer of industrial ALTI-type Karavella for precise microprocessing of electronic component materials remains for today The Istok Co. (Fryazino, Moscow region), which is the only manufacturer of industrial high-performance and long-life industrial sealed-off pulsed AE pulsed CVL. A number of other leading research institutes and enterprises (see Chapter 1) actively participate in the research of pulsed CVLs and their application in technology: Bauman Moscow State Technological University, VELIT (Istra, MO), Lebedev Physical Institute and the Prokhorov General Physics Institute, Research Institute of Precision Engineering (TM) and Mekhatron, State Polytechnic University of St. Petersburg, St. Petersburg National Research University of Information Technologies, Mechanics and Optics, Laser Technology Center" (St. Petersburg), Lasers and Instrumentation, The Kurchatov Institute, Medical Sterilization Systems, Institute of Semiconductor Physics (Novosibirsk), Institution of the Russian Academy of Sciences Joint Institute for High Temperature Russian Academy of Sciences (Moscow), Tomsk State University (TSU) and Institute of Atmospheric Optics, Siberian Division of the RAS, Precision Electromechanics (Moscow), Chistye Tekhnologii (Izhevsk).

In the foremost foreign countries there is also a noticeable interest in the development and application of technological equipment based on pulse CVLs for productive and high-quality microprocessing of metallic and non-metallic materials. The large volume of foreign scientific and technical publications was devoted to the use of CVL for microprocessing purposes. The main part of the first papers was generalized in [20]. Corresponding works are carried out in scientific centres and firms of the USA, Great Britain, France, Australia, Japan,

Germany, Italy, etc. (detailed in Chapter 1). For example, at Oxford Lasers, a laser technological installation of the MP200X model based on a pulsed CVLS operating according to the MO–PA scheme with an average radiation power of up to 60 W with a 10 kHz PIC and a working laser field of 200 × 200 mm was developed. The main and significant disadvantage of the MP200X installation is that the AE in CVLS operate in the neon buffer gas pumping mode, that is, the technology of their manufacture does not allow operating in the sealed-off mode. But the operating mode of operation, as is well known, worsens the stability of the output radiation parameters and lowers the life of the laser. The domestic industrial AEs of the CVLS of the series Kulon and Kristall effectively operate in the sealed-off mode, have 3–4 times more guaranteed operating time, high stability and reproducibility of output parameters, which is predetermined by thorough scientific and technological design study, as well as the quality of materials used and nodes.

8.5. Conclusions and results for Chapter 8

1. Numerous examples of the use of the automated industrial technological equipment of ALTI Karavella on the basis of pulsed CVLs and CVLSs for the manufacture of precision parts of the electronic devices are presented.

2. Precision parts are made at industrial ALTI Karavella with high productivity and quality, both from refractory (Mo, W, Ta, Re) and high-conductivity (Cu, Al, Ag, Au) and other metals (Ni, Ti, Zr, Fe, Pb, Zr), their alloys, a large range of semiconductors and dielectrics (Si, Ge, GaAs, Al_2O_3, 22XC, sapphire, artificial diamond, glass, quartz), graphite, composite materials.

3. It is shown that the laser method of microprocessing with pulsed radiation of CVL with nanosecond duration has a number of significant advantages over the known traditional methods for manufacturing precision parts, including electrospark machining, especially complex configurations: a wide range of processed structural materials, a reduction in the number of operations and transitions several times, materials in the manufacture and cutting of blanks from the sheet, an order of magnitude and higher productivity, high efficiency in R&D, single and small-scale deliveries and mass production, machining without mechanical pressure, efficient processing in air atmosphere (without blowing of process gas), accuracy of manufacturing parts with tolerances in the range of 4–10

μm, small surface roughness of the cut (≤1–2 μm) and the heat-affected zone (≤10–20 μm), absence of delamination and microcracks of metal (molybdenum, tungsten), high percentage of yield of suitable parts (up to 100%), low laboriousness of manufacturing and versatility of the tooling, presence of a computer database on the modes of laser micromachining of materials, process control from the PC, significantly reducing the impact of human factors on the reproducibility of the process.

4. Eight directions for using pulsed nanosecond laser radiation in the technology of laser microprocessing of electronic component materials have been clearly outlined: contour precision cutting, drilling of microholes with the method of direct piercing, scribing, cutting of solders, processing of film coatings, surface cleaning of parts, formation of images in the volume of transparent media, marking and engraving with high resolution.

5. At present, tens of hundreds of types of flat and light-weight steel plates are manufactured in the automated mode at the industrial ALTIs Karavella-1, Karavella-1M, Karavella-2 and Karavella-2M in accordance with the programs for working drawings together with volume precision parts of simple and complex configurations in providing R&D, single, small-scale and serial production of microwave devices in the Istok Co.

Conclusion

1. A comparative analysis of the properties of the radiation of a copper vapour laser (CVL) operating in the single-generator mode and a copper vapour laser system (CVLS) operating according to the master oscillator (MO)–power amplifier (PA) scheme with known technological lasers has shown that the CVL and CVLS with emission wavelengths of 510.6 and 578.2 nm, short pulses (20–40 ns), high pulse repetition rates (5–30 kHz) and low pulsed energy (0.1–10 mJ) remain today the most powerful pulsed sources of coherent radiation in the visible spectral region and are ideal instruments with a light diameter of the light spots of 5–20 μm for effective micromachining materials for electronic device components.

2. A new generation of industrial, with high efficiency, reliability and quality of radiation of sealed-off laser AEs on copper, CVL and CVLS vapours with an average radiation power of 1–20 W and 30–100 W has been developed and investigated on the basis of a set of scientific, technical, technological and circuit solutions, surpassing foreign analogs in the removal of radiation power from a unit of active medium 2–4 times and a minimum operating time 4–5 times.

3. The regularities of the dynamics of the formation of pulsed CVL radiation are investigated when operating in the mode of a generator with optical resonators. It is shown for the first time that the output radiation of CVL has a multibeam structure: two always present superradiance beams and several beams formed directly by the resonator, the number of which (usually 2–4) is determined by the time of existence of population inversion in the active medium (20–40 ns). Each beam has its spatial, temporal and energy characteristics and uneven intensity distribution in the focusing plane, which is unacceptable for high-quality microprocessing.

4. It has been established that for effective microprocessing of materials in the evaporation regime, pulsed radiation of CVL requires the formation of single-beam radiation with diffraction divergence

and peak power density in a focused light spot with a diameter of 5–20 μm within 10^9-10^{12} W/cm^2.

This radiation quality is achieved when new highly selective optical systems are used in CVL and CVLS: with an unstable resonator with two convex mirrors and a telescopic resonator in combination with a spatial collimator filter.

5. The physical properties of the active medium of a pulsed CVL were investigated and it was established for the first time that, in relation to the intrinsic radiation, the active medium has four characteristic time zones, one after another and repeating from pulse to pulse, of weak absorption (τ = 20–30 ns), amplification (τ = 30–40 ns), total absorption (τ > 1000 ns) and maximum transparency (τ > 1000 ns). Based on these properties, methods and devices have been developed for the automated control of laser microprocessing regimes by changing the population inversion.

6. It is shown that the radiation of CVLs can efficiently process materials with a thickness of not more than 1 mm. And it is even more effective to use CVL at thicknesses less than 0.6 mm, when the processing speed is one and tens of mm/s, which is an order of magnitude greater than in electroerosion processing.

7. The quality of the surface of the laser cutting and the structure of the HAZ for high-heat-conducting and refractory materials were studied in processing by CVL radiation with a light spot diameter of 10–20 μm and a density of peak power of 10^{10}–10^{11} W/cm^2 and it was established that:

– the surface of the laser cut has a multichannel structure with a period of repetition of the channels of 10–15 μm and depth (roughness) to 3–4 μm;

– increasing the speed and number of processing passes leads to a decrease in the roughness of the cutting surface to values less than 1 μm;

– in high-heat-conducting metals, hardening or softening of the material does not occur in the zone adjacent to the laser cut (10–20 μm);

– in refractory metals with lower thermal conductivity insignificant hardening takes place in the zone adjacent to the laser cut (10–20 μm).

8. Modern ALTI Karavella-1 and Karavella-1M with a light spot diameter of 10–20 μmicrons and a density of peak power of 10^9–10^{12} W/cm^2 at 14 kHz PRF for high-performance precision microprocessing of thin-sheet materials of electronic components

0.3–0.5 mm and 0.6–1 mm. The ALTI contains industrial CVLS Kulon-10 with an average output power of 10-15 W and Kulon-20 with a power of 20–25 W, new highly selective optical systems and methods for controlling radiation parameters and a precision XYZ three-axis table with a working field of the XY horizontal table 150 × 150 mm in size and the accuracy of positioning along the axes of ±2 μm.

The ALTIs Karavella-2 and Karavella-2M with a light spot diameter of 10–20 μm and a peak power density of 10^9–10^{11} W/cm^2 at PRF of 15 kHz are developed for high-performance and high-quality precision microprocessing of foil materials of electronic components with a thickness of 0.01–0.2 mm. The ALTI employs an industrial CVL Kulon-06 with an average radiation power of 5–8 W, a new highly selective optical system and methods for controlling radiation parameters and precision XYZ three-axis tables with a working field of a horizontal XY table 100 × 100 mm and 200 × 200 mm and precision positioning along the axes of ±2 μm.

10. It has been shown that high heat conductivity (Cu, Al, Ag, Au) and high-melting (Mo, W, Ta, Re) and other (Ni, Ti, Zr, Fe, Pb) metals, their alloys, a large number of semiconductors and dielectrics (Si, Ge, GaAs, Al_2O_3, 22KhS, sapphire, artificial diamond, glass, quartz), graphite and composite materials 0.01–1 mm thick can be processed efficiently by the radiation of CVL with nanosecond pulse duration.

11. It has been established that the laser method of microprocessing by CVL radiation with nanosecond pulse duration has a number of significant advantages over the known traditional methods for manufacturing precision parts, including electroerosion processing, especially details of complex configurations:

– a wide range of processed structural materials, machining without mechanical pressure, efficient processing in the air atmosphere (without blowing of process gas), saving materials in the manufacture and cutting of blanks from the sheet, low labor input and versatility of the tooling;

– an order of magnitude and higher productivity, a reduction in the number of operations and transitions by several times, high efficiency in R&D and in single, small and batch production;

– small roughness of the cutting surface (1–2 μm) and the heat-affect zone (5–10 μm), absence of delamination and microcracks of the metal (molybdenum, tungsten);

– availability of a computer database for laser microprocessing of materials, control of technological processes from the PC, a significant reduction in the influence of the human factor on the reproducibility of the technological process, a high percentage of yield of suitable parts (up to 100%).

12. The most promising technological directions of application of pulsed nanosecond laser radiation for laser microprocessing of materials have been established: contour precision cutting, drilling of microholes with the method of direct piercing, scribing, cutting of solders, processing of film coatings, surface cleaning of parts, imaging in the volume of transparent media, marking and engraving with high resolution.

References

1. Fowles G.R., Silfast W.I., Appl. Phys. Lett. 1965. V. 6, No. 12. P. 236–237.
2. Piltch M., et al., Appl. Phys. Lett. 1965. V. 7, No. 11. P. 309–310.
3. Piltch M., et al., Bull. Amer. Phys. Soc. 1966. V. 11, No. 1. P. 113.
4. Piltch M., Gould G., Rev. Sci. Instr. 1966. V. 37. P. 925–927.
5. Walter W.T., Bull. Amer. Phys. Soc. 1967. V. 12, No. 1. P. 90.
6. Walter W.T., IEEE J. Quant. Electron. 1968. V. 4, No. 5. P. 355–356.
7. Leonard D.A., IEEE J. Quant. Electron. 1967. V. 3, No. 9. P. 380.
8. Petrash G.G., Usp. Fiz. Nauk. 1971. V. 105, No. 4. P. 645–676.
9. Isaev A.A., Kazaryan M.A., Petrash G.G., Pis'ma v Zh. Eksp. Teor. Fiz. 1972. P. 16. P. 40–42.
10. Isaev A.A., Petrash G.G., in: Pulsed gas-discharge lasers on transitions of atoms and molecules: Tr. FIAN. V. 81. Moscow: Nauka, 1975. P. 3–87.
11. Petrash G.G., Lasers on metal vapors: Handbook of lasers: In 2 vols. /Ed. A.M. Prokhorov. V. 1. Ch. 8. Moscow: Sov. radio, 1978. P. 183–197.
12. Buzhinsky O.I., Evolution of copper laser research and the possibilities of its practical application: Review. Moscow: IAE I.V. Kurchatov, 1983.
13. Effective gas-discharge vapor-metal lasers: Proceedings, Ed. PA Bohan. Tomsk: IOA SB RAS, 1978..
14. Soldatov A.I., Solomonov V.I., Gas-discharge lasers based on self-terminating transitions in metal vapors. Novosibirsk: Nauka, 1985.
15. Pulsed Metal Vapour Lasers: Proc. NATO Advansed Research Workshop on Pulsed Metal Vapour Lasers – Physics and Emerging Applicaitions in Industry, Medical and Science, St. Andrews (UK), Aug. 6–10, 1995/Ed. C. E. Little, N.V. Sabotinov. Dordrecht, 1996. 479 p.
16. Grigor'yants A.G., et al., Copper vapor lasers: design, characteristics and applications. Moscow: Fizmatlit, 2005. – 312 p.
17. Batenin VM, Buchanov VV, Kazaryan MA et al. Lasers on self-limited transitions of metal atoms. Moscow: Nauchnaya kniga, 1998. 544 p.
18. Batenin V.M., et al., Lasers on Self-Terminating Transitions of Metal Atoms. In 2 vols. Vol. 1/Ed. V.M. Batenin. Moscow: Fizmatlit, 2009. 544 p.
19. Batenin V.M., et al., Lasers on Self-Terminating Transitions of Metal Atoms. 2. In 2 vols. Vol. 2 / Ed. V.M. Batenin. Moscow: Fizmatlit, 2009. 616 p.
20. Little C.E., Metal Vapor Lasers: Physics, Engineering and Applications. Chichester (UK): J. Wiley and Sons Ltd., 1999. 620 p.
21. Aikhler Yu., Eichler G.I., Lasers. Execution, management, application. Moscow: Tekhnosfera, 2008. 440 p.
22. Bokhan P.A., et al., Laser separation of isotopes in atomic vapor. Moscow: Fizmatlit, 2004. 208 pp.

23. Bokhan P.A., et al. Optical and laser-chemical separation of isotopes in atomic vapor. Moscow: Fizmatlit, 2010. 208 pp.
24. Pasmanik G.A., Zemskov K.I., Kazaryan M.A., Optical systems with luminance amplifiers. Gorky: IPF Academy of Sciences of the USSR, 1988. 173 p.
25. Grigor'yants A.G., et al., Technological processes of laser processing: Proc. Manual for universities / Ed. A.G. Grigor'yants. Moscow: Bauman MSTU, 2006. 664 p.
26. Lyabin N.A., Development and investigation of industrial sealed copper vapor lasers with a power of 10–50 W for technological and medical applications: Dissertation. Fryazino, 2002. 244 p.
27. Ambartsumyan R.V., et al., Kratkie soobshcheniya po fizike. 1988. No. 8. P. 35–37.
28. Cherepnin N.V., Physical concepts of the laser annealing mechanism of implanted semiconductor structures: A review of electronic engineering. Ser. Technology, production organization and equipment. Issue 8, Moscow, 1981. 25 p.
29. Zuev, VE, Propagation of laser radiation in the atmosphere. Moscow: Radio and Communication, 1981. – 287 p.
30. Zuev V.Ee., Naats I.E.. Inverse problems of laser sounding of the atmosphere. Novosibirsk: Nauka, 1982. 260 p.
31. Kazaryan M.A., et al., Kvant. Elektronika. 1998. V. 25, No. 9. P. 773–774.
32. Physico-chemical processes in the selection of atoms and molecules: Proc. 4th All-Russia. (International) Scientific. Conf., October 4–8. 1999, Zvenigorod, Ed. V.Yu. Baranov, Yu. A. Kolesnikov. M., 1999. 271 p.
33. Zharikov V.M.,et al., Laser-Inform: Information Bulletin. Laser Association. 1999. No. 9 (168). P. 2–8.
34. Korolev A.N., et al., Laser technological installation Karavella-1 for precision microprocessing of thin-sheet materials of electronics products, Elektronnaya promst.. 2006. No. 3. P. 61–74.
35. Petrash G.G., Izv. VUZ. Fizika. 1999. V. 42, No. 8. P. 18–22.
36. Soldatov A.N., Izv. VUZ. Fizika. 1999. V. 42, No. 8. P. 23–36.
37. Yakovlenko S.I., Izv. VUZ. Fizika. 1999. V. 42, No. 8. P. 82–87.
38. Evtushenko G.S., Izv. VUZ. Fizika. 1999. V. 42, No. 8. P. 88–95.
39. Gorelik V.S., Nauka proizvodstvu. 2000. No. 6 (31). P. 2–6.
40. Nasibov A.S., et al., ibid, 2000. No. 6 (31). P. 20–21.
41. Domanov M.S., et al., ibid, 2000. No. 6 (31). P. 55–57.
42. Lyabin N.A., et al., Kvant. elektronika. 2001. T. 31, No. 3. Pp. 191–202.
43. Aleinikov V.S,, et al., Elektronaya promst. 1984. Issue. 10. pp. 10–12.
44. Armichev A.V., et al., Elektronaya promst. 1984. Issue. 10. P. 32–35.
45. Aleinikov V.S., et al., Elektronaya promst. 1984. Issue. 10. P. 32–35.
46. Devyatkov N.D., et al., Khirurgiya. 1986. No. 4. P. 116–121.
47. Lasers Use in Oncology: CIS Selected Papers/Ed. by A.V. Ivanov et al., Proc. SPIE 1996. No. 2728.
48. Ponomarev I.V., Application of lasers on metal vapors in medicine. Moscow: Lebedev Physical Institute. Russian Academy of Sciences, 1997.
49. Ponomarev I.V., Laser surgery of vascular and pigmented skin defects. Moscow: Lebedev Physical Institute. Russian Academy of Sciences,, 2001. 48 p.
50. Treatment of epithelial skin formations with a laser medical apparatus on copper vapor Yakhroma-Med. Methodological recommendations. – SPb .: SPbGMA im. Mechnikova, 2004. – 52 p.
51. Klyuchareva S.V., et al., Eksper. Klinicheskaya dermatokosmitologiya. 2008. No. 6. P. 50–55.
52. Lepekhin N.M., et al., Lazernaya meditsina. 2004. V. 8, no. 3. P. 186.

53. Sokolov V.V., et al., in: Proc. of the scientific-practical conference Modern achievements of laser medicine and their application in practical public health. Ed. A.V. Heynits. October 5–6, 2006, Moscow. P. 200–201.
54. Photodynamic therapy: Materials of the III All-Russian Symposium, October 11–12. 1999, Moscow/Ed. E.F. Stranadko. Moscow:GNTsLM, 1999. 208 p.
55. Stranadko E.F.,et al., Photodynamic therapy in the treatment of malignant neoplasms of various localizations: A manual for physicians, ed. E.F. Stranadko. Moscow: GNTsLM, 1999. .
56. Evtushenko V.A., et al., Izv. VUZ. Fizika. 1999. Vl. 42, No. 8. P. 109–118.
57. Voronko Yu.K., et al., Neorganicheskie materialy. 2009. V. 45, No. 5. P. 619–625.
58. Laser equipment for processing material, Electronic resource: http://www.laserarp.com (reference date: 9.09.2011).
59. Saprykin L.G., Kudryavtseva A.L., Metalloobrabotka i stankostroenie. 2008. No. 10. P. 39.
60. Kudryavtseva A.L., Metalloobrabotka i stankostroenie. 2008. No. 11–12. C. 27.
61. Kudryavtseva A.L., Metalloobrabotka i stankostroenie. 2009. No. 2. 28.
62. Industrial fiber lasers. – Advertising IRE-Polus/Electronic resource: http://www.ntoire-polus.ru (reference date: 19.08.2011).
63. Industrial fiber lasers – the choice of the day. – Advertising of NTO IRE-Pole/Ritm. 2007. No. 3. P. 78–79.
64. Holidays and everyday life of fiber lasers. – Advertising of NTO IRE-Pole/Ritm. 2010. No. 2. 64.
65. Excimer lase, Electronic resource: http://www.wikipedia.org/wiki/Excimer laser (date of circulation: 19.08.2011).
66. Excimer laser, Electronic resource: http://www.triniti.ru/Excimer. htm (date of circulation: 19.08.2011).
67. Laser diode, Electronic resource: http://ru.wikipedia.org/wiki (date of circulation: 19.08.2011).
68. Diode lasers/Electronic resource: http:,//www.jenoptic-components.ru (date of reference: 19.08.2011).
69. Overton G., et al., Laser markets develop despite 'counter winds' in the global economy, Laser-Inform: Information Bulletin. Laser Association. 2013. No. 3 (498). Pp. 1–6.
70. Our word famous copper vapor lasers, Electronic resource: http://www.oxfordlasers.com (reference date 12.09.2011).
71. Karnakis D., et al., High-quality laser milling of ceramics, dielectrics and metals using nanosecond and picosecond lasers, Photonic West 2006, LASE – Laser Applications in Microelectronic Manufacturing XI, San Jose, January 2006, USA.
72. Coutts D.W., et al., Generation of Flat-top Focussed Beams for Percussion Drilling of Ceramic and Metal, ICALEO 2002, Orlando Florida, USA.
73. Rabinowitz P., Chimenti R., J. Opt. Soc. Am. 1970. V. 60. P. 1577–1578.
74. Chimenti R., Walter, W.T., Bull. Am. Phys. Soc. 1971. V. 16. P. 41–2.
75. Kazaryan M.A., Investigation of pulsed lasers on metal vapors: Dissertation, Moscow, 1974. – 151 p.
76. Petrash G.G., Pulsed gas-discharge lasers: Dissertation, Moscow, 1972. – 392 p.
77. Isaev A.A., Limmerman G.Yu., Kvant. elektronika. 1977. V. 4. P. 1413–1417.
78. Kazaryan M.A., Optical systems with image brightness amplifiers: Dissertation, Moscow, 1988. – 310 p.
79. Isaev A.A., Effective pulsed-periodic lasers on copper vapor: Dissertation, Moscow, 1988. – 339 p.

80. Isaev A.A., et al., Quantum electronics: Proceedings, Ed. N.G. Basov. – Moscow Radio, 1972. No. 5 (11). C. 100.
81. Isaev A.A., et al., Zh. Prikl. Spektr. 1973. P. 18. P. 483–484.
82. Isaev A.A., et al., in: Lasers based on complex organic compounds: Mater. All-Union. Conf. Minsk, 1975. P. 10.
83. Isaev A.A., Kazaryan M.A. Pulsed laser on lead vapor with repetition rate up to 20 kHz, Kr. Soobshch. po fiz. FIAN. 1976. No. 10. P. 29–30.
84. Isaev A.A., et al., ibid, 1976. No. 10. P. 3–5.
85. Isaev A.A., et al., Quantum electronics: Collectio,. 1973. No. 4 (10). P. 123–126.
86. Kazaryan M.A., et al., Kvant. elektronika. 1975. V. 2. P. 503–507.
87. Isaev A.A., et al., Kvant. elektronika. 1979. V. 6. P. 1942–1947.
88. Isaev A.A., Lemmerman G.Yu., Kvant. elektronika. 1985. No. 12. P. 68–73.
89. Isaev A.A., et al., et al., Tr. FIAN. 1987. V. 181. P. 3–17.
90. Isaev A.A., et al., Kvant. elektronika. 1976. V. 3. P. 1802–1805.
91. Zemskov K.I., et al., Kvant. elektronika. 1974. V. 1. P. 863–869.
92. Zemskov K.I., et al., in: 2nd Symp. on the Physics of gas lasers: – Novosibirsk, 1974. P. 141.
93. Isaev A.A., et al., Kvant. elektronika. 1974. V. 1, No. 6. P. 1379–1388.
94. Isaev A.A., et al., Kvant. elektronika. 1975. Vol. 2, No. 6. P. 1125–1137.
95. Zemskov K.I., et al., Kvant. elektronika. 1998. TV 25, No. 7. P. 616–618.
96. Jones Q.R., et al., IEEE J. Quant. Electron. 1995. V. 31, No. 4. P. 747–753.
97. Le Guyadec E. et al. 280-W average power Cu-Ne-HBr laser amplifier, ibid. 1999. V. 35. P. 1616–1622.
98. Zemskov K.I., et al., Optical Systems with Brightness Amplifiers, Proceedings of P.N. Lebedev FIAN. 1991. V. 206. P. 63–100.
99. Lyabin N.A., et al., in: Physico-chemical processes in the selection of atoms and molecules: Proc. 6th International Scientific. Conf. Zvenigorod, October 1–5. 2001. pp. 123–131.
100. Kazaryan M.A., et al., in: Physicochemical processes in the selection of atoms and molecules: Proc. VII International Scientific. Conf. Zvenigorod, September 30-Oct 4 2002. P. 240–248.
101. Lyabin N.A., et al., in: Innovative technologies in science, engineering and education: Proc. of the International scientific and technical conference. Taba, Egypt, 14–21 November. 2009
102. Kazaryan M.A., et al., Proc. SPIE. 1997. V. 3317. P. 275–278.
103. Gas and plasma lasers: Encyclopedia of low-temperature plasma. ed. S.I. Yakovlenko. T. XI-4. Moscow: Fizmatlit, 2005. P. 381–405.
104. Kazaryan M.A., et al., Iav. RAN. Ser. Fiz.1999. V. 63, No. 6. P. 1190–1191.
105. Averyushkin A.S., et al., Kratkie Soobshcheniya po Fizike. 2006. No. 10. P. 42–46.
106. Grigoryan G.V., et al., in: Lasers on metal vapors, Proc. Symp. Loo, September 22–26, 2008. P. 25.
107. Grigoryan G.V., et al., in: Lasers on metal vapors: Proc. Symp. Loo, September 20–24, 2010. P. 35.
108. Biryukov A.S., et al., in: Physico-chemical processes in the selection of atoms and molecules: Proc.. XIV International Scientific. Conf. Zvenigorod, 4–5 October 2010. P. 14.
109. Averyushkin A.S., et al., in: Lasers and laser technologies, Proceedings of the Youth School-Conference with international participation, ed. A.N. Soldatov. Tomsk, November 22–27, 2010. P. 134–135.
110. Biryukov A.S., et al., in: Lasers and laser technology, ibid, pp. 163–164.

111. Kazaryan M.A., et al., in: Power Laser Ablation: Proc. SPIE. 2002. V. 4900. P. 1094–1098.
112. Kazaryan M.A., et al., Journal of Russian Laser Research. 2004. V. 25, No. 3. P. 267–297.
113. Grigorianz A.G., et al., Laser micromachining in self-conjugated projection microscope, Techical Digest ICONO Minsk, Belorus, May 28–June 1. 2007. LO2–22 .
114. Bokhan P.A., et al., in: Atomic and Molecular Pulsed Lasers, Abstracts of the 10th International Conference Tomsk, Russia, September 12–16, 2011. P. 26.
115. Kazaryan M.A., et al., ibid, P. 74–75.
116. Grigoryan G.V., et al., ibid, P. 84.
117. Combination dispersion – 80 years of research: Collective monograph, e.d. V.S. Garelik. Moscow: P.N. Lebedev Physical Institute. 2008. 604 p.
118. Boson-Verdura F., et al., Kvant. elektronika. 2003. V. 33, No. 8. P. 714–720.
119. Izgaliev A.T., et al., Kvant. elektronika. 2004. V. 34, No. 1. P. 47–50.
120. Shafeev G.A., Kvant. elektronika. 2010. V. 40, No. 11. P. 941–953.
121. Voron'ko Yu.K., et al., Neorganicheskie materialy. 2009. V. 45, No. 5. P. 619–625.
122. Soldatov A.N., Yancharina A.M., Izv. VUZ. Fizika. 1999. Vol. 42, No. 8. Pp. 4–13.
123. Bokhan P.A., et al., Kvant. elektronika. 1975. V. 2. P. 159.
124. Bohan P.A., et al., Kvant. elektronika. 1978. V. 5. P. 198.
125. Vlasov G.Ya., et al., Kvant. elektronika. 1979. V. 6. P. 1359.
126. Bokhan P.A., et al., Kvant. elektronika. 1980. V. 7, No. 6. P. 1264–1269.
127. Soldatov A.N., Fedorov V.F., Izv. VUZ. Fizika. 1983. No. 9. P. 80.
128. Shiyanov D.V.,et al., Optika atmosfery i okeana. 2000. V. 13, No. 3. P. 254–257.
129. Batenin V.M., et al., Teplofizika vysokikh temperatur. 1979. V. 17, No. 3. P. 483–489.
130. Klimovskii I.I., Morozov A.V., Kvant. elektronika. 1986. V. 13, No. 4. P. 828–930.
131. Batenin V.M., et al., Teplofizika vysokikh temperatur. 1978. V. 16, No. 6. P. 1145–1151.
132. Batenin V.M., Kvant. elektronika. 1977. V. 4, No. 7. P. 1572–1575.
133. Batenin V.M., Teplofizika vysokikh temperatur. 1976. V. 14, No. 6. P. 1316–1319.
134. Vokhmin P.A., Klimovskii I.I., Teplofizika vysokikh temperatur. 1978. V. 16, No. 5. 1080–1085.
135. Batenin V.M., et al., Teplofizika vysokikh temperatur. 1982. V. 20, No. 1. Pp. 177–180.
136. Klimovskii I.I., Lasers on self-terminating transitions of metal atoms: Dissertation, Moscow, 1991.
137. Karpukhin V.T., Malikov M.M., Kvant. elektronika. 2003. V. 33, No. 5. P. 411–415.
138. Karpukhin V.T., Malikov M.M., Kvant. elektronika. 2003. V. 33, No. 5. P. 416–418.
139. Karpukhin V.T., Malikov M.M., Zh. Teor. Fiz. 2000. V. 70, No. 4. P. 87–89.
140. Batenin V.M., et al., Kvant. elektronika. 2005. V. 35, No. 9. P. 844–848.
141. Batenin V.M., et al., in: Lasers in Science, Technology, and Medicine: Proceedings, Ed. V.A. Petrov. Moscow: MNTORES, 2008. P. 14–18.
142. Batenin V.M., Prikl. Fizika. 2009. No. 4. P. 128–132.
143. Batenin V.M., Kvant. elektronika. 2009. V. 39, No. 5. P. 405–409.
144. Kondratenko V.S., et al., Vestnik MGUPI. 2009. No. 17. P. 124–131.
145. Lepekhin N.M., Kvant. elektronika. 2007. V. 37, No. 8. P. 765–769.
146. Fogelson T.B., et al., Pulsed hydrogen thyratrons. Moscow: Sov. radio, 1974. 212 p.
147. Skripnichenko A.S., Glikin L.S., –Author Cert.. 986683. USSR, M. Cl.^{3} B 23 K 26/00. Device for processing materials with a laser beam. No. 2609152/25-27; claimed April 21, 1978, publ. 07.01.83, Bulletin. No. 1.

148. Glikin L.S., et al., Author. Cert. No. 980364. USSR, Cl. B 23 K 26/00. Device for laser projection processing. No. 3614019/25–27; claimed. 04.07.83, publ. 23.04.90, Bulletin. No. 15.
149. Glikin L.S., et al., International application No. WO 81/02951, Cl. H01S 3/00. Device for laser processing objects with visual control on the lumen. No. 4227277/24-10; claimed 16.03.87, publ. 07.08.90, Bulletin No. 29.
150. Physico-chemical processes in the selection of atoms and molecules: Proc. IX-th National (international) Scientific. Conf., October 4–8. 2004, Zvenigorod, Ed. Yu.A. Kolesnikov. Moscow: TsNIIatominform, Troitsk, SSC RF TRINITI 2004. 320 p.
151. Physico-chemical processes in the selection of atoms and molecules: Proc. XII-th National (international) Scientific. Conf., March 32 – April 4. 2008, Zvenigorod, Ed. V.E. Cherkovtsev. Moscow: TsNIIatominform, Troitsk, SSC RF TRINITI 2008. 417 p.
152. Physico-chemical processes in the selection of atoms and molecules, Perspective materials, special issue (10). Moscow: Interkontakt Nauka, February 2011. 349 p.
153. Kazaryan M.A., et al. in: Proc. 3rd All-Russian Sci. Conf. Zvenigorod, October 5–9, 1998. P. 44–45.
154. Kazaryan M.A.,et al., in: Combinational scattering – 70 years of research: Proc.. Intern. Conf. Moscow, Nov. 16–18. 1998. P. 446–448.
155. Gradoboev Yu.G., et al., Kvant. elektronika. 2004. V. 34, No. 12. P. 1133–1137.
156. Guiragossian Z.G., et al., Color laser system, Lasers'99: Proc. Intern. Conf. Quebec, Canada, December 13–16, 1999. P. 685–687.
157. Lepekhin N.M., Prikl. Fizika. 2001. No. 5. P. 46–49.
158. Domanov M.S., et al., in: Lasers on metal vapors: Proc. Symposium Loo, September 24–26, 2002. P. 40.
159. Lepekhin N.M., et al., Prikl. Fizika. 2006. No. 1. P. 8–14.
160. Lepekhin N.M., et al., Patent 2226022 The Russian Federation. IPC^7 H 01 S 3/0975. A generator of nanosecond pulses for excitation of lasers on self-limited transitions of metal atoms/Applicants and patent holders. No. 2002118259; claimed 10.07.2002; publ. 20.03.2004, Bulletin No. 8.
161. Lepekhin N.M., et al., Patent No. 2251 799. Russian Federation. IPC^7 H 01 S $3/09^7$. The method of excitation of pulsed lasers on self-terminating transitions of metal atoms operating in self-heating mode and the device for its implementation/Applicants and patent holders. No. 2003120867/28 filed 11.07.2003; publ. 27.04.2005, Buletin No. 12.
162. Lepekhin N.M., et al., Patent 2226030 Russian Federation. IPC^7 H 03 K 3/53. Pulsed submodulator of a nanosecond pulse generator. The applicant and the patent owner of LLC NPP VELIT. No. 2002118261/09; claimed 10.07.2002; publ. 03/20/2004, CD-ROM: MIMOSA RFD 2004/003 MRFD2004003.
163. Lepekhin N.M., et al., Patent 2226740 Russian Federation. IPC^7 H 03 K 3/53. A method for regulating the voltage on a capacitor accumulator of a nanosecond pulse generator. The applicant and the patent owner of LLC NPP VELIT. No. 2002118260/09; claimed 10.07.2002; publ. 10.04.2004, CD-ROM: MIMOSA RFD 2004/004 MRFD2004004.
164. Lepekhin N.M., et al., Prikl. Fizika. 2005. No. 1. P. 110–115.
165. Kondratenko V.S., et al., Pribory. 2009. No. 2 (109). P. 49–54.
166. Lyabin N.A., et al., in: Lasers in Science and Technology: Proceedings, Ed. V.A. Petrov. Moscow: A.S. Popov MNTORES, 2005. P. 98–105.
167. Lepekhin N.M., et al., in: Lasers in Science and Technology: Proceedings, Ed. V.A. Petrov. Moscow: A.S. Popov MNTORES, 2006. P. 92–95.

168. Aliyeva E.A, et al., in: Lasers in Science and Technology: Proceedings, Ed. V.A. Petrov. Moscow: A.S. Popov MNTORES, 2007. P. 70–75.
169. Priseko Yu.S., Kondratenko V.S., in: Information technologies in science, engineering and education: Proceedings. Intern. scientific and technical conf. Turkey November 16-19, Moscow: MGUPI. 2008. P. 87–93.
170. Lepekhin N.M., et al., in: Information technologies in science, engineering and education: Proceedings. Intern. scientific and technical conf. Turkey November 16-19, Moscow: MGUPI. 2008. P. 113–122.
171. Lepekhin N.M., et al., in: Lasers in Science and Technology: Proceedings, Ed. V.A. Petrov. Moscow: A.S. Popov MNTORES, 2007. P. 39–42.
172. Chissov V.I., et al., Lazernaya meditsina. 2011. V. 15, No. 4. P. 40–47.
173. Vasil'ev N.E., Romanov V.N., ibid, 2002. Vol. 6, No. 1. P. 30–32.
174. Lepekhin N.M., et al., in: Lasers in Science and Technology: Proceedings, Ed. V.A. Petrov. Moscow: A.S. Popov MNTORES, 2003. P. 72–74.
175. Lepekhin N.M., et al., in: Lasers in Science and Technology: Proceedings, Ed. V.A. Petrov. Moscow: A.S. Popov MNTORES, 2004. P. 18–20.
176. Lepekhin N.M., et al., in: Lasers in Science and Technology: Proceedings, Ed. V.A. Petrov. Moscow: A.S. Popov MNTORES, 2004. P. 20–23.
177. Lepekhin N.M., et al., in: Lasers in Science and Technology: Proceedings, Ed. V.A. Petrov. Moscow: A.S. Popov MNTORES, 2004. P. 23–25.
178. Lepekhin N.M., et al., Elektrotekhnika 2010: Proc. VIII Symp. Moscow, VEI, May 24–26, 2005. P. 153.
179. Lepekhin N.M., Elektrotekhnika 2030: Proc. IX Symp. Moscow, VEI, May 29–31 2007. P. 277.
180. Lepekhin N.M., et al., in: Lasers on Metal Vapours: Proc. Symp. Loo, September 21–23, 2004. P. 50.
181. Lepekhin N,M,, et al., ibid. 21–23, 2004. P. 51.
182. Soldatov A.N., et al., ibid, 21–23, 2004. P. 88.
183. Lepekhin N.M., et al., ibid, 25–29, 2006. P. 7.
184. Laser technology and technology: Textbook for universities: In 7 volumes, ed. A.G. Grigoryants. Book 3: Grigoryants A.G., Safonov A.N., Methods of surface treatment. – Moscow: Vysshaya shkola. 1987. – 191 p.
185. Laser technology and technology: Textbook for universities: In 7 volumes, ed. A.G. Grigoryants. Book 5: Grigoryants A.G., Shiganov I.N., Laser welding of metals. – Moscow Vysshaya shkola, 1988. – 207 p.
186. Grigoryants A.G., Svar. Proizvod. 1996. No. 8. P. 2–4.
187. Grigoryants A.G., Zharikov V.M., Vestn. MSTU im. N.E. Baumana, Ser. Mashinostroenie. 2000. No. 2.
188. Betina L.L., et al., Lasers in Engineering. 2003. V. 12, No. 3. P. 167–185.
189. Betina L.L., et al., Lasers in Engineering. 2003. V. 13, No. 3. P. 167–185.
190. Lyabin N.A., et al., in: Lasers on metal vapors: Proc. Symp. Loo, September 21–23, 2004. P. 46–47.
191. Grigoryants A.G., ibid, 21–23, 2004. P. 49.
192. Grigoryants A.G., et al., Patent 2288845 Russian Federation IPC B44C 5/08 (2006.01), CO3C 23/00 (2006.01). A high-resolution imaging device inside a transparent or slightly transparent solid/Applicants and patent holders. No. 2005111797/12; claimed 21.04.2005; publ. 10.12.2006, Bulletin. No. 34.
193. Ivanov I.A., Development and investigation of the method of separation of transparent brittle dielectric materials by radiation of a copper vapor laser: Dissertation, Moscow, 2006. 157 p.

194. Fahlen T.S., J. Appl. Phys. 1974. V. 45, No. 9. P. 4132–4133.
195. Decker C.D., et al., ibid. 1975. V. 46, No. 5. P. 2308–2309.
196. Anderson R.S.S., IEEE J. Quant. Electron. 1975. V. 11. P. 172–174.
197. Anderson R.S.S., et al., ibid. 1975. V. 11. P. 560–570.
198. Warner B.E., Proc. SPIE. 1987. V. 737. P. 2–6.
199. Hackel R.P., Warner B.E., ibid. 1993. V. 1859. P. 120–129.
200. Mishin V.A., Isotopes: properties, production, application, Moscow: IzdAT, 2000. P. 308.
201. Grant B., Lasers improve uranium enrichment, Proton. Spectra. 1997. V. 31, No. 10. P. 46.
202. Kiernan V., Laser Focus World. 1997. V. 33, No. 10. P. 78–79.
203. Bettinger A., et al., Proc. SPIE. 1993. V. 1859. P. 108–116.
204. Camarcat N. et al., ibid. 1993. V. 1859. P. 14–23.
205. Shirayama S., et al., ibid. 1990. V. 1225. P. 279–288.
206. Tabata Y., et al., Proc. SPIE. 1992. V. 1628. P. 32–43.
207. Morioka N., ibid. 1993. V. 1859. P. 2–13.
208. Konagai C., et al., Development of highpower copper vapor laser amplifier, Proc.15th. Ann. Meeting Laser Soc. Jpn. 1995. V. 15. P. 112.
209. Iseki Y., et al., Jpn. J. Appl. Phys. 1994. V. 33. P. 860–862.
210. Kimura H., et al., J. Nucl. Sci. Technol. 1994. V. 31(1). P. 34–47.
211. Chang I.I., et al., Proc. SPIE. 1994. V. 2118. P. 2–8.
212. MP200X Micro-machining System, http:, www. oxfordlasers.com
213. Forrest G.T., Laser Focus. 1986. V. 22(4). P. 23–30.
214. Smilanski I., et al., Opt. Commun. 1979. V. 30(1). P. 70–74.
215. Stanco A., Proc. SPIE. 1987. V. 737. P. 7–9.
216. Postdecline News, Laser Focus. 1988. V. 24(4). P. 14.
217. Carman R.J., et al., Proc. SPIE. 1997. V. 3092. P. 68–71.
218. Withford M.J., et al., Opt. Commun. 1997. V. 135. P. 164–170.
219. Marasov O., et al., J. Phys. E: Sci. Instrum. 1984. V. 17. P. 127–130.
220. Marasov O., Konstadinov I., ibid. 1989. V. 22. P. 441–445.
221. Rewiew and forecast of laser markets – Part 1, Laser Focus World. 1998. V. 34, No. 1. P. 78–98.
222. Burmakin V.A., et al., Author Cert. 555776. USSR, MKI3 H 01 S 3/22. Active element of an optical quantum generator on vapors of chemical substances. No. 2003845/25; claimed 03/11/1974.
223. Burmakin V.A., et al., Kvant. elektronika. 1978. Vol. 5, No. 8. P. 1000–1004.
224. Burmakin V.M., Kvant. elektronika. 1979. Vol. 6, No. 7. P. 1589–1590.
225. Zubov V.V., et al., Kvant. elektronika. 1983. Vol. 10, No. 9. P. 1908–1910.
226. Zharikov V.M., et al., Kvant. elektronika. 1984. Vol. 11, No. 5. P. 918–923.
227. Zubov V.V., et al., Elektronnaya promst.. 1984. No. 10. P. 28–30.
228. Belyaev V.P.,et al., Kvant. elektronika. 1985. Vol. 12, No. 1. P. 74–79.
229. Grigoryants A.G., Vasil'tsov V.V., Inzh. Zhurna;: nauka i innovatsiya. 2012. No. 6 (6). P. 1.
230. Lyabin N.A., Proc. conference. TsNIIE. Ser. 11. Laser technology and optoelectronics. Issue. 3 (237). Moscow, 1986. P. 15–16.
231. Zubov V.V., et al., Kvant. elektronika. 1986. V. 13, No. 12. P. 2431–2436.
232. Zubov V.V., et al., Kvant. elektronika. 1988. V. 15, No. 10. S. 1947–1954.
233. Lyabin N.A., Kvant. elektronika. 1989. V. 16, No. 4. P. 652–657.
234. Armichev A.V., et al., in: Laser technology and its application in medicine: Mater. All-Union. school. Mosc, 1984. P. 19.

235. Armichev A.V., et al., in: Prospective directions of laser medicine: Mater. Intern. Conf. Moscw, 1992. P. 440.
236. Armichev A.V., et al., in :New directions of laser medicine: Mater. Intern. Conf. Moscow, 1996. P. 353.
237. Armichev A.V., Laser on solutions of dyes pumped by a laser on copper vapor and a phototherapeutic device based on them: Dissertation. Moscow, 1999. – 26 p.
238. Stranadko E.F., Laser Market. 1993. No. 7–8. P. 22–24.
239. Ponomarev I.V., Lasers on copper and gold vapor in medicine. – Moscow: P.N. Lebedev FIAN. 1998. 56 p.
240. Ivanov A.V., Current trends in the development of methods of photodynamic therapy of tumors, Laser-Inform: Information Bulletin. Laser Association. 2000. No. 23–24. P. 1–4.
241. Armichev A.V., et al., Biomeditsinskaya Elektronika. 2000. No. 11. P.24–28.
242. Aleinikov V.S., et al., in: Laser Technology and Technology: Proc.. Branch scientific-technical seminar. Bryansk, 1991. P. 28.
243. Zharikov V.M., et al., ibid, Bryansk, 1991. P. 66.
244. Zharikov V.M., Zubov V.V., Pis'ma Zh. Teor. Fiz. 1992. V. 18, No. 7. P. 67–68.
245. Konkov N.V., et al., Elektronnaya tekhnika. Ser. SVCh-tekhnika. 1993. No. 4 (458). P. 29–33.
246. Dickmann K., et al. Laser und Optoelektronik. 1996. No. 28 (1). P. 52–57.
247. Zharikov V.M., et al. J. Moscow Phys. Soc. 1997. No. 7. P. 339–350.
248. Betina L.L., et al., Application of copper vapor for processing of microelectronics articles, Lasers'98: Proc. Intern. Conf. Tucson, Arizona, Desember 7–11, 1998. P. 367–370.
249. Zharikov V.M., et al., Influence of the resonant parameters and radiation characteristics on the luminescence potential, Lasers'99: Proc. Intern. Conf. Quebec, Canada, December 13–16, 1999. P. 690–694.
250. Zharikov V.M. Investigation of physical processes of interaction of laser radiation on copper vapor with materials of electronic engineering and development of technology for their precision processing: author's abstract of thesis. Dissertation. Fryazino, 1999.
251. Zharikov V.M., et al., Distictive features of processing transparent dielectrics by copper vapor lasers emission, Lasers 2000: Proc. Intern. Conf. Albuquerque, New Mexico, December 4-8, 2000. P. 911–914.
252. Guiragossian Z.G., et al., Color laser system, Lasers 2000: Proc. Intern. Conf. Albuquerque, New Mexico, December 4–8, 2000. P. 685–687.
253. Lyabin N.A., et al., Kvant. elektronika. 1990. V. 17, No. 1. Pp. 28–31.
254. Zubov V.V., et al., Proc. SPIE. 1993. V. 2110. P. 78–89.
255. Lyabin N.A., et al., J. of Russian Laser Reseach. 1996. V. 17, No. 4. P. 346–355.
256. Zubov V.V., et al., in: Fizprom-96: Proc.. Intern. Conf. Golitsyno, Moscow. 22–26 Sept., 1996. P. 195–196.
257. Lyabin N.A., Kazaryan M.A., in: Physico-chemical processes in selection of atoms and molecules: Proc. doc. 2-nd All-Russia. Sci. Conf. Zvenigorod, September 29–October 3, 1997. P. 74–77.
258. Lyabin N.A., et al., in: Lasers in Science, Technology, Medicine: Proc. VIII Int. scientific-techn. Conf. Pushkinskie Gory, 1997. P. 73.
259. Lyabin, N.A., et al., Lasers'98: Proc. Intern. Conf. Tucson, Arizona, December 7–11, 1998. P. 359-366.
260. Lyabin N. A., Kazaryan M.A. Spatial and temporal characteristics of sealed-off copper vapor laser radiation, Ibid. P. 695–700.

261. Lyabin N. A., et al., Lasers 2000: Proc. Intern. Conf. Albuquerque, New Mexico, December 4–8, 2000. P. 915–922.
262. Kazaryan M.A., et al., Power Laser Ablation: Proc. SPIE. 2000. V. 4065. P. 719–727.
263. Kazaryan M.A., Lyabin N.A., in: Physico-chemical processes in the selection of atoms and molecules: Proc. 4th All-Russia. Sci. Conf. Zvenigorod, 4–8 October. 1999. P. 110–114.
264. Kazaryan M.A., et al. in: Physico-chemical processes in the selection of atoms and molecules: Proc. 5th All-Russia. Sci. Conf. Zvenigorod, 2–6 October 2000. P. 118–119.
265. Shiyanov D.V., et al ., in: Lasers on metal vapors: Proc. Symposium. Lazarevskoe, September 25–29 2000. P. 9.
266. Bohan P.A., et al., ibid, P. 13.
267. Lyabin N.A., et al., ibid. P. 48–49.
268. Chursin A.D., et al., ibid. P. 51.
269. Lyabin N.A.,et al., Izv. VUZ. Fizika. 1999. V. 42, No. 8. P. 67–73.
270. Lyabin N.A., Optika atmosfery i okeana. 2000. V. 13, No. 3. P. 258–264.
271. Lyabin N.A., Development and investigation of industrial sealed copper vapor lasers with a power of 10–50 W for technological and medical applications: Dissertation. Fryazino, 2002. 244 p.
272. Batygin V.N., et al., Vacuum-dense ceramics and its fittings with metals. Moscow: Energiya, 1973. 410 p.
273. Andrianov N.T., Lukin E.S., Thermal ageing of ceramics. Moscow: Metallurgiya, 1979. 100 p.
274. Metelkin I.I., et al., Welding of ceramics with metals. Moscow: Metallurgiya, 1977. 158 p.
275. Gladkov A.S., et al., Metals and alloys for electrovacuum devices. Moscow: Energiya, 1969. 600 p.
276. Belskii E.I., et al., New materials in engineering. Minsk: Belarus, 1971. 270 p.
277. Physical and chemical properties of oxides: Handbook. Moscow: Mashinostroenie, 1969. 455 p.
278. Savitsky E.M., et al., Rhenium alloys. Moscow: Nauka, 1965. .
279. Savitskiy E.M., Burkhanov K.B., Metal science of refractory metals and alloys. Moscow: Nauka, 1967. 324 p.
280. Smittels C.G. Tunsgten, translated from the English; Moscow: Metallurgizdat, 1958. 414 p.
281. Kudintseva G.A., et al. Thermoelectronic cathodes. Moscow: Energiya, 1966. 368 p.
282. Kiselev A.B., Metal oxide cathodes of electronic devices. Moscow: Publishing House of MFTI, 2002. 240 p.
283. Palatnik L.S., Dokl. AN SSSR. 1953. V. 89, No. 3. P. 34–36.
284. Gusev G.V., Zh. Teor. Fiz. 1955. V. 25, no. 4. P. 1672–1687.
285. Zipgerman A.S., Fiz. Met. Metalloved. 1957. V. 5, no. 1. P. 243–251.
286. Samsonov G.V., Verkhoturov A.D., Elektronnaya obrabotka materialov. 1969. No. 1. P. 25–29.
287. Samsonov G.V., et al., ibid, 1966. No. 1. P. 23–32.
288. Orgel L., Introduction to the chemistry of transition metals, translated from the English; Moscow: Mir, 1964. 210 p.
289. Chang Y.Y., et al., IEEE Trans. Plasma Sci. 1997. V. 25. P. 392–399.
290. Springer L.W., IEEE Intern. Pulsed Power Conf. 1976. V. 9. P. 1–6.
291. Dul'nev G.N., Zarichnyak Yu.P., Thermal conductivity of mixtures and composite materials. Leningrad: Energiya, 1974. 264 p.

292. Kharlamov A.G., Thermal conductivity of high-temperature heat insulators. Moscow: Atomizdat, 1979. 100 p.
293. Krasulin Yu.L., Porous structural ceramics. Moscow: Metallurgiya, 1980. – 100 p.
294. Isachenko V.P., et al., Heat transfer. Moscow: Energiya, 1975. 417 p.
295. Kondakova L.V., Mikhailova V.A., Glass-metal housings for semiconductor and vacuum devices. Moscow: Energiya, 1979. 97 p.
296. Smilanski I., et al., Opt. Commun. 1979. V. 30. No. 1. P. 70–74.
297. McDaniel I., Collision Processes in Ionized Gases [in Russian], Trans. from the English; Ed. L. A. Artsimovich. Moscow: Mir, 1967. 832 p.
298. Kovalenko V.F., Thermophysical processes and electrovacuum devices. Moscow: Sov. radio, 1975. 215 p.
299. Ageev V.P., et al., Zh. Teor. Fiz. 1986. V. 56. P. 1387–1390.
300. Buzhinsky O.I., et al., Kvant. elektronika. 1980. V. 8, No. 12. P. 2644–2646.
301. Bakiev A.M., Valiev S.Kh., Kvant. elektronika. 1989. V. 16, No. 12. P. 2489–2492.
302. Alipov D.T., et al. Zh. Teor. Fiz. 1990. V. 60. P. 97–105.
303. Ananyev Yu.A., et al., Zh. Teor. Eksper. Fiz. 1968. V. 55. P. 130–137.
304. Grove R.E., et al., IEEE J. Quant. Electron. 1981. V. 17, No. 12. P. 51.
305. Kalugin M.M., et al., Kvant. elektronika. 1981. V. 8, No. 5. Pp. 1085–1089.
306. Kazaryan M.A., et al., The generator-amplifier system based on a copper vapor laser: USSR Academy of Sciences. Preprint/FIAN No. 163. Moscow, 1982.
307. Pod'yapolsky B.A., Popov V.K., Pulse modulator lamps. Moscow: Sov. radio, 1967. 64 p.
308. Volkov I.V., Ganchenko L.M., Tekhn. elektrodynamika. 2000. No. 3. P. 23–27.
309. Evtushenko G.S., et al., Optika atmosfery i okeana. 2000. V. 13, No. 3. P. 265–266.
310. Bohan P.A., et al., Kvant. elektronika. 2002. V. 32, No. 7. Pp. 570–585.
311. Kolokolov I.S., et al., Prikl. fizika. 2003. No. 3. Pp. 84–89.
312. Lyabin N.A., et al., Elektronaya tekhnika. Ser. SVCh-tekhnika. 2003. Issue. 2 (482). P. 17–35.
313. Pleshanov S.A., et al., Patent 1565320 Russian Federation. MKI[5] H 01 S 3/08. Pulsed laser. Applicant and patent holder of FGUP SPE Istok. No. 4423717/24-25; claimed. 05/13/1988.
314. Lyabin N.A.. Author Cert. 1438549. USSR, MKI[4] H 01 S 3/10. Non-cavity pulsed laser. No. 4211869/24–25; claimed 19.03.87.
315. Lyabin N.A.,et al., Patent 1813307 A3 of the USSR, MKI[5] H 01 S 3/041. Metal vapor laser. Applicant and patent holder FGUP SPE Istok. No. 4861801/25; claimed 24.08.90.
316. Lyabin N.A., et al., Patent 20617 Russian Federation, MKI[7] H 01 S 3/00, B23 K26/00. Pulsed metal vapor laser. Applicant and patent holder FGUP SPE Istok. No. 2001111795/20; claimed 03.05.2001; publ. 10.11.2001, Bulletin No. 31.
317. Lyabin N.A., Chursin A.D., Patent 2191452 Russian Federation, MKI7 H 01 S 3/03, 3/227. A discharge tube of a metal vapor laser. Applicant and patent holder FGUP SPE Istok. No. 2000112859/28; claimed. 23.05.2000; publ. 20.10.2002, Bulletin No. 29.
318. Lyabin N.A., et al., Patent 35177 The Russian Federation. MKI[7] H 01 S 3/03. A discharge tube of a metal vapor laser. Applicant and patent holder FGUP SPE Istok. No. 2003119771/20; claimed 07/01/2003; publ. 27.12.2003; Bulletin No. 36.
319. Chursin A.D., Patent 2023334 Russian Federation. IPC[5] H 01 S 3/22. Active element on chemical vapor. Applicant and patent holder FGUP SPE Istok. No. 4866118/25; claimed 09.07.90; publ. 15.11.94, Bulletin No. 21.
320. Lyabin N.A., et al., Patent for utility model 30468 Russian Federation, MKI[7] H 01 S

3/03. Active element of a metal vapor laser. Applicant and patent holder FGUP SPE Istok. No. 2002116609/20; claimed 24.06.2002; publ. 27.06.2003; Bulletin No. 18.
321. Chursin A.D., Zubov V.V., Author Cert. for invention No. 1572367. USSR, Cl. H 01 S 3/03. Active element of the laser on chemical vapors. Applicant and patent holder FGUP SPE Istok. No. 4486293/24–25; claimed 09/26/1988.
322. Lyabin N.A., et al., in: Lasers in science and technology: Collection of scientific papers ed. V.A. Petrova. Moscow: A.S. Popov MNTORES, 2006. pp. 129–131.
323. Kolokolov I.S., et al., in: Lasers on metal vapor: Proc. Symp. Loo, September 24–26, 2002. P. 32.
324. Kolokolov I.S., et al., in: Physico-chemical processes in the selection of atoms and molecules: Proc. VII All-Russia. (international) Scientific. Conf. Zvenigorod, September 30-Oct 4 2002. P. 249–250.
325. Bergmann H.W., et al. Micromachining of metals with copper vapor lasers, CLEO'92: Texch. Dig.-Opt. Soc. Am. Washington, DC. 1992. P. 320–321.
326. Nikonchuk M.O., Proc. SPIE. 1991. V. 1412. P. 38–49.
327. Gorny S.G., et al., Vestn. mashinostr. 1999. No. 2. Pp. 30–32.
328. Kautz D.D., et al. Drilling with fiber-transmitted, visible lasers, Lasers'93: Proc. Int. Conf. – STS Press: McLean, VA, 1994. P. 30–36.
329. Coutts D.W., et al. UV micromachining using copper vapour lasers, Pulsed Metals Vapour Lasers – Physics and Emerging Applications in Industry, Medicine and Seince. Ed. by C. E. E. Little, N.V. Sabotinov. – Kluwer Acad. Publishers: Dortrecht, 1996. P. 365–370.
330. Kim H.S., et al., Appl. Phys. Lett. 1989. V. 55 (8). P. 726–728.
331. Kearsley A., Errey K., Opto & Laser Europa. 1995. V. 26. P. 24–26.
332. Lach, J.S., Gilgenbach R.M., Sci. Instrum. 1993. V. 64 (11). P. 3308–3313.
333. Grigoryants A.G., Fundamentals of laser processing of materials. Moscow: Mashinostroenie, 1989. 304 p.
334. Knowles M., Webb C.E., Physics World. 1995. V. 8 (5). P. 41–44.
335. Dickman K., Laser Optoelectron. 1996. V. 28 (1). P. 52–57.
336. Laser technology and technology: Textbook for universities: in 7 volumes, Ed. A.G. Grigoryants. Book 7: Grigoriants A.G., Sokolov A.A., Laser cutting of metals. Moscow, Vysshaya shkola, 1988. 127 p.
337. Nikonchuk M.O., et al., Proc. SPIE. 1990. V. 1225. P. 419–430.
338. Vejko V.P., Laser processing of film elements. Lenigrad: Mashinostroenie, 1986. 248 p.
339. Warner B.E., et al., Industrial applications of high-power copper vapor lasers, Pulsed Metal Vapour Lasers – Physics and Emerging Applications in Industry, Medicine and Science/Ed. by C. E. Little, N.V. Sabotinov. Kluwer Acad. Publichers. Dortrecht, 1996. P. 331–346.
340. Hang J.J., Warner B.E., Appl. Phys. Lett. 1996. V. 69(4). P. 473–475.
341. Grigoryants A.G. Basics of Laser Material Processing, CRC Press, Boca Raton (USA), 1994. 312 p.
342. Knowles M. R. H., Foster-Turner R., Kearsley A. J. Processing of ceramics with copper lasers, CLEO/Europe'94: Tech. Dig. – IEEE. Piscataway, NJ, 1994. P. 294–295.
343. Laser technology and technology: Textbook for universities: in 7 volumes, Ed. A.G. Grigoryants. Book 4: Grigoryants A.G., Sokolov A.A., Laser processing of nonmetallic materials. Moscow: Vysshaya shkola, 1988. 191 p.
344. Grigoryants A.G., "Modeling of the process of laser thermoscaling of tubular glassware," Vestn. MGTU in N.E. Bauman. Ser. Mashinostroenie. 1999. No. 2. P. 28–35.
345. Pini R., et al., Improve drilling of crystals and other optical materials by a diffraction

limited copper vapor laser, Lasers'93: Proc. Int. Conf. – STS Press: McLean, VA, 1994. P. 364–368.

346. Pini R., et al., Imaging and mechanism analysis of high quality drilling by a copper vapor laser, CLEO/Europe'94: Tech. Dig. – IEEE, Piscataway, NJ, 1994. P. 295–296.
347. Pini R., et al., High-quality drilling with copper vapour lasers, Opt. Quant. Elektron. 1995. V. 27. P. 1243–1256.
348. Pini R., et al., Appl. Phys. 1995. V. 61(5). P. 505–510.
349. Pini R., Drilling and cutting transparent substates with copper vapour lasers, Pulsed Metal Vapour Lasers. Physics and Emerging Applications in Industry, Medicine and Scince. Ed. by C. E. Little, N.V. Sabotinov. – Kluwer Acad. Publishers. Dortricht, 1996. P. 359–364.
350. Precision machining with copper vapour lasers, ibid. P. 317–330.
351. Riva R., et al., Proc. SPIE. 1996. V. 2789. P. 345–351.
352. Konagai C., et al., Underwater direct metal processing by high-power copper vapour lasers, Pulsed Metal Vapour Lasers. Physics and Emerging Applications in Industry, Medicine and Science, Ed. by C. E. Little, N.V. Sabotinov. – Kluwer Acad. Publishers: Dortricht, 1996. P. 371–376.
353. Opacohko I.I., Shimon L.L., Proc. SPIE. 1993. V. 2120. P. 250–267.
354. Glover C.J., et al., IEEE J. Sel. Topics Quantum Electron. 1995. V. 1(3). P. 830–836.
355. Illy E.K., et al., J.A. Ablation characteristics for high-speed micro-machining of polymers with UV-copper vapour lasers, IQEC'96: Tech. Dig. – Opt. Soc. Am. Washington, DC, 1996. P. 207–208.
356. Grigoriants A. G., Bazanova N.I. Photochemical technologies for the modification of polymers of medical purpose, Vestn. MGTU im. N.E. Bauman. Ser. Estestv. nauki. 2000. No. 1. P. 97–106.
357. Glover A.C.J., et al., Progress in high-speed UV micromachining with high repetition rate frequency-doubled copper vapour lasers, Laser Material Processing: Proc. CLEO'94. Laser Inst. Am. 1994. V. 79. P. 343–351.
358. Zemskov K.I., et al., Kvant. elektronika. 1984. V. 11, No. 2. P. 418–420.
359. Mildren R.P., Piper J.A., Opt. Lett. 2003. V. 28. P. 1936–1938.
360. Kazaryan M.A., et al., Laser Phys. 2002. V. 12, No. 10. P. 1281–1285.
361. Kazaryan M.A., Lyabin N.A., Effective lasers for the visible, near infrared and ultraviolet spectral regions, International Conference on Luminescence. Proc. Moscow, October 17–19, 2001. P. 245.
362. Lyabin N.A., et al., in: Physico-chemical processes in the selection of atoms and molecules. Proc. IX All-Russian (international) Scientific. Conf. Zvenigorod, 4–8 October. 2004. P. 208–210.
363. Azizbekyan G.A., et al. ibid, P. 45.
364. Lyabin N.A., et al., in: Physico-chemical processes in the selection of atoms and molecules. Proc. XII International Sci. Conf. Zvenigorod, March 31–April 4 2008. P. 250–254.
365. Lyabin N.A., et al., Patent No. 2264011. Russian Federation. IPC^7 H 01 S 3/02, 3/09. Method of excitation of radiation pulses of laser systems on self-organic transitions (variants). Applicant and patent holder FGUP SPE Istok. No. 2004108765/28; claimed 24.03.2004; publ. 10.11.2005, Bulletin. No. 31.
366. Korolev A.N., et al., Elektronaya tekhnika. Ser. 1. SVCh-tekhnika. 2009. Issue 2 (501). P. 45–52.
367. Zharikov V.M., et al., in: Lasers in Science and Technology: Abstracts. XII Int. Scientific-techn. Conf. (Sochi, 2001). Moscow: Bauman MGTU, 2001. P. 92–93.

368. Lyabin N.A., et al., ibid, 2001. P. 24–25.
369. Lyabin N.A., et al., in: Lasers in science and technology: Collection of scientific papers. Moscow: A.S. Popov MNTORES, 2005. P. 149–154.
370. Lyabin N.A., et al., ibid, Moscow: A.S. Popov MNTORES, 2007. P. 32–39.
371. Lyabin N.A., et al., Patent 2432652 Russian Federation, IPC H 01 S 3/03,/05. Pulsed laser. Applicant and patent holder FSUE SPE Istok. No. 2010133034; claimed 05.08.2010; publ. 10/27/2011, Bulletin No. 30.
372. Lyabin N.A., et al., in: Innovative technologies in science, engineering and education: Collection of works, T. 1. Moscow: MGUPI, 2008. P. 103–112.
373. Lyabin N.A., et al., ibid, T. 1. Moscow: MGUPI, 2009. P. 73–80.
374. Lyabin N.A., Kazaryan M.A., Lasers on atomic metal vapor. New technologies – the 21st century. Moscow, 'Vlad', 2009. No. 6. P. 14–15.
375. Lyabin N.A., et al., Lasers in science and technology: Collection of scientific papers. Moscow: A.S. Popov MNTORES,, 2003. P. 74–75.
376. Kazaryan M.A., et al., in: Lasers on metal vapors: Proc. Symp. Loo, September 21–23, 2004. P. 10.
377. Kazaryan M.A., et al., in: Lasers on metal vapors: Proc. Symp. Loo, September 21–23, , 2004. P. 31.
378. Lyabin N.A., et al., in in: Lasers on metal vapors: Proc. Symp. Loo, September 21–23, 2004. P. 45.
379. Gradoboev Yu.V., et al., in: Lasers on metal vapors: Proc. Symp. Loo, September 21–23, 2004. P. 48.
380. Manucharyan R.G., in: Lasers on metal vapors: Proc. Symp. Loo, September 21–23, 2006. P. 14.
381. Lyabin N.A., et al., in: Lasers on metal vapors: Proc. Symp. Loo, September 21–23, 2006. P. 47.
382. Kazaryan M.A., et al., in: Lasers on metal vapors: Proc. Symp. Loo, September 21–23, 2006. P. 48.
383. Lyabin N.A., et al., in: Lasers on metal vapors: Proc. Symp. Loo, September 21–23, 2006. P. 49.
384. Averyushkin A.S., et al., in: in: Lasers on metal vapors: Proc. Symp. Loo, September 21–23, 2010. P. 3.
385. Biryukov A.S., et al., in: Lasers on metal vapors: Proc. Symp. Loo, September 21–23, 2010. P. 9.
386. Domanov M.S., et al., in: Atomic and Molecular Pulsed Lasers: Abstracts The 6-th International Conference Tomsk, Russia, September 15–19, 2003. P. 19.
387. Kolokolov I.S., et al., in: Atomic and Molecular Pulsed Lasers: Abstracts The 6-th International Conference Tomsk, Russia, September 15–19, 2003 – P. 68.
388. Gradoboev Yu.G., et al., in: Atomic and Molecular Pulsed Lasers: Abstracts The 6-th International Conference Tomsk, Russia, September 15–19, 2003 – P. 94.
389. Grigoriants A.G., et al., Al'ternativnaya energetika i ekologiya. 2013. V. 129, No. 7. P. 31–43.
390. Grigoryants A.G., et al., Tekhnologiya mashinostroeniya. 2005. No. 10. P. 60–64.
391. Biryukov A.S., et al., Perspektivnya materialy. 2011. No. 10. P. 148–152.
392. Lyabin N.A., et al., Elektronnaya tekhnika. Ser. 1. SVCh-tekhnika. 2013. Vol. 3 (518). P. 211–220.
393. Bokhan P.A., et al., Kvant. elektronika, 2013, 43 (8), 715–719.
394. Figurovsky D.K., et al., Naukoemkie teknologii v mashinostroenii, Moscow: Mashinostronie, 2013. No. 8. P. 26–30.
395. Lyabin N.A., et al., Nauka i obrazovanie. 2014. No. 8. P. 30–62; DOI:

10.7463/0814.0720903.

396. Lyabin N.A., ibid, 2014. No. 6. Pp. 1–15: DOI: 10.7463/0614.0717060.
397. Lyabin N.A., ibid, 2014. No. 7. Pp. 20–35. DOI: 10.7463/0714.0717617.
398. Lyabin N.A., et al., Naukoemkie teknologii v mashinostroenii, Moscow: Mashinostronie, 2014. No. 9. Pp. 19–26.
399. Lyabin N.A., et al., ibid, 2014. No. 10. Pp. 41–48.
400. Lyabin N.A., et al., Elektronnaya tekhnika. Ser. 1. SVCh-tekhnika. 2014. No. 3 (536). P. 211–220.
401. Directory of the company Vicon-Standa. – Opto-Mechanical Products, 2012/2013. .
402. Veiko V.P., A basic summary of lectures on the course 'Physical and technical fundamentals of laser technologies'. Section: Laser microprocessing. – St Petersburg: SPb GU ITMO, 2005. 110 p.
403. Libenson M.N., et al., Interaction of laser radiation with matter (power optics). Lecture notes. Part I. Absorption of laser radiation in matter. ed V.P. Veiko – St Petersburg: SPb GU ITMO, 2008. 141 p.
404. Veiko V.P., et al., Laser cleaning in mechanical engineering and instrument making. St Petersburg: SPb GU ITMO, 2013. 103 p.
405. Vejko V.P., A basic summary of lectures on the course 'Physical and technical fundamentals of laser technologies'. Section: Technological lasers and laser radiation. St Petersburg: SPb GU ITMO, 2007. 52 p.
406. Vejko V.P., A basic summary of the lectures 'Laser micro- and nanotechnologies in microelectronics'. St Petersburg: SPb GU ITMO, 2012. 141 p.

Index

A

A.N. Prokhorov Institute of General Physics xviii, 160
Atzevus 24
AVLIS program 18, 20, 21, 22, 23, 73, 159, 205
AVLIS technology xiii, xviii, 19, 21, 160

B

Bauman Moscow State Technical University xviii, 37
Bombay Atomic Research Center 23

C

Chistye Tekhnologii xviii, 39, 394
crystals
  BBO xiii, 9, 10, 13, 15

D

DKDP xiii, 9, 10, 15

E

electropolishing 353, 354
excimer xi, xv, 42, 48, 51, 66, 67

H

High Density Interconnect Board 364

I

Institute of Optics and Atmosphere of the SB RAS xviii
Institute of Semiconductor Physics xviii, 10, 37, 160, 230, 294, 394
Istok xviii, 6, 7, 9, 10, 11, 12, 14, 15, 16, 26, 27, 33, 34, 35, 37, 38, 39, 5
3, 54, 55, 67, 70, 76, 78, 80, 84, 86, 104, 108, 137, 156, 158, 159, 2
30, 243, 244, 268, 289, 294, 351, 377, 389, 394, 396

K

Kareliya 30, 31, 32, 54, 55, 56, 67, 76, 251, 284
Kazaryan M.A. 3, 6
KDP, xiii
Kulon-Med xx, 15, 36, 39
Kulon–Med xiii
Kurchatov Institute xviii, 10, 37, 38, 115, 160, 230, 293, 295, 298, 394

L

laser
barium vapour laser 6
CO2 lasers xiii, 24, 42, 48
continuous-wave lasers 44
CuBr laser 8, 38, 121
CVL with enhanced kinetics 4, 25
diode lasers xv, 42, 66
Excimer gas lasers xv
fiber xiv, xv, 16, 17, 22, 24, 31, 32, 42, 48, 51, 66, 89, 133, 308
gold vapour laser 6, 12, 137, 149
Kriostat laser 243
metal vapour lasers 1
Nd:YAG laser xiv, 13
Rofin-Sinar Laser xiv
TRUMPF xiv
YAG:Nd laser 49
Yb:YAG xiv, 24
Lawrence Livermore National Laboratory xviii, 18, 37, 38, 44, 73, 159, 205
Lazery i apparatura TM 16

M

Macquarie University xviii, 37
Mashinoexport 26, 37, 157, 158
MP200X installation 23, 67, 395

O

optical quantum amplifier 27
Oxford Lasers xviii, 22, 23, 37, 44, 51, 52, 53, 67, 156, 157, 158, 395

P

P.N. Lebedev Institute of Physics 3, 7, 11, 26
P.N. Lebedev Physical Institute xviii, 70

Q

Q-factor 44

R

r–m junctions xii, 7

S

scattering
  hyper-Raman 6
  hyper-Rayleigh 6
  Raman 6, 7
Spectronica 26, 37
St. Petersburg Technical University 44, 160
system
  motion and control system 304, 318, 333

T

titanium sapphire xiii

V

VELIT Research and Production Enterprise 14

Y

Yakhroma-Med xx, 11, 32, 36, 39
Yakhroma–Med xiii

Printed in the United States
by Baker & Taylor Publisher Services